Stephen Barnett
University of Essex

Matrices
Methods and Applications

CLARENDON PRESS · OXFORD

Oxford University Press, Walton Street, Oxford OX2 6DP

Oxford New York Toronto
Delhi Bombay Calcutta Madras Karachi
Petaling Jaya Singapore Hong Kong Tokyo
Nairobi Dar es Salaam Cape Town
Melbourne Auckland

and associated companies in
Berlin Ibadan

Oxford is a trade mark of Oxford University Press

Published in the United States
by Oxford University Press, New York

© Stephen Barnett, 1990
First published 1990
Reprinted 1992

British Library Cataloguing in Publication Data
Barnett, Stephen, 1938–
Matrices: methods and applications.
1. Algebra. Matrices
512.9'434
ISBN 0-19-859665-0
ISBN 0-19-859680-4 (Pbk)

Library of Congress Cataloging in Publication Data
Barnett, S.
Matrices: methods and applications / Stephen Barnett.
(Oxford applied mathematics and computing science series)
1. Matrices. I. Title. II. Series.
QA188.B36 1990 512.9'434—dc20 89-23942
ISBN 0-19-859665-0
ISBN 0-19-859680-4 (pbk.)

Printed and bound in
Great Britain by Biddles Ltd,
Guildford and King's Lynn

Oxford Applied Mathematics and Computing Science Series

General Editors
R. F. Churchouse, W. F. MacColl, and A. B. Tayler

OXFORD APPLIED MATHEMATICS AND COMPUTING SCIENCE SERIES

I. Anderson: *A First Course in Combinatorial Mathematics* (*Second Edition*)

D. W. Jordan and P. Smith: *Nonlinear Ordinary Differential Equations* (*Second Edition*)

D. S. Jones: *Elementary Information Theory*

B. Carré: *Graphs and Networks*

A. J. Davies: *The Finite Element Method*

W. E. Williams: *Partial Differential Equations*

R. G. Garside: *The Architecture of Digital Computers*

J. C. Newby: *Mathematics for the Biological Sciences*

G. D. Smith: *Numerical Solution of Partial Differential Equations* (*Third Edition*)

J. R. Ullmann: *A Pascal Database Book*

S. Barnett and R. G. Cameron: *Introduction to Mathematical Control Theory* (*Second Edition*)

A. B. Tayler: *Mathematical Models in Applied Mechanics*

R. Hill: *A First Course in Coding Theory*

P. Baxandall and H. Liebeck: *Vector Calculus*

D. C. Ince: *An Introduction to Discrete Mathematics and Formal System Specification*

P. Thomas, H. Robinson, and J. Emms: *Abstract Data Types: Their Specification, Representation, and Use*

D. R. Bland: *Wave Theory and Applications*

R. P. Whittington: *Database Systems Engineering*

P. Gibbons: *Logic with Prolog*

J. J. Modi: *Parallel Algorithms and Matrix Computation*

D. J. Acheson: *Elementary Fluid Dynamics*

S. Barnett: *Matrices: Methods and Applications*

P. Grindrod: *Patterns and Waves: The Theory and Applications of Reaction–Diffusion Equations*

J. Kondo: *Integral Equations*

Preface

Matrix methods are an essential tool for virtually all users of mathematics, to whom this book is primarily addressed. Moreover, many students find the subject a welcome relief from the intricacies of calculus and analysis. In order to offer a distinctive contribution to the extensive range of textbooks available on matrices and linear algebra, I have adopted an approach which reflects my own experience and interests. These have developed during a period of over twenty-five years active research and teaching in matrix theory and applications. There are in consequence two specific features of the book. Firstly, my treatment of the material is non-abstract and relatively informal; I have aimed at clarity of explanations, so that results are often justified using simple illustrative examples, rather than by formal proofs. Secondly, I have placed a strong emphasis on topics which are of value for applications in engineering and physical, biological, and social sciences. Despite the vast increase in publications in matrix theory over the last decade, most non-mathematicians remain more interested in how matrix methods can be used than in the detailed derivation of theorems, elegant though these may be.

The book falls naturally into two halves. The first seven chapters provide a first course on matrices, commencing with an introduction showing how the concept of a matrix arises in various applications. This is followed by three chapters on the basic algebra of matrices, solution of linear equations, and determinant and inverse. The next three chapters develop the subject further with the concept of rank and non-unique solution of equations, eigenvalues and eigenvectors, and quadratic forms. This material is mainly standard but I have included some interesting and useful topics (for example, companion form and Kronecker product) which are often omitted from elementary books, and relevant applications (e.g. error-correcting codes in Chapter 5) are described throughout. The second half of the book consists of more advanced topics suitable for a second course, where there are few textbooks appropriate for the audience I am aiming at. A chapter on canonical forms is followed by chapters on matrix functions, generalized inverses, polynomials and stability, polynomial matrices, and patterned matrices (including circulant and Toeplitz forms). The book ends with a collection of some other topics, including matrix equations, non-negative matrices, and norms. The total spread of material included is wider than in any other applications-oriented books currently available.

My coverage of numerical and computational aspects is deliberately limited, since this is now a well established area in its own right with an expanding list of well-written textbooks (e.g. Golub and Van Loan 1989; Hager 1988). However, a computational theme which recurs throughout the book is the simple but powerful method of gaussian elimination. In various guises this is used for the solution of linear equations (Chapters 3 and 5), triangular decomposition (Chapter 3), evaluation of determinant and inverse (Chapter 4) and of rank (Chapter 5), testing the sign nature of a quadratic form (Chapter 7), computation of generalized inverse (Chapter 10), and determination of the greatest common divisor of two polynomials (Chapter 11).

The exercises in the text are essential for understanding the subject, and should be attempted as they occur, or at the very least read through. As with learning to play a musical instrument, practice is indispensable in order to become a successful user of mathematics. The problems at the end of each chapter are also important but are generally more difficult than the exercises, so need not all be tried at a first reading. Again, they are worth scanning at least, since I have used the problems to introduce further developments and applications. An extensive list of textbooks for further reading is also included.

Chapters 1 to 7 and Chapter 9 are based on the elementary book *Matrix methods for scientists and engineers,* published by McGraw-Hill in 1979, where I acknowledged my debt to such classic texts as Mirsky (1963) for stimulating my interest in matrices. It was therefore a particular source of satisfaction that the late Leon Mirsky wrote in a review of my earlier work that 'The appropriate criterion by which such a book must be judged is its usefulness to students. And on these terms it passes the test very well and can be readily commended' (*Linear and Multilinear Algebra,* **9,** 77 (1980)). I can only hope that this new volume would have similarly met with his approval, and that it succeeds in conveying some of the power and fascination of the subject.

I am grateful to Dr M. J. C. Gover who has made many helpful suggestions for improvements, and also to Professors R. E. Hartwig, E. I. Jury, and P. C. Parks who have read and commented on some of the chapters. I also wish to thank Mrs Valerie Hunter for her usual superlative efforts on typing yet another of my books, and the University of Bradford Graphics Unit for producing the figures.

Bradford S. B.

July 1989

Contents

Notation

$\bar{z}$	conjugate of complex number z
$|z|$	modulus of complex number a
$\mathrm{Re}(z)$	real part of complex number z
$\mathrm{sgn}(z)$	sign function, $= +1$, $\mathrm{Re}(z) > 0$; -1, $Re(z) < 0$
$A = [a_{ij}]$	matrix having element a_{ij} in row i, column j
$a = [a_1, a_2, \ldots, a_n]$	row n-vector having a_i as ith component
A^{T}	transpose of A, $= [a_{ji}]$
$b = [b_1, b_2, \ldots, b_n]^{\mathsf{T}}$	column n-vector
$\langle a, b \rangle$, $\bar{a}b$	scalar (or inner) product of a and b, $= \bar{a}_1 b_1 + \cdots + \bar{a}_n b_n$
$\bar{A}$	complex conjugate of A, $= [\bar{a}_{ij}]$
A^*	conjugate transpose of A, $= [\bar{a}_{ji}]$
$\mathrm{diag}[d_{11}, d_{22}, \ldots, d_{nn}]$	$n \times n$ diagonal matrix, $d_{ij} = 0$, $i \neq j$
$\mathrm{diag}[D_1, \ldots, D_n]$	block diagonal matrix
I_n	$n \times n$ unit matrix, $= \mathrm{diag}[1, 1, \ldots, 1]$
K_n	$n \times n$ reverse unit matrix
$\mathrm{tr}(A)$	trace of $n \times n$ matrix A, $= a_{11} + a_{22} + \cdots + a_{nn}$
$A \otimes B$	Kronecker product of A and B
$A \circ B$	Hadamard (Schur) product of A and B
$\det A$, $|A|$	determinant of square matrix A
PDi	property of determinants, number i (Section 4.1.2)
$\mathrm{adj}\, A$	adjoint of A
A^{-1}	inverse of non-singular A
$R(A)$	rank of A
$\rho(A)$	spectral radius of A
$\dot{x}$	$\mathrm{d}x/\mathrm{d}t = [\mathrm{d}x_1/\mathrm{d}t, \ldots, \mathrm{d}x_n/\mathrm{d}t]$
$\ddot{x}$	$\mathrm{d}^2 x/\mathrm{d}t^2$
(Ri), (Cj), (Lk)	row i, column j, line k (of A)

$(Li) \leftrightarrow (Lj)$	interchange lines i and j
$(Li) \times k$	multiply line i by a scalar k
$(Li) + p(Lj)$	add p times line j to line i
$k(\lambda)$	polynomial of degree n, $= \lambda^n + k_1\lambda^{n-1} + \cdots + k_n$
C_n	companion matrix for $k(\lambda)$
(U/W)	Schur complement of W in U
$v(A)$	column vector of rows of A [vec(A)]
J_i	Jordan block of order k_i
$\mathscr{J} = \mathrm{diag}[J_1, J_2, \ldots, J_q]$	Jordan canonical form
$\mathrm{sgn}(A)$	matrix sign function
A^+	Moore–Penrose inverse of A
$\mathrm{In}(A)$	inertia of A, $= (\pi_A, \nu_A, \delta_A)$
$A(\lambda)$	polynomial matrix, $= A_0\lambda^N + A_1\lambda^{N-1} + \cdots + A_N$
$d_k(\lambda)$	kth determinantal divisor of $A(\lambda)$
$i_k(\lambda)$	kth invariant factor of $A(\lambda)$
$S(\lambda)$	Smith normal form of $A(\lambda)$
$\mathrm{circ}(c_1, c_2, \ldots, c_n)$	circulant matrix of order n
$\|x\|, \|x\|_2$	euclidean norm of vector x, $= (\sum |x_i|^2)^{1/2}$
$\|A\|, \|A\|_e$	euclidean norm of A, $= (\sum\sum |a_{ij}|^2)^{1/2}$
$\|x\|_p$	p-norm of vector x, $= (\sum |x_i|^p)^{1/p}$
$\|A\|_p$	p-norm of A, $= \max \|Ax\|_p/\|x\|_p$, $x \neq 0$
$\binom{k}{r}$	binomial coefficient, $= k!/r!(k-r)!$
$\mathscr{C} = [B, AB, A^2B, \ldots, A^{n-1}B]$	controllability matrix for pair A, B
(7.38)	refers to equation number (7.38) in Chapter 7

1 How matrices arise

In this chapter we describe a few of the very many areas of applications in which matrices are used, the choice of topics being deliberately diverse.

Example 1.1 Suppose that the prices (in some monetary units) of four kinds of canned foods F_1, F_2, F_3, F_4 at three different supermarkets S_1, S_2, S_3 are as given in Table 1.1. Thus for example the price of item F_3 in supermarket S_2 is 13. The total cost of buying one can of each food at S_1 is $17 + 7 + 11 + 21 = 56$, and similarly 56 at S_2, 60 at S_3.

Table 1.1

	F_1	F_2	F_3	F_4
S_1	17	7	11	21
S_2	15	9	13	19
S_3	18	8	15	19

The rectangular array of numbers in Table 1.1 is called a *matrix,* in this case having three *rows* and four *columns*. An array like this will arise whenever there are two sets (above, the foods and the supermarkets) whose members are linked by a set of numbers (above, the prices).

Example 1.2 Table 1.2 gives distances (in miles) between four American cities, and is an example of a familiar feature of many road maps.

Table 1.2

	Chicago	New York City	San Francisco	Washington DC
Chicago	0	841	2212	704
New York City	841	0	3033	224
San Francisco	2212	3033	0	2835
Washington DC	704	224	2835	0

The array here has the same number of rows as columns, and is called *square*. The line of zeros forms the *principal diagonal* of this square array. Notice also that the numbers in Table 1.2 are symmetric with respect to this diagonal, e.g. the distance from New York to San Francisco is given either by the number in row 2, column 3, or by the number in row 3, column 2. We shall see that such *symmetric* matrices have interesting properties.

Example 1.3 Figure 1.1, called a *network*, represents connections between two airports A_1, A_2 in one country with airports B_1, B_2, B_3 in a second country. The number on each linking line gives the number of different airlines flying on that route, e.g. there are two airlines offering flights from A_2 to B_2. In tabular form the information can be presented as follows:

$$\begin{array}{c} \\ A_1 \\ A_2 \end{array} \begin{array}{ccc} B_1 & B_2 & B_3 \\ \left[\begin{array}{ccc} 4 & 1 & 3 \\ 0 & 2 & 2 \end{array}\right] \end{array}. \tag{1.1}$$

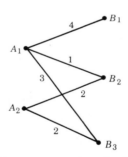

Fig. 1.1 Airport connections for Example 1.3.

This time we have enclosed the array within square brackets, and this is the standard notation for matrices.

Example 1.4 It can be shown that the currents i_1, i_2, i_3 in the electric circuit represented in Fig. 1.2 satisfy the equations

$$(R_1 + R_4 + R_5)i_1 - R_4 i_2 - R_5 i_3 = E_1$$

$$-R_4 i_1 + (R_2 + R_4 + R_6)i_2 - R_6 i_3 = E_2 \tag{1.2}$$

$$-R_5 i_1 - R_6 i_2 + (R_3 + R_5 + R_6)i_3 = E_3.$$

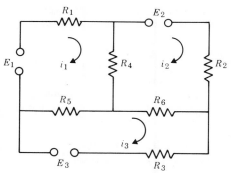

Fig. 1.2 Electric circuit for Example 1.4.

These equations are to be solved for the three unknowns i_1, i_2, i_3, the values of the other parameters being known. Equations (1.2) are called *linear* because no powers or products of the i's are involved. The information in (1.2) can be presented in a more concise way using matrix notation, as follows:

$$\begin{bmatrix} (R_1 + R_4 + R_5) & -R_4 & -R_5 \\ -R_4 & (R_2 + R_4 + R_6) & -R_6 \\ -R_5 & -R_6 & (R_3 + R_5 + R_6) \end{bmatrix} \begin{bmatrix} i_1 \\ i_2 \\ i_3 \end{bmatrix} = \begin{bmatrix} E_1 \\ E_2 \\ E_3 \end{bmatrix} \quad (1.3)$$

At this stage we regard (1.3) merely as an alternative way of *writing* (1.2), having some advantage in economy of symbols since the coefficients in the equations have been recorded separately from the unknowns. However we shall see in Chapter 2 that a fundamental significance can be attached to this representation. Sets of linear equations arise in many applications and their solution forms a major theme of this book.

Example 1.5 An oil refinery makes two grades of petrol, 'supreme' and 'terrific', from two crude oils c_1 and c_2. Two possible blending processes can be used, and the inputs and outputs for a single production run are shown in Table 1.3. Thus if x_1, x_2 are the numbers of

Table 1.3

	Input		Output	
	crude c_1	crude c_2	supreme	terrific
Process 1	3	4	4	6
Process 2	5	2	5	3

production runs of processes 1 and 2 respectively, then the amount of crude c_1 used is $3x_1 + 5x_2$ and the amount of supreme petrol produced is $4x_1 + 5x_2$, and so on.

The maximum amounts available of crudes c_1, c_2 are 170 and 200 units respectively. The profits per run for the two processes are 2.1 and 2.3 units. It is required to produce at least 150 units of supreme petrol and at least 110 units of terrific petrol so as to maximize the total profit $2.1x_1 + 2.3x_2$. In mathematical terms this means that we must determine $x_1 \geqslant 0$, $x_2 \geqslant 0$ satisfying

$$\left. \begin{array}{l} 3x_1 + 5x_2 \leqslant 170 \\ 4x_1 + 2x_2 \leqslant 200 \end{array} \right\} \quad \begin{array}{l} \text{constraints on availability} \\ \text{of crude oils} \end{array}$$

$$\left. \begin{array}{l} 4x_1 + 5x_2 \geqslant 150 \\ 6x_1 + 3x_2 \geqslant 110 \end{array} \right\} \quad \text{production requirements.} \tag{1.4}$$

This is a simple example of a *linear programming* (LP) problem. The linear inequalities can easily be converted into equations. For example, if we define the amount of unused crude c_1 as x_3, then

$$x_3 = 170 - 3x_1 - 5x_2$$

and the first inequality in (1.4) implies $x_3 \geqslant 0$. Therefore this inequality can be replaced by the equation

$$3x_1 + 5x_2 + x_3 = 170, \tag{1.5}$$

in which, like x_1 and x_2, the new variable x_3 must be non-negative.

Similarly, the other three constraints for the problem become

$$4x_1 + 2x_2 + x_4 = 200$$
$$4x_1 + 5x_2 - x_5 = 150 \tag{1.6}$$
$$6x_1 + 3x_2 - x_6 = 110$$

with x_4, x_5, x_6 all to be non-negative. In (1.5) and (1.6) there are now more unknowns (six) than equations (four). This is a typical feature of LP problems, which constitute an important field of application of matrices.

In matrix form (1.5) and (1.6) can be written

$$\begin{bmatrix} 3 & 5 & 1 & 0 & 0 & 0 \\ 4 & 2 & 0 & 1 & 0 & 0 \\ 4 & 5 & 0 & 0 & -1 & 0 \\ 6 & 3 & 0 & 0 & 0 & -1 \end{bmatrix} \begin{bmatrix} x_1 \\ x_2 \\ x_3 \\ x_4 \\ x_5 \\ x_6 \end{bmatrix} = \begin{bmatrix} 170 \\ 200 \\ 150 \\ 110 \end{bmatrix}.$$

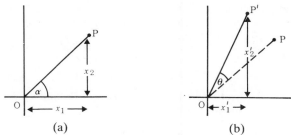

(a)   (b)

Fig. 1.3 (a) Point P in Example 1.6. (b) Point P' in Example 1.6.

Example 1.6 Let P be a point in the plane having cartesian coordinates x_1 and x_2, and let O be the origin, as shown in Fig. 1.3(a). Suppose that OP is rotated in an anticlockwise direction through an angle θ so that P moves to $P' \equiv (x_1', x_2')$ as in Fig. 1.3(b). Since OP = OP'($= r$, say) we have $x_1 = r \cos \alpha$, $x_2 = r \sin \alpha$ and

$$x_1' = r \cos(\theta + \alpha)$$
$$= r \cos \theta \cos \alpha - r \sin \theta \sin \alpha$$
$$= (\cos \theta)x_1 - (\sin \theta)x_2$$

and similarly

$$x_2' = (\sin \theta)x_1 + (\cos \theta)x_2$$

or in matrix terms

$$\begin{bmatrix} x_1' \\ x_2' \end{bmatrix} = \begin{bmatrix} \cos \theta & -\sin \theta \\ \sin \theta & \cos \theta \end{bmatrix} \begin{bmatrix} x_1 \\ x_2 \end{bmatrix}. \tag{1.7}$$

The change of coordinates from x_1, x_2 to x_1', x_2' is an example of a *transformation* (or *mapping*). These are useful in geometry.

Example 1.7 Consider a mechanical system composed of two masses lying on a smooth table, connected to each other and to a fixed support by springs and dampers as shown in Fig. 1.4. The displacements of the masses from equilibrium are x_1 and x_2, and their velocities are x_3 and x_4

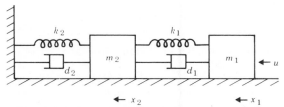

Fig. 1.4 Mass-spring system for Example 1.7.

respectively. Assuming that the springs obey Hooke's law, and the dampers exert forces proportional to velocity, Newton's equations of motion can be written (ignoring u):

$$\left.\begin{array}{l} \text{For } m_1: \ m_1\ddot{x}_3 = -k_1(x_1 - x_2) - d_1(x_3 - x_4) \\ \text{For } m_2: \ m_2\ddot{x}_4 = k_1(x_1 - x_2) + d_1(x_3 - x_4) - k_2 x_2 - d_2 x_4 \end{array}\right\}, \quad (1.8)$$

where (˙) (a dot over a character) denotes $d(\ )/dt$, and k_1, k_2, d_1, d_2 are the spring and damping coefficients respectively. Since

$$\dot{x}_1 = x_3, \qquad \dot{x}_2 = x_4, \qquad (1.9)$$

we can write (1.8) and (1.9) in the combined form

$$\begin{bmatrix} \dot{x}_1 \\ \dot{x}_2 \\ \dot{x}_3 \\ \dot{x}_4 \end{bmatrix} = \begin{bmatrix} 0 & 0 & 1 & 0 \\ 0 & 0 & 0 & 1 \\ -k_1/m_1 & k_1/m_1 & -d_1/m_1 & d_1/m_1 \\ k_1/m_2 & -(k_1+k_2)/m_2 & d_1/m_2 & -(d_1+d_2)/m_2 \end{bmatrix} \begin{bmatrix} x_1 \\ x_2 \\ x_3 \\ x_4 \end{bmatrix},$$

$$(1.10)$$

constituting a set of linear *differential* equations.

If the right-hand mass were given an initial push then the system would perform unforced oscillations. Alternatively, it may be required to determine how a force u should be applied to m_1 so as to make the masses move in a desired fashion—this a problem of *control theory*. In both areas matrices play a very important role, some aspects of which will be dealt with in later chapters.

Example 1.8 Suppose that the female population of a certain species of bird is classified into juveniles and adults, and only the latter lay eggs. A yearly count of birds reveals that there are x_n juvenile birds and y_n adult females in year n, beginning with $n = 0$ when the initial totals are x_0 and y_0. It is assumed that:

(i)   a proportion α of juvenile females in year n survives to become adults in the spring of year $n + 1$;

(ii) each surviving adult female lays eggs in spring to produce an average of k juveniles by the next spring;

(iii) adult birds die, a proportion β surviving from one spring to the next.

In view of (ii), the number of juveniles in year $n + 1$ is

$$x_{n+1} = ky_n, \qquad n = 0, 1, 2, \ldots. \qquad (1.11)$$

The other assumptions mean that the number of adults in year $n + 1$ is

$$y_{n+1} = \alpha x_n + \beta y_n, \qquad n = 0, 1, 2, \ldots. \qquad (1.12)$$

Equations (1.11) and (1.12) are a pair of linear *difference* equations, and can be written in the combined form

$$\begin{bmatrix} x_{n+1} \\ y_{n+1} \end{bmatrix} = \begin{bmatrix} 0 & k \\ \alpha & \beta \end{bmatrix} \begin{bmatrix} x_n \\ y_n \end{bmatrix}. \tag{1.13}$$

Such models are used widely in population studies, and a general definition of the matrices which arise is given in Chapter 13.

Example 1.9 A car rental company has depots in three cities c_1, c_2, c_3. Customers can rent cars in any city and return to any other. On the basis of past experience it is found that 70 per cent of customers renting from c_1 return the car to the same location, whereas 20 per cent return to c_2 and 10 per cent return to c_3. These can be regarded as the probabilities 0.7, 0.2, 0.1 that a customer renting from c_1 will return to c_1, c_2, or c_3 respectively. The complete set of such probabilities can be recorded in the following matrix:

renting from city

$$\begin{array}{ccc} c_1 & c_2 & c_3 \end{array}$$
$$\begin{bmatrix} 0.7 & 0.1 & 0.2 \\ 0.2 & 0.8 & 0.3 \\ 0.1 & 0.1 & 0.5 \end{bmatrix} \begin{array}{l} c_1 \\ c_2 \\ c_3 \end{array} \quad \text{returning to city.}$$

Notice that since it is assumed that all cars are returned, each column of the matrix sums to one. Matrices having this form are called *stochastic*, and will be studied in Chapter 14.

Simple discussions of a wide variety of applications of matrices are given in the books listed at the end of this chapter, and many more texts on applications are included in the Bibliography at the end of the book.

Problems

1.1 The network in Fig. 1.5 represents roads connecting three cities in country D to three cities in country E, and then to two cities in country F. The connections between the cities in countries D and E can be described by the following matrix:

$$\begin{array}{c} \\ d_1 \\ d_2 \\ d_3 \end{array} \begin{array}{ccc} e_1 & e_2 & e_3 \\ \begin{bmatrix} 1 & 1 & 0 \\ 1 & 0 & 1 \\ 1 & 1 & 0 \end{bmatrix} \end{array}$$

where the 1's indicate pairs of cities which are connected, and the zeros pairs which are not. Write down the matrix for road connections between countries E and F.

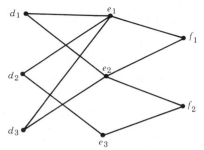

Fig. 1.5 Road network for Problem 1.1.

1.2 Suppose that for the network in Fig. 1.1, onward flights from country B to country C are as shown in Fig. 1.6.

   (a) Write down the matrix giving information on flights between B and C.

   (b) By combining Figs. 1.1 and 1.6, obtain the matrix giving the numbers of different ways of flying from airports in A to those in C.

1.3 Two people, A, B, play a game in which they each toss a coin simultaneously. The rules are as follows:

   if both coins come up heads A pays B five pounds; if both come up tails B pays A seven pounds; if A's coin shows a head and B's coin a tail then B pays A two pounds; and if A's shows a tail and B's a head then A pays B three pounds.

   Represent this information in matrix form. How would the rules have to be modified so as to make this matrix symmetric?

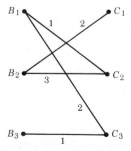

Fig. 1.6 Onward airport connections for Problem 1.2.

Table 1.4

	Proteins	Fats	Carbohydrates
Food 1	2	1	16
Food 2	8	2	0
Food 3	0	28	0
Food 4	2	9	1

1.4 The minimum daily requiremnts of an adult person are 2000 units of proteins, 2500 units of fats, and 8500 units of carbohydrates. The contents of certain foods are given in Table 1.4 in units per gram.

Let x_i, $i = 1, 2, 3, 4$, represent the weights bought of each food. Determine the inequalities that represent the condition that the minimum daily requirements be met or exceeded. Convert these inequalities into linear equations, and hence express the problem in matrix terms.

If in addition the daily requirements are to be met at minimum cost, then this is an LP problem.

1.5 Using the notation of Example 1.6, what do the following transformations represent in geometrical terms?

(a) $\begin{bmatrix} x_1' \\ x_2' \end{bmatrix} = \begin{bmatrix} 1 & 0 \\ 0 & -1 \end{bmatrix} \begin{bmatrix} x_1 \\ x_2 \end{bmatrix}$;

(b) $\begin{bmatrix} x_1' \\ x_2' \end{bmatrix} = \begin{bmatrix} 2 & 0 \\ 0 & 2 \end{bmatrix} \begin{bmatrix} x_1 \\ x_2 \end{bmatrix}$,

(c) $\begin{bmatrix} x_1' \\ x_2' \\ x_3' \end{bmatrix} = \begin{bmatrix} \cos\theta & -\sin\theta & 0 \\ \sin\theta & \cos\theta & 0 \\ 0 & 0 & 1 \end{bmatrix} \begin{bmatrix} x_1 \\ x_2 \\ x_3 \end{bmatrix}$.

1.6 The female population of a certain species of beetle is observed to obey the following rules:

(i) half the newly-born beetles survive into their second year;

(ii) one-third of two-year-old beetles survive into their third year;

(iii) no beetle lives longer than three years;

(iv) offspring are only produced in the third year of life, at an average rate of six per beetle.

Denote the numbers of three-year-old, two-year-old and one-year-old beetles in year n ($n = 0, 1, 2, 3, \ldots$) by x_n, y_n and z_n respectively, and obtain the 3×3 matrix A relating the numbers in year $n + 1$ to those in year n.

How matrices arise

1.7 A taxi-driver operates in two neighbouring cities B and L. On the basis of past experience he knows that 40 per cent of passengers picked up in B have a destination in the same city, and 60 per cent wish to travel to L. Similarly, when in L the percentages travelling within L or to B are 70 per cent and 30 per cent respectively. Write down the matrix showing the probabilities applying to passengers.

References

Bradley and Meek (1986), Fletcher (1972), Rorres and Anton (1984).

2 Basic algebra of matrices

2.1 Definitions

The first step in the development of matrix algebra is to denote an array by a single letter–almost invariably an upper case ('capital') letter is used, for example

$$A = \begin{bmatrix} 1 & 7 & 5 \\ 3 & -1 & 2 \end{bmatrix}. \tag{2.1}$$

In some books bold type, $\mathbf{A}$, is used for matrices. The numbers in the array are called the *elements* of A. The standard notation for elements is to use the same letter but lower case, and with two suffices which describe the position of an element. For example, in eqn (2.1) $a_{11} = 1$, $a_{12} = 7$, $a_{21} = 3$, etc. Generally, a_{ij} denotes the element in *row i* and *column j* of the array, and is called the *i, j* element: thus

$$A = \begin{bmatrix} a_{11} & a_{12} & a_{13} & \cdot & \cdot & \cdot & a_{1n} \\ a_{21} & a_{22} & a_{23} & \cdot & \cdot & \cdot & a_{2n} \\ \vdots & \vdots & \vdots & \vdots & a_{ij} & \vdots & \vdots \\ a_{m1} & a_{m2} & a_{m3} & \cdot & \cdot & \cdot & a_{mn} \end{bmatrix} \quad \leftarrow \text{row } i. \tag{2.2}$$

$$\uparrow$$
$$\text{column } j$$

The matrix A in (2.2) has m rows and n columns, and is said to have *dimensions* $m \times n$, or simply to be an $m \times n$ matrix. A useful shorthand for (2.2) is

$$A = [a_{ij}], \qquad i = 1, 2, \ldots, m; \qquad j = 1, 2, \ldots, n. \tag{2.3}$$

When $m = n$ the matrix is called *square* of *order n*; otherwise A is called *rectangular*. In this book the elements will be mainly real or complex numbers. When all the a_{ij} are real then A is called a *real* matrix; if some or all of the a_{ij} are complex then A is a *complex* matrix. For a square matrix, the diagonal from the top left corner to the bottom right corner is called the *principal* diagonal.

Two matrices A and B are *equal* if they have the same dimensions and all their elements in corresponding positions are identical, i.e. $a_{ij} = b_{ij}$, for all possible i and j.

One of the crucial advantages of matrix notation is that very often properties and theorems can be obtained by direct manipulation of matrices denoted by single letters, without involving use of the actual elements in the arrays.

Exercise 2.1   Consider the matrix

$$B = \begin{bmatrix} 1 & 3 \\ 7 & -1 \\ 5 & 4 \\ 2 & -6 \end{bmatrix}. \tag{2.4}$$

What are: (a) the dimensions of B; (b) the elements b_{12}, b_{31}, b_{13}?

Exercise 2.2   Write down the matrix A in (2.3) when (a) $m = 2$, $n = 3$, $a_{ij} = 2i - j$; (b) $m = 3$, $n = 3$, $a_{ij} = |i - j|$.

2.2   Basic operations

2.2.1   Addition

Two matrices A and B can be added together only if they have the same dimensions, and then the elements in corresponding positions are added, i.e. if $A + B = C$ then

$$c_{ij} = a_{ij} + b_{ij}. \tag{2.5}$$

Clearly the order of addition does not matter, i.e. $A + B = B + A$, so addition is *commutative*.

2.2.2   Multiplication by a scalar

If k is a constant then the product kA is formed by multiplying every element of A by k. In our shorthand notation this can be written

$$k[a_{ij}] = [ka_{ij}]. \tag{2.6}$$

The two rules (2.5) and (2.6) can be combined together. For

example, taking $k = -1$ in (2.6), the rule for subtraction is that the i, j element of $A - B = A + (-1)B$ is

$$a_{ij} + (-1)b_{ij} = a_{ij} - b_{ij}, \tag{2.7}$$

that is, corresponding elements are subtracted.

Example 2.1 If A is the matrix in (2.1) and

$$B = \begin{bmatrix} 2 & 4 & 2 \\ 0 & 3 & -1 \end{bmatrix}, \tag{2.8}$$

then

$$A + B = \begin{bmatrix} (1+2) & (7+4) & (5+2) \\ (3+0) & (-1+3) & (2-1) \end{bmatrix}$$

$$= \begin{bmatrix} 3 & 11 & 7 \\ 3 & 2 & 1 \end{bmatrix}$$

$$2A = \begin{bmatrix} 2 & 14 & 10 \\ 6 & -2 & 4 \end{bmatrix}$$

$$3B = \begin{bmatrix} 6 & 12 & 6 \\ 0 & 9 & -3 \end{bmatrix}$$

$$2A - 3B = \begin{bmatrix} (2-6) & (14-12) & (10-6) \\ (6-0) & (-2-9) & (4+3) \end{bmatrix}$$

$$= \begin{bmatrix} -4 & 2 & 4 \\ 6 & -11 & 7 \end{bmatrix}.$$

After a little practice intermediate steps can be missed out.

Exercise 2.3 If

$$A = \begin{bmatrix} 1 & -1 \\ 0 & 4 \\ 4 & 2 \\ 3 & -8 \end{bmatrix} \tag{2.9}$$

and B is the matrix in (2.4), determine (a) $A + B$, (b) $A - B$, (c) $2B$, (d) $2B - 4A$.

It is obvious that for A in (2.1)

$$A - A = \begin{bmatrix} 0 & 0 & 0 \\ 0 & 0 & 0 \end{bmatrix} \tag{2.10}$$

and this will clearly be true for any matrix. A matrix like that in

(2.10) having all its elements zero is called, naturally enough, a *zero matrix* (sometimes, a *null matrix*).

Notice also that (2.6) can be applied in reverse so as to remove a factor common to all the elements of a matrix. For example,

$$\begin{bmatrix} 3 & 9 \\ -6 & 12 \end{bmatrix} = 3 \begin{bmatrix} 1 & 3 \\ -2 & 4 \end{bmatrix}.$$

The rules presented so far can be extended to more than two matrices in an obvious fashion. For example, if A, B, C are three matrices having the same dimensions then their sum is

$$A + B + C = A + (B + C)$$
$$= (A + B) + C, \text{ etc.}$$

This states that the result of adding A, B, and C is independent of the order in which it is done. In formal terms, matrix addition is thus *associative*. Similarly, $2A + 3B + 5C = (2A + 3B) + 5C$, etc.

Exercise 2.4 If A, B, C are three matrices having the same dimensions, and

$$A + C = B + C$$

show by considering elements on both sides of the equation that $A = B$.

Example 2.2 Consider an equation

$$X + A = B, \tag{2.11}$$

which is to be solved for the matrix X, given A and B. Add $-A$ on to both sides of (2.11) to give

$$X + A + (-A) = B + (-A)$$
$$X + (A - A) = B - A$$
$$X + 0 = B - A,$$

where 0 denotes the zero matrix, so

$$X = B - A \tag{2.12}$$

since $X + 0 = X$. Thus, although A, B, X stand for arrays of numbers, we have been able to manipulate them as single entities. In this case the solution (2.12) of (2.11) is the same as would be obtained if A, B, X were scalars. In the next section we begin to break away from ordinary algebra.

2.2.3 Multiplication of two matrices

Example 2.3 Return to the supermarket problem described in Example 1.1. Let x_i be the number of cans purchased of food F_i, $i = 1, 2, 3, 4$. Then from Table 1.1 the total cost at supermarket S_1 would be $17x_1 + 7x_2 + 11x_3 + 21x_4$, and similarly for S_2 and S_3. We can write these products in the following form, as in Example 1.4:

$$\begin{bmatrix} 17 & 7 & 11 & 21 \\ 15 & 9 & 13 & 19 \\ 18 & 8 & 15 & 19 \end{bmatrix} \begin{bmatrix} x_1 \\ x_2 \\ x_3 \\ x_4 \end{bmatrix} = \begin{bmatrix} 17x_1 + 7x_2 + 11x_3 + 21x_4 \\ 15x_1 + 9x_2 + 13x_3 + 19x_4 \\ 18x_1 + 8x_2 + 15x_3 + 19x_4 \end{bmatrix}. \quad (2.13)$$
$$\quad\quad A \quad\quad\quad\quad X$$

The elements of the matrix on the right-hand side of (2.13) give the costs of the purchases at S_1, S_2, and S_3 respectively. It therefore seems natural to define the *product* of A and X as in (2.13): each of the rows of A is multiplied term-by-term with the elements of X. For this to work it is clear that the number of *columns* of A must be the same as the number of *rows* in X.

Exercise 2.5 Three types of food f_1, f_2, f_3 have vitamin content in units per kilogram given in Table 2.1. Express the vitamin content of 5 kg of f_1, 3 kg of f_2, and 7 kg of f_3 as a matrix product, and evaluate it. If the costs per kilogram of the three foods are 75, 90 and 80 units respectively, express the total cost as a matrix product, and evaluate it.

If the second matrix in the product has more than one column we simply multiply each of its columns in turn by A, using the same rule. For example, if

$$\begin{bmatrix} a_{11} & a_{12} \\ a_{21} & a_{22} \end{bmatrix} \begin{bmatrix} b_{11} & b_{12} \\ b_{21} & b_{22} \end{bmatrix} = \begin{bmatrix} (a_{11}b_{11} + a_{12}b_{21}) & (a_{11}b_{12} + a_{12}b_{22}) \\ (a_{21}b_{11} + a_{22}b_{21}) & (a_{21}b_{12} + a_{22}b_{22}) \end{bmatrix},$$
$$\quad A \quad\quad\quad B \quad\quad\quad\quad\quad\quad C \quad\quad\quad\quad\quad (2.14)$$

then we write $AB = C$. The *first* column of C in (2.14) is obtained

Table 2.1

	f_1	f_2	f_3
Vit. A	3	2	4
Vit. B	5	7	9

by multiplying the *first* column of B by A, as in (2.13); the *second* column of C is obtained by multiplying the *second* column of B by A.

Example 2.4   If

$$A = \begin{bmatrix} 1 & 2 \\ 3 & 4 \end{bmatrix}, \qquad B = \begin{bmatrix} -1 & 4 \\ 2 & -3 \end{bmatrix} \tag{2.15}$$

then

$$AB = \begin{bmatrix} \{1 \times (-1) + 2 \times 2\} & \{1 \times 4 + 2 \times (-3)\} \\ \{3 \times (-1) + 4 \times 2\} & \{3 \times 4 + 4 \times (-3)\} \end{bmatrix}$$

$$= \begin{bmatrix} 3 & -2 \\ 5 & 0 \end{bmatrix}. \tag{2.16}$$

The general rule is as follows:

If $A = [a_{ij}]$ is $m \times n$ and $B = [b_{ij}]$ is $n \times p$ then $C = AB$ is $m \times p$ and

$$c_{ij} = \text{term-by-term product of the } i\text{th row of } A$$
$$\text{with the } j\text{th column of } B$$

$$= [a_{i1}a_{i2} \cdots a_{in}] \begin{bmatrix} b_{1j} \\ b_{2j} \\ \vdots \\ b_{nj} \end{bmatrix} \tag{2.17}$$

$$= a_{i1}b_{1j} + a_{i2}b_{2j} + \cdots + a_{in}b_{nj} \tag{2.18}$$

$$= \sum_{k=1}^{n} a_{ik}b_{kj}. \tag{2.19}$$

Thus AB is constructed by multiplying the first column of B by each of the rows of A in turn, giving the first column of AB; this is repeated with the second column of B, and so on. Notice again that the product AB is defined only if the number of *columns* of A is equal to the number of *rows* of B, and A and B are then said to be *conformable* for multiplication. The dimensions of the resulting product can be found by the simple rule:

$$\begin{array}{ccc} A & B & = & C \\ (m \times n) & (n \times p) & & (m \times p) \end{array} \tag{2.20}$$

Example 2.4 (continued) For A and B in (2.15) we have

$$BA = \begin{bmatrix} \{(-1) \times 1 + 4 \times 3\} & \{(-1) \times 2 + 4 \times 4\} \\ \{2 \times 1 + (-3) \times 3\} & \{2 \times 2 + (-3) \times 4\} \end{bmatrix}$$

$$= \begin{bmatrix} 11 & 14 \\ -7 & -8 \end{bmatrix},$$

which has no elements in common with AB in (2.16). This illustrates the general fact that $AB \neq BA$, i.e. matrix multiplication is *not* commutative—a fundamental departure from ordinary algebra.

In AB the matrix B is said to be multiplied *on the left*, or *premultiplied* by A; in BA the matrix B is multiplied *on the right*, or *postmultiplied* by A. In general we shall have to specify which is meant—the expression 'the product of A and B' is too vague.

Example 2.5 If A is the matrix in (2.1) and

$$B = \begin{bmatrix} -1 & 6 & 7 \\ 4 & 5 & 3 \\ 3 & 0 & 4 \end{bmatrix},$$

then using the rule for multiplication

$$AB = \begin{bmatrix} \{1 \times (-1) + 7 \times 4 + 5 \times 3\} & (1 \times 6 + 7 \times 5 + 5 \times 0) \\ \{3 \times (-1) + (-1) \times 4 + 2 \times 3\} & \{3 \times 6 + (-1) \times 5 + 2 \times 0\} \end{bmatrix}$$

$$\begin{matrix} (1 \times 7 + 7 \times 3 + 5 \times 4) \\ \{3 \times 7 + (-1) \times 3 + 2 \times 4\} \end{matrix}\bigg]$$

$$= \begin{bmatrix} 42 & 41 & 48 \\ -1 & 13 & 26 \end{bmatrix}$$

and the dimensions of AB can be obtained from (2.20): $(2 \times 3)(3 \times 3)$. Notice that BA does not exist: consider $(3 \times \underline{3})(\underline{2} \times 3)$—the two underlined dimensions are not equal.

Exercise 2.6 Determine where possible $A + B$, AB, and BA for each of the following pairs:

(a) $\qquad A = \begin{bmatrix} 0 & 1 \\ 1 & 1 \end{bmatrix}, \qquad B = \begin{bmatrix} 0 & -1 \\ 1 & 0 \end{bmatrix};$

(b) $\qquad A = \begin{bmatrix} 1 & 2 & 3 \\ 1 & 3 & 6 \end{bmatrix}, \qquad B = \begin{bmatrix} 1 & 1 & 1 \\ 1 & 2 & 3 \end{bmatrix};$

(c) $\qquad A = \begin{bmatrix} 2 & -1 \\ 1 & 0 \\ -3 & 4 \end{bmatrix}, \qquad B = \begin{bmatrix} 1 & -2 & -5 \\ 3 & 4 & 0 \end{bmatrix}.$

Exercise 2.7 If A has m rows and $m + 5$ columns, B has n rows and $11 - n$ columns, and both AB and BA exist, what are the values of m and n?

Example 2.6 An important special matrix can now be introduced. Let

$$I = \begin{bmatrix} 1 & 0 \\ 0 & 1 \end{bmatrix} \tag{2.21}$$

and let A be the arbitrary 2×2 matrix in (2.14). Then with $B = I$ in (2.14) it is easy to verify that

$$AI = A \tag{2.22}$$

and similarly

$$IA = A. \tag{2.23}$$

Since I behaves like the number 1 in ordinary algebra it is called the *unit matrix* (or *identity* matrix), being defined in general as the $n \times n$ matrix having 1's along its principal diagonal and zeros everywhere else. The results in eqns (2.22) and (2.23) then hold for any $n \times n$ matrix A. We shall often write I_n to emphasize that the order is n.

Example 2.7 It can turn out that AB does equal BA, as is easily verified when

$$A = \begin{bmatrix} 1 & 2 \\ -2 & 1 \end{bmatrix}, \qquad B = \begin{bmatrix} 3 & 4 \\ -4 & 3 \end{bmatrix}.$$

In this case A and B are said to *commute* with each other. In view of eqns (2.22) and (2.23), I_n is a matrix which commutes with *all* $n \times n$ matrices.

Powers of a matrix are defined in an obvious fashion:

$$A^2 = AA, \qquad A^3 = AA^2, \qquad A^4 = AA^3, \ldots . \tag{2.24}$$

It is important to realize that the dimension requirement in (2.20) implies that (2.24) is meaningful only if A is square. Since

$$A^3 = AAA = A^2 A$$

and so on, it follows that all powers of A commute with each other, and $A^k A^l = A^{k+l}$ for positive integers k and l.

Exercise 2.8 If

$$A = \begin{bmatrix} 1 & -1 & 1 \\ 2 & -1 & 0 \\ 1 & 0 & 0 \end{bmatrix},$$

calculate A^2 and verify that $A^2 A = AA^2 = I_3$.

Exercise 2.9 If

$$A = \begin{bmatrix} 3 & -1 & -1 \\ 1 & 1 & -1 \\ 1 & -1 & 1 \end{bmatrix} \qquad (2.25)$$

calculate A^2 and verify that $A^2 - 3A + 2I_3 \equiv 0$.

Notice that it easily follows from the definition that matrix multiplication is *associative*, i.e.

$$A(BC) = (AB)C, \qquad (2.26)$$

so we can simply write ABC without brackets. The dimensions of this product are shown below:

$$\begin{array}{cccc} A & B & C & = & D \\ (m \times n) & (n \times p) & (p \times q) & (m \times q) \end{array}.$$

It is also easy to show that mutiplication is *distributive* with respect to addition, i.e.

$$A(B + C) = AB + AC. \qquad (2.27)$$

The results (2.26) and (2.27) mean that we can deal with brackets in the same way as for ordinary algebra.

Exercise 2.10 If A is $a_1 \times a_2$, B is $b_1 \times b_2, \ldots, Z$ is $z_1 \times z_2$, what are the conditions for the product $ABC \ldots YZ$ to exist, and what are its dimensions?

Exercise 2.11 Test (2.26) with A, B as in Exercise 2.6(c) and

$$C = \begin{bmatrix} 2 & 1 \\ 0 & -4 \\ 1 & 2 \end{bmatrix}.$$

We now consider some further ways in which matrix multipication differs from multiplication of scalars.

Exercise 2.12 If

$$A = \begin{bmatrix} 2 & -3 & -5 \\ -1 & 4 & 5 \\ 1 & -3 & -4 \end{bmatrix}, \qquad B = \begin{bmatrix} -1 & 3 & 5 \\ 1 & -3 & -5 \\ -1 & 3 & 5 \end{bmatrix},$$

$$C = \begin{bmatrix} 2 & -2 & -4 \\ -1 & 3 & 4 \\ 1 & -2 & -3 \end{bmatrix},$$

show that (a) $AB = BA = 0$; (b) $AC = A$, $CA = C$, and hence show that $ACB = CBA$.

Part (a) of Exercise 2.12 illustrates the fact that a product AB can equal the zero matrix even if neither A nor B is itself zero. This is in sharp contrast to ordinary algebra, where the result that a product can only be zero if one of the factors is zero is applied almost instinctively. For example, to solve a quadratic equation like

$$x^2 - 3x + 2 = 0 \qquad (2.28)$$

we factorize as

$$(x - 1)(x - 2) = 0 \qquad (2.29)$$

and then conclude that one of the factors in (2.29) must be zero, so the solution of (2.28) is $x = 1$ or $x = 2$. However, in attempting to solve a corresponding matrix equation

$$X^2 - 3X + 2I = 0, \qquad (2.30)$$

where X is an $n \times n$ matrix, then although this factorizes into

$$(X - I)(X - 2I) = 0$$

(verify this!) we cannot assume that the only solutions of (2.30) are $X = I$ or $X = 2I$. Indeed, when X is a 3×3 matrix then a solution of (2.30) is the matrix in (2.25).

Another aspect of the preceding remarks is that if $AC = AD$, it cannot necessarily be inferred that $C = D$ (i.e. the A cannot be 'cancelled') for it is possible to have

$$AC - AD = A(C - D) = 0$$

even though $C - D$ may not equal zero. This is illustrated by part (b) of Exercise 2.12, where $AC = A = AI_3$ but $C \neq I_3$.

Exercise 2.13 If A and B are two $n \times n$ matrices, expand the product $(A - B)(A + B)$. Under what conditions is this equal to $A^2 - B^2$?

Exercise 2.14 Expand $(A + B)^2$, $(A + B)^3$ for two arbitrary $n \times n$ matrices A and B. What is the condition for the usual binomial theorem expressions to be obtained?

Exercise 2.15 Give an example of a 2×3 matrix A and a 3×2 matrix B for which $AB = I_2$, but $BA \neq I_3$.

We close this section with some more applications of matrix multiplication, emphasizing further that the definition is not a mathematical abstraction.

Example 2.8 Consider the network in Fig. 1.5, showing roads connecting three sets of cities. The matrices representing connections between countries D and E, and E and F are respectively

$$
A = \begin{array}{c} \\ \end{array}
\begin{matrix} e_1 & e_2 & e_3 \\ \begin{bmatrix} 1 & 1 & 0 \\ 1 & 0 & 1 \\ 1 & 1 & 0 \end{bmatrix} & \begin{matrix} d_1 \\ d_2, \\ d_3 \end{matrix} \end{matrix}
\qquad
B = \begin{matrix} f_1 & f_2 \\ \begin{bmatrix} 1 & 0 \\ 1 & 1 \\ 0 & 1 \end{bmatrix} & \begin{matrix} e_1 \\ e_2, \\ e_3 \end{matrix} \end{matrix}
\qquad (2.31)
$$

where $a_{ij} = 1$ if d_i is connected to e_j, and $a_{ij} = 0$ otherwise, and similarly for matrix B. Consulting Fig. 1.5, it is easily seen that the number of routes connecting d_1 and f_1 is 2 (via e_1 or via e_2). Proceeding in this way, the matrix giving numbers of routes between D and F is found to be

$$
\begin{matrix} f_1 & f_2 \\ \begin{bmatrix} 2 & 1 \\ 1 & 1 \\ 2 & 1 \end{bmatrix} & \begin{matrix} d_1 \\ d_2 \\ d_3 \end{matrix} \end{matrix}
\qquad (2.32)
$$

The reader can verify that the product AB of the matrices in (2.31) gives precisely the matrix in (2.32).

Example 2.9 If $f(x, y)$ is a function of two independent variables and a transformation into $f(u, v)$ is made, then the theory of partial differentiation gives the derivatives of f with respect to the new variables in the form

$$
\frac{\partial f}{\partial u} = \frac{\partial f}{\partial x} \frac{\partial x}{\partial u} + \frac{\partial f}{\partial y} \frac{\partial y}{\partial u}
$$

$$
\frac{\partial f}{\partial v} = \frac{\partial f}{\partial x} \frac{\partial x}{\partial v} + \frac{\partial f}{\partial y} \frac{\partial y}{\partial v}.
$$

In matrix terms this becomes

$$
\begin{bmatrix} f_u \\ f_v \end{bmatrix} = \begin{bmatrix} x_u & y_u \\ x_v & y_v \end{bmatrix} \begin{bmatrix} f_x \\ f_y \end{bmatrix},
$$

where x_u denotes $\partial x / \partial u$, etc.

Example 2.10 Let $z_1 = a_1 + ib_1$ be an arbitrary complex number with a_1 and b_1 real and $i^2 = -1$. Form the matrix

$$
A^{(1)} = \begin{bmatrix} a_1 & b_1 \\ -b_1 & a_1 \end{bmatrix}
\qquad (2.33)
$$

and denote the relationship by $A^{(1)} \sim z_1$. It turns out that properties of z_1 can be interpreted in terms of properties of $A^{(1)}$. To see this, let z_2 be a second complex number with $A^{(2)} \sim z_2$. Then

$$A^{(1)} + A^{(2)} = \begin{bmatrix} (a_1 + a_2) & (b_1 + b_2) \\ -(b_1 + b_2) & (a_1 + a_2) \end{bmatrix}$$

$$\sim (a_1 + a_2) + \mathrm{i}(b_1 + b_2) \tag{2.34}$$

and the number in (2.34) is just $z_1 + z_2$, which means that addition of complex numbers can be done by adding matrices in the form of that in (2.33).

This also applies for multiplication:

$$A^{(1)}A^{(2)} = \begin{bmatrix} a_1 & b_1 \\ -b_1 & a_1 \end{bmatrix}\begin{bmatrix} a_2 & b_2 \\ -b_2 & a_2 \end{bmatrix}$$

$$= \begin{bmatrix} (a_1 a_2 - b_1 b_2) & (a_1 b_2 + b_1 a_2) \\ -(a_1 b_2 + b_1 a_2) & (a_1 a_2 - b_1 b_2) \end{bmatrix}$$

$$\sim (a_1 a_2 - b_1 b_2) + \mathrm{i}(a_1 b_2 + b_1 a_2) \tag{2.35}$$

and the complex number in (2.35) is precisely $z_1 z_2$.

Further aspects of this representation of complex numbers via 2×2 matrices in the form of (2.33) will be developed in Problem 2.9, Exercises 4.1 and 6.3, and in Section 8.1.

Exercise 2.16 A square $n \times n$ matrix $D = [d_{ij}]$ is called *diagonal* if all elements off the principal diagonal are zero, i.e. $d_{ij} = 0$, $i \neq j$. This is written

$$D = \mathrm{diag}[d_{11}, d_{22}, \ldots, d_{nn}]. \tag{2.36}$$

Prove that

$$D^2 = \mathrm{diag}[d_{11}^2, d_{22}^2, \ldots, d_{nn}^2]$$

and obtain the expression for D^k, where k is a positive integer. Show also that any two $n \times n$ diagonal matrices commute with each other.

2.3 Transpose

2.3.1 Definition and properties

If A is the general $m \times n$ matrix in (2.2) then the $n \times m$ matrix obtained from A by interchanging the rows and columns is called the *transpose* of A, written A^{T} (in some books A').

Example 2.11 If A is the matrix in (2.9) then

$$A^T = \begin{bmatrix} 1 & 0 & 4 & 3 \\ -1 & 4 & 2 & -8 \end{bmatrix} \tag{2.37}$$

since the first row in (2.9) becomes the first column in (2.37), the second row becomes the second column, and so on. (Similarly, the first row of (2.37) is the first column of (2.9), etc.)

In general, the i, j element a_{ij} of A becomes the j, i element of A^T. For example, the 2, 3 element of (2.37) is the 3, 2 element of (2.9). In particular when A is square, the elements on the principal diagonal of A^T are a_{ii}, the same as those on the principal diagonal of A.

It is obvious that transposing twice in succession returns any matrix to itself, i.e.

$$(A^T)^T = A. \tag{2.38}$$

Similarly it follows immediately from the definition that

$$(A + B)^T = A^T + B^T \tag{2.39}$$

$$(kA)^T = kA^T, \text{ for any scalar } k. \tag{2.40}$$

It is not quite so easy to prove that, if A and B are conformable for multiplication, then

$$(AB)^T = B^T A^T. \tag{2.41}$$

Note the reversal of order between the two sides of (2.41).

Exercise 2.17 Determine $B^T A^T$ for the matrices in (2.15), and hence test the validity of (2.41).

To see how (2.41) is proved in general, suppose A and B are both 3×3 and let $C = AB$. The i, j element of C^T is c_{ji}, which from (2.18) with i and j interchanged is

$$c_{ji} = a_{j1}b_{1i} + a_{j2}b_{2i} + a_{j3}b_{3i}. \tag{2.42}$$

The i, j element of $B^T A^T$ is the term-by-term product of the ith row of B^T with the jth column of A^T. However, the ith row of B^T is equal to the ith *column* of B and the jth *column* of A^T is equal to the jth *row* of A. Thus the i, j element of $B^T A^T$ is

$$b_{1i}a_{j1} + b_{2i}a_{j2} + b_{3i}a_{j3},$$

which is identical to (2.42), showing $C^T = B^T A^T$ as required. The argument is easily extended to general dimensions.

If

$$x = \begin{bmatrix} x_1 \\ x_2 \\ \vdots \\ x_n \end{bmatrix}, \qquad y = \begin{bmatrix} y_1 \\ y_2 \\ \vdots \\ y_n \end{bmatrix}, \qquad (2.43)$$

these $n \times 1$ matrices are called *column n-vectors*, and x^T, y^T are *row n-vectors*. The elements x_i, y_i are called *components* of the vectors. Again, bold type, **x**, **y**, is often used for vectors. The product

$$x^\mathsf{T} y = x_1 y_1 + x_2 y_2 + \cdots + x_n y_n \qquad (2.44)$$

is the *scalar* (or *inner*) product of x and y, and the notation $\langle x, y \rangle$ is also used.

Exercise 2.18   Prove, using (2.41), that $(ABC)^\mathsf{T} = C^\mathsf{T} B^\mathsf{T} A^\mathsf{T}$.

Exercise 2.19   Let e_i denote the ith row of I_n. For example, if $n = 3$ then $e_2 = [0, 1, 0]$. If A is an arbitrary $n \times n$ matrix, what do the products $e_i A$ and $A e_i^\mathsf{T}$ represent?

If some or all the a_{ij} are complex numbers, then the *complex conjugate* $\bar{A}$ of A is the matrix obtained by replacing each a_{ij} by its conjugate, $\bar{a}_{ij}$. The *conjugate transpose* of A is

$$A^* = (\bar{A})^\mathsf{T} \qquad (2.45)$$

and the order of the operations in (2.45) doesn't matter, i.e.

$$A^* = (\overline{A^\mathsf{T}}).$$

The results of (2.38), (2.39), and (2.41) still hold for conjugate transpose with superscript T replaced by *. Similarly, (2.40) holds if k is real, but if k is complex we have

$$(kA)^* = \bar{k} A^*. \qquad (2.46)$$

Example 2.12

$$A = \begin{bmatrix} 2 + 3i & 1 + i \\ 2 - i & 4i \end{bmatrix}, \qquad A^* = \begin{bmatrix} 2 - 3i & 2 + i \\ 1 - i & -4i \end{bmatrix}. \qquad (2.47)$$

Notice that any complex matrix can be written as

$$A = A_1 + iA_2, \qquad (2.48)$$

where the elements of A_1 and A_2 are purely real. For example, for (2.47)

$$A = \begin{bmatrix} 2 & 1 \\ 2 & 0 \end{bmatrix} + i \begin{bmatrix} 3 & 1 \\ -1 & 4 \end{bmatrix}.$$

2.3.2 Symmetric and hermitian matrices

We encountered the idea of a *symmetric* square matrix in Example 1.2. This can now be defined by

$$A^{\mathsf{T}} = A, \tag{2.49}$$

which implies that $a_{ij} = a_{ji}$, for all i and j. Similarly, A is *skew symmetric* if

$$A^{\mathsf{T}} = -A, \tag{2.50}$$

which implies $a_{ij} = -a_{ji}$, for all i and j, so in particular all the diagonal elements a_{ii} are zero.

Example 2.13 The matrices

$$A = \begin{bmatrix} 1 & 3 & 7 \\ 3 & 4 & 2 \\ 7 & 2 & 0 \end{bmatrix}, \qquad B = \begin{bmatrix} 0 & 3 & 7 \\ -3 & 0 & 2 \\ -7 & -2 & 0 \end{bmatrix} \tag{2.51}$$

are respectively symmetric and skew symmetric.

An interesting fact is that any square matrix A can be expressed *uniquely* as

$$A = M + S, \tag{2.52}$$

where M is symmetric and S is skew symmetric. To show this, transpose both sides of (2.52):

$$\begin{aligned} A^{\mathsf{T}} &= (M + S)^{\mathsf{T}} \\ &= M^{\mathsf{T}} + S^{\mathsf{T}} \\ &= M - S. \end{aligned} \tag{2.53}$$

Adding (2.52) and (2.53) gives the *symmetric part* of A

$$M = \frac{1}{2}(A + A^{\mathsf{T}}) \tag{2.54}$$

and similarly by subtraction the *skew symmetric part* is

$$S = \frac{1}{2}(A - A^{\mathsf{T}}). \tag{2.55}$$

Exercise 2.20   Determine M and S for the matrix A in (2.25).

Exercise 2.21   Verify that the matrices of (2.54) and (2.55) are indeed symmetric and skew symmetric.

Exercise 2.22   If A is any symmetric $n \times n$ matrix and P is an arbitrary $m \times n$ matrix, prove that PAP^{T} is symmetric.

Exercise 2.23   Prove that the maximum number of different elements in an $n \times n$ symmetric matrix is $\frac{1}{2}n(n + 1)$. What is the maximum number for a skew symmetric matrix (ignoring signs and zeros)?

When A has complex elements, two further important definitions are used: If

$$A^* = A, \tag{2.56}$$

then A is called *hermitian* (after the French mathematician Hermite), and (2.56) implies $a_{ij} = \bar{a}_{ji}$; and if

$$A^* = -A, \tag{2.57}$$

then A is called *skew hermitian*, with $a_{ij} = -\bar{a}_{ji}$.

Example 2.14   The matrices

$$
\begin{bmatrix}
2 & 1+i & 5-i \\
1-i & 7 & i \\
5+i & -i & -1
\end{bmatrix},
\qquad
\begin{bmatrix}
2i & 1+i & 5-i \\
-1+i & 7i & i \\
-5-i & i & -i
\end{bmatrix}
$$

are respectively hermitian and skew hermitian.

This example illustrates the general fact that for a hermitian matrix $a_{ii} = \bar{a}_{ii}$, so all a_{ii} are purely real; and similarly, all a_{ii} are purely imaginary for a skew hermitian matrix.

Important applications of the matrices introduced in this section will be studied in Chapter 7.

Exercise 2.24   If A is any rectangular matrix prove that A^*A and AA^* are both hermitian.

Exercise 2.25   Consider the case when A in (2.48) is hermitian. (a) Show that A_1 is symmetric and A_2 is skew symmetric. (b) Determine the condition to be satisfied by A_1 and A_2 for A^*A to be a real matrix.

Exercise 2.26   If A is any skew hermitian matrix prove that iA and $-iA$ are both hermitian.

Exercise 2.27   Obtain the generalization of eqns (2.52), (2.54), and (2.55) when A is complex.

2.4 Partitioning and submatrices

It is often convenient to subdivide or *partition* a matrix into smaller blocks of elements.

Example 2.15 The following 3×5 matrix is partitioned into four blocks

$$A = \begin{bmatrix} 1 & 0 & 2 & \vdots & 3 & 5 \\ 2 & 1 & 4 & \vdots & 3 & 0 \\ \hdashline 5 & 7 & 1 & \vdots & 1 & 4 \end{bmatrix} = \begin{bmatrix} \overset{3}{B} & \overset{2}{C} \\ D & E \end{bmatrix}\overset{2}{\underset{1}{}}, \tag{2.58}$$

where B, C, D, E are the arrays indicated by the dashed lines. The dimensions can be marked as shown in (2.58).

More generally, if some rows and columns of any matrix A are deleted then the resulting matrix is called a *submatrix* of A. For example, B in (2.58) is the submatrix obtained by deleting row 3 and columns 4 and 5 of A. It is a convention that A can be regarded as a submatrix of itself.

Example 2.15 (continued) Deleting row 2 and columns 2, 4, 5 of the matrix A in (2.58) gives the following 2×2 submatrix

$$\begin{bmatrix} 1 & 2 \\ 5 & 1 \end{bmatrix}.$$

When A is square, an important special type of submatrix is obtained by building up square arrays, starting in the top left corner, and finishing with A itself. These are called the *leading principal submatrices* of A.

Example 2.16 The leading principal submatrices of A in (2.51) are

$$[1], \qquad \begin{bmatrix} 1 & 3 \\ 3 & 4 \end{bmatrix}, \qquad A,$$

which are built up as indicated below:

$$\begin{array}{ccc} 1 & 3 & 7 \\ 3 & 4 & 2 \\ 7 & 2 & 0 \end{array} \tag{2.59}$$

It is clear from this example that the principal diagonal of each leading principal submatrix is part (or all) of the principal diagonal of A. More generally, any square submatrix of A whose principal diagonal satisfies this property is called simply a *principal submatrix* of A. Thus for A in (2.59), principal submatrices are

$$\begin{bmatrix} 4 & 2 \\ 2 & 0 \end{bmatrix}, \qquad \begin{bmatrix} 1 & 7 \\ 7 & 0 \end{bmatrix}. \tag{2.60}$$

Exercise 2.28 What rows and columns of A in (2.59) are deleted to give (2.60)?

Partitioning is useful when applied to large matrices since manipulations can be carried out on the smaller blocks. For example, if A_1 is a 3×5 matrix partitioned into blocks B_1, C_1, D_1, E_1 in the same way as in (2.58) then

$$A + A_1 = \begin{bmatrix} (B + B_1) & (C + C_1) \\ (D + D_1) & (E + E_1) \end{bmatrix}.$$

More importantly, when multiplying matrices in partitioned form the basic rule can be applied to the blocks as though they were single elements. For example, if

$$X = \begin{bmatrix} X_1 \\ X_2 \end{bmatrix}_2^3, \tag{2.61}$$

then with A in (2.58) we obtain

$$AX = \begin{bmatrix} B & C \\ D & E \end{bmatrix} \begin{bmatrix} X_1 \\ X_2 \end{bmatrix}$$
$$= \begin{bmatrix} BX_1 + CX_2 \\ DX_1 + EX_2 \end{bmatrix}. \tag{2.62}$$

The only restriction is that the blocks must be conformable for multiplication, so all the products BX_1, CX_2, etc. in (2.62) exist. This requires that in a product AX the number of *columns* in each block of A must equal the number of *rows* in the corresponding block of X (as illustrated in (2.58) and (2.61)).

Exercise 2.29 Complete the partitioning for each of the following products so that each matrix is divided into four submatrices which are conformable for multiplication.

(a) $\begin{bmatrix} \times & \times & \times & \times \\ \times & \times & \times & \times \\ \times & \times & \times & \times \\ \times & \times & \times & \times \end{bmatrix} \begin{bmatrix} \times & \times & \times & \times \\ \times & \times & \times & \times \\ \times & \times & \times & \times \\ \times & \times & \times & \times \end{bmatrix} = \begin{bmatrix} \times & \times & \times & \times \\ \times & \times & \times & \times \\ \times & \times & \times & \times \\ \times & \times & \times & \times \end{bmatrix}$

(b) $\begin{bmatrix} \times & \times & \times & \times \\ \times & \times & \times & \times \\ \times & \times & \times & \times \\ \times & \times & \times & \times \end{bmatrix} \begin{bmatrix} \times & \times & \times & \times \\ \times & \times & \times & \times \\ \times & \times & \times & \times \\ \times & \times & \times & \times \end{bmatrix} = \begin{bmatrix} \times & \times & \times & \times \\ \times & \times & \times & \times \\ \times & \times & \times & \times \\ \times & \times & \times & \times \end{bmatrix}$

Exercise 2.30 Choose arbitrary numbers for the elements of X_1 and X_2 in (2.61) and hence evaluate (2.62). Also evaluate AX directly without partitioning.

The rule for transposing a partitioned matrix is best explained by applying it to (2.58). We obtain

$$A^{\mathsf{T}} = \begin{bmatrix} \overset{2}{B^{\mathsf{T}}} & \overset{1}{D^{\mathsf{T}}} \\ C^{\mathsf{T}} & E^{\mathsf{T}} \end{bmatrix} \begin{matrix} 3 \\ 2 \end{matrix}$$

showing that the rows and columns of blocks are interchanged, and in addition each submatrix is itself transposed.

If A is square and its only nonzero elements can be partitioned as principal submatrices, then it is called *block diagonal*.

Example 2.17 The matrix

$$A = \begin{bmatrix} 1 & 2 & \vdots & 0 \\ 3 & 4 & \vdots & 0 \\ \hdashline 0 & 0 & \vdots & 2 \end{bmatrix} \tag{2.63}$$

is block diagonal. A convenient notation which generalizes (2.36) is to write (2.63) as

$$A = \mathrm{diag}[A_1, A_2], \tag{2.64}$$

where

$$A_1 = \begin{bmatrix} 1 & 2 \\ 3 & 4 \end{bmatrix}, \qquad A_2 = [2].$$

The expression (2.64) is also written as

$$A = A_1 \oplus A_2,$$

where $\oplus$ denotes *direct sum*.

Exercise 2.31 If A_1 and A_2 in (2.64) are square matrices having arbitrary dimensions, obtain A^{T} and A^2 in partitioned form.

2.5 Kronecker and Hadamard products

The reader should appreciate that we adopted the definition of multiplication in Section 2.2.3 because it arose in a natural way, and has properties which are used in very many applications. However, in certain applications other definitions are useful, and in this section we introduce two important cases. The first is the *Kronecker product* (sometimes called the *direct* product). Let A be the $m \times n$ matrix in (2.2) and let B be another arbitrary matrix having dimensions $p \times q$. Then $A \otimes B$ is defined to be the partitioned matrix

$$A \otimes B = \begin{matrix} q & q & \cdots & q \\ \begin{bmatrix} a_{11}B & a_{12}B & \cdots & a_{1n}B \\ \vdots & \vdots & & \vdots \\ a_{m1}B & a_{m2}B & \cdots & a_{mn}B \end{bmatrix} & & & \end{matrix} \begin{matrix} p \\ \vdots \\ p \end{matrix}. \qquad (2.65)$$

Each submatrix in (2.65) has dimensions $p \times q$, so $A \otimes B$ has dimensions $(mp) \times (nq)$.

Example 2.18   Let A and B be the 2×2 matrices in (2.15). Then

$$A \otimes B = \begin{bmatrix} B & 2B \\ 3B & 4B \end{bmatrix} = \begin{bmatrix} -1 & 4 & -2 & 8 \\ 2 & -3 & 4 & -6 \\ -3 & 12 & -4 & 16 \\ 6 & -9 & 8 & -12 \end{bmatrix}, \qquad (2.66)$$

$$B \otimes A = \begin{bmatrix} -A & 4A \\ 2A & -3A \end{bmatrix} = \begin{bmatrix} -1 & -2 & 4 & 8 \\ -3 & -4 & 12 & 16 \\ 2 & 4 & -3 & -6 \\ 6 & 8 & -9 & -12 \end{bmatrix}. \qquad (2.67)$$

Notice that $A \otimes B \neq B \otimes A$, as for 'ordinary' multiplication. However, if in (2.67) the second and third columns are interchanged, and the second and third rows are interchanged, then (2.66) is obtained.

It can be shown (see Section 13.4.5) that the above result holds in general: the elements of $B \otimes A$ are simply a rearrangement of those of $A \otimes B$, so $B \otimes A$ can be transformed into $A \otimes B$ by suitable row and column interchanges. Thus the difficulty of non-commutativity encountered with ordinary multiplication has to some extent been overcome. In other results also the Kronecker product has advantages, for example

$$(A \otimes B)^{\mathsf{T}} = A^{\mathsf{T}} \otimes B^{\mathsf{T}}, \qquad (2.68)$$

so unlike (2.41) there is no reversal of order (see also Problem 2.13). The Kronecker product is not just an interesting mathematical idea, as it has applications in statistics, numerical analysis, communications theory, and elsewhere.

Exercise 2.32 Verify (2.68) by comparing the right-hand side with the transpose of (2.65). Similarly, show that $(A \otimes B)^* = A^* \otimes B^*$.

Exercise 2.33 Show that the Kronecker product is distributive and associative, i.e.

$$A \otimes (B + C) = A \otimes B + A \otimes C; \qquad A \otimes (B \otimes C) = (A \otimes B) \otimes C.$$

The second special product, which is particularly useful in some areas of statistics, is called the *Hadamard* (or *Schur*) *product*. If A is again the $m \times n$ matrix in (2.2) and $B = [b_{ij}]$ has the same dimensions, then $A \circ B$ is obtained simply by multiplying together corresponding elements in each matrix, so that

$$A \circ B = \begin{bmatrix} a_{11}b_{11} & a_{12}b_{12} & \cdots & a_{1n}b_{1n} \\ a_{21}b_{21} & a_{22}b_{22} & \cdots & a_{2n}b_{2n} \\ . & . & \cdots & . \\ a_{m1}b_{m1} & a_{m2}b_{m2} & \cdots & a_{mn}b_{mn} \end{bmatrix}. \qquad (2.69)$$

It is obvious that in this case multiplication is always commutative, that is $A \circ B = B \circ A$ for all matrices A and B having the same dimensions.

Example 2.19 Again consider the 2×2 matrices in (2.15). Then

$$A \circ B = \begin{bmatrix} (1)(-1) & (2)(4) \\ (3)(2) & (4)(-3) \end{bmatrix} = \begin{bmatrix} -1 & 8 \\ 6 & -12 \end{bmatrix}.$$

Notice that $A \circ B$ is the submatrix obtained from $A \otimes B$ by taking rows 1, 4 and columns 1, 4 in (2.66).

Example 2.20 Consider the situation of the taxi-driver described in Problem 1.7. The matrix of probabilities was found to be

$$\begin{array}{cc} & \text{picked up} \\ & \text{in city} \\ & B \quad L \end{array}$$
$$A = \begin{bmatrix} 0.4 & 0.3 \\ 0.6 & 0.7 \end{bmatrix} \begin{matrix} B \\ L \end{matrix} \begin{matrix} \text{dropped off} \\ \text{in city} \end{matrix}.$$

Suppose that the average fare within city B is 6, within city L is 9 and between the two cities is 16 (in some monetary units). These fares can be represented in matrix form as

$$F = \begin{bmatrix} 6 & 16 \\ 16 & 9 \end{bmatrix}.$$

For example, the expected reward for a journey within city B is therefore $6 \times 0.4 = 2.4$. The matrix of expected rewards is

$$A \circ F = \begin{bmatrix} 2.4 & 4.8 \\ 9.6 & 6.3 \end{bmatrix}.$$

It is easy to see by comparing (2.65) and (2.69) that in general $A \circ B$ is a submatrix of $A \otimes B$, obtained by selecting rows 1, $p + 1$, $2p + 1$, ..., $(m - 1)p + 1$ and columns 1, $q + 1$, $2q + 1$, ..., $(n - 1)q + 1$ in (2.65).

Exercise 2.34 Show that the Hadamard product is distributive and associative, i.e.

$$A \circ (B + C) = A \circ B + A \circ C, \qquad A \circ (B \circ C) = (A \circ B) \circ C.$$

Exercise 2.35 Show that

$$(A \circ B)^T = A^T \circ B^T, \qquad (A \circ B)^* = A^* \circ B^*.$$

Exercise 2.36 Show that

(a) if A and B are hermitian so is $A \circ B$;

(b) if A and B are skew hermitian then $A \circ B$ is hermitian;

(c) if A is hermitian and B is skew hermitian then $A \circ B$ is skew hermitian.

2.6 Derivative of a matrix

If the elements $a_{ij}(t)$ of A are differentiable functions of t then the *derivative* of A, denoted by dA/dt or $\dot{A}$, is the matrix whose elements are da_{ij}/dt. The usual rules for differentiation still hold, for example:

$$\frac{d}{dt}(A + B) = \frac{dA}{dt} + \frac{dB}{dt} \tag{2.70}$$

$$\frac{d}{dt}(AB) = \left[\frac{dA}{dt}\right]B + A\left[\frac{dB}{dt}\right]. \tag{2.71}$$

Note that when applying the 'product rule' (2.71) the order of the terms must be preserved, e.g. $\dot{A}B \neq B\dot{A}$.

Example 2.21 If

$$A = \begin{bmatrix} t^2 & t^3 \\ t & \sin t \end{bmatrix}, \qquad B = \begin{bmatrix} e^{2t} & 0 \\ \cos t & 4 \end{bmatrix}, \tag{2.72}$$

then

$$\frac{dA}{dt} = \begin{bmatrix} 2t & 3t^2 \\ 1 & \cos t \end{bmatrix}, \qquad \frac{dB}{dt} = \begin{bmatrix} 2e^{2t} & 0 \\ -\sin t & 0 \end{bmatrix}. \qquad (2.73)$$

Exercise 2.37 Evaluate $d(AB)/dt$ for the two matrices in (2.72): (a) by first determining AB; (b) by using (2.71).

Exercise 2.38 Deduce, using A in (2.72), that in general $A\dot{A} \neq \dot{A}A$. Hence obtain expressions for $d(A^2)/dt$ and $d(A^3)/dt$ for an arbitrary square matrix A.

The matrix result corresponding to the 'quotient rule' must be delayed until we have defined the inverse of a matrix in Chapter 4 (see Problem 4.19).

The *integral* of a matrix whose elements are integrable functions of t is defined similarly as the matrix obtained by integrating each element, i.e. the i, j element of $\int A \, dt$ is $\int a_{ij} \, dt$.

Problems

2.1 Fill in the missing elements in the following product

$$\begin{bmatrix} 2 & 0 & 0 \\ \cdot & 2 & 0 \\ 0 & \cdot & 2 \end{bmatrix} \begin{bmatrix} \cdot & \cdot & 0 \\ 0 & \cdot & \cdot \\ 0 & 0 & 2 \end{bmatrix} = \begin{bmatrix} 4 & -2 & 0 \\ -2 & 5 & -2 \\ 0 & -2 & 5 \end{bmatrix}.$$

2.2 The relationship between the input current i_1 and voltage v_1 and the output current i_2 and voltage v_2 for the four-terminal network in Fig. 2.1 is given by

$$\begin{bmatrix} v_1 \\ i_1 \end{bmatrix} = \begin{bmatrix} 1 + i\omega CR & R(2 + i\omega CR) \\ i\omega C & 1 + i\omega CR \end{bmatrix} \begin{bmatrix} v_2 \\ i_2 \end{bmatrix}.$$

Find v_1 and i_1 in terms of v_3 and i_3 for the network shown in Fig. 2.2.

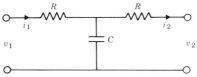

Fig. 2.1 Four-terminal network for Problem 2.2.

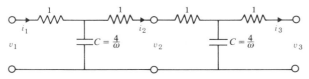

Fig. 2.2 Second network for Problem 2.2

2.3 Consider again the airlines problem described in Example 1.3 and Problem 1.2. Verify that the product of the matrix of (1.1) and the matrix you found in part (a) of Problem 1.2 is equal to the matrix of part (b), giving flight information from country A to country C.

2.4 For a square matrix, the diagonal from the top right corner to bottom left corner is called the *secondary* diagonal, and is perpendicular to the principal diagonal.

Define K_n to be the $n \times n$ *reverse unit matrix* having 1's along the secondary diagonal and zeros everywhere else, e.g.

$$K_3 = \begin{bmatrix} 0 & 0 & 1 \\ 0 & 1 & 0 \\ 1 & 0 & 0 \end{bmatrix}.$$

The terms *exchange matrix* and *flip matrix* are also used.

(a) Prove $K_n^2 = I_n$.

(b) If A is an arbitrary $n \times n$ matrix what are the products KA and AK?

2.5 The generalization of the linear mapping in (1.7) can be written $x' = Tx$, where x and x' are column n-vectors and T is an $n \times n$ matrix, and represents a change from coordinates $x_1, \ldots, x_n$ to $x_1', \ldots, x_n'$. If $x_1' = 2x_1 + 3x_2$, $x_2' = 4x_3$, $x_3' = x_1' - 7x_2 + 3x_3$, write down T.

2.6 Use (2.41) to prove that $(A^T)^2 = (A^2)^T$ and similarly show that $(A^T)^k = (A^k)^T$ for any positive integer k and square matrix A.

2.7 The *trace* of an $n \times n$ matrix A is defined as the sum of the elements on the principal diagonal, i.e.

$$\text{tr}(A) = a_{11} + a_{22} + \cdots a_{nn} = \sum_{i=1}^{n} a_{ii}. \tag{2.74}$$

Prove that: (a) $\text{tr}(A + B) = \text{tr}(A) + \text{tr}(B)$;

(b) $\text{tr}(AB) = \text{tr}(BA)$;

(c) $\text{tr}(AA^T) = \sum_{i=1}^{n} \sum_{j=1}^{n} a_{ij}^2$

(use (2.19) for (b) and (c)).

2.8   A square matrix is called *upper triangular* if all the elements below the principal diagonal are zero. (a) Prove that the product of two upper triangular matrices is also upper triangular. (b) If A is a 3×3 upper triangular matrix with all $a_{ii} = 0$, prove that $A^3 = 0$.

2.9   Consider the matrix representation of complex numbers described in Example 2.10. Show that (a) $i^2 \sim -I_2$, (b) $\bar{z}_1 \sim (A^{(1)})^{\mathsf{T}}$, where $A^{(1)}$ is the matrix in (2.33).

2.10   An $n \times n$ matrix is called *normal* if $A^*A = AA^*$. For example, a real symmetric matrix is normal since $A^*A = A^{\mathsf{T}}A = AA = AA^*$. What other matrices are normal?

2.11   The *Fibonacci numbers* are $1, 1, 2, 3, 5, 8, 13, \ldots$, each number being obtained as the sum of the preceding two. These arise in many applications. If x_k denotes the kth number, then

$$x_{k+2} = x_{k+1} + x_k, \qquad k = 1, 2, 3, \ldots$$

with $x_1 = 1$, $x_2 = 1$. To obtain a matrix form define new variables $X_1(k) = x_k$, $X_2(k) = x_{k+1}$ so that $X_1(k + 1) = X_2(k)$ and $X_2(k + 1) = X_2(k) + X_1(k)$, and hence

$$\begin{bmatrix} X_1(k+1) \\ X_2(k+1) \end{bmatrix} = \begin{bmatrix} 0 & 1 \\ 1 & 1 \end{bmatrix} \begin{bmatrix} X_1(k) \\ X_2(k) \end{bmatrix} = A \begin{bmatrix} X_1(k) \\ X_2(k) \end{bmatrix}, \qquad k = 1, 2, 3, \ldots$$

(2.75)

The equations (2.75) represent a pair of linear *difference* equations. Show that

$$\begin{bmatrix} X_1(8) \\ X_2(8) \end{bmatrix} = A^7 \begin{bmatrix} 1 \\ 1 \end{bmatrix} = A^5 \begin{bmatrix} 2 \\ 3 \end{bmatrix}$$

and hence calculate x_8 and x_9.

2.12   If C, D are arbitrary matrices having dimensions $n \times r$ and $q \times s$ respectively, and A and B are as in (2.65), show that the *ordinary* product of $A \otimes B$ and $C \otimes D$ satisfies the relationship

$$(A \otimes B)(C \otimes D) = AC \otimes BD \qquad (2.76)$$

by comparing terms on both sides of (2.76).

2.13   Define the *Kronecker power* of A by

$$A^{[2]} = A \otimes A, \ A^{[3]} = A \otimes A \otimes A = A \otimes A^{[2]}, \text{ etc.}$$

Use (2.76) to prove that $(AC)^{[2]}$ is equal to the ordinary product $A^{[2]}C^{[2]}$, and hence show that $(AC)^{[k]} = A^{[k]}C^{[k]}$ for any positive integer k. (A is $m \times n$, C is $n \times r$.)
    Under what conditions on A and C does $(AC)^k = A^k C^k$?

2.14 The idea of Example 2.10 can be extended to the complex matrix A in (2.48), using the real partitioned matrix

$$D = \begin{bmatrix} A_1 & A_2 \\ -A_2 & A_1 \end{bmatrix}^n_n \sim A_1 + iA_2 = A. \qquad (2.77)$$

(a) Using the result of Exercise 2.25(a) prove that if A is hermitian then D is symmetric.

(b) If $E \sim B_1 + iB_2 = B$, prove that $DE \sim AB$.

2.15 Suppose that in Problem 1.6 the initial numbers of one-, two- and three-year-old beetles are equal (i.e. $x_0 = y_0 = z_0 = c$). Compute A^3, and hence show that the numbers fluctuate in a three-year cycle. Show also that if the initial numbers are in the ratios $1:3:6$ then these remain unaltered subsequently.

2.16 Let A be an $m \times n$ matrix whose elements a_{ij} represent statistical data. Define an $m \times n$ matrix $\Delta = (\delta_{ij})$ with

$$\delta_{ij} = \begin{cases} 1, & \text{if } a_{ij} \text{ is known} \\ 0, & \text{if } a_{ij} \text{ is unknown.} \end{cases}$$

Show that

$$A = (U - \Delta) \circ A + \Delta \circ A$$
$$= A_u + A_k,$$

where U is an $m \times n$ matrix all of whose elements are one, and A_u and A_k are the unknown and known parts of A respectively.

Decompose in this way the data matrix

$$A = \begin{bmatrix} 9 & 1 & 3 \\ * & 2 & 4 \\ 5 & * & 7 \\ 1 & 6 & * \end{bmatrix},$$

where the asterisks denote unknown elements.

3 Unique solution of linear equations

In this chapter we shall study the solution of n simultaneous linear equations in n unknowns $x_1, x_2, \ldots, x_n$ in the form

$$\left.\begin{aligned}
a_{11}x_1 + a_{12}x_2 + \cdots + a_{1n}x_n &= b_1 \\
a_{21}x_1 + a_{22}x_2 + \cdots + a_{2n}x_n &= b_2 \\
&\vdots \\
a_{n1}x_1 + a_{n2}x_2 + \cdots + a_{nn}x_n &= b_n
\end{aligned}\right\}, \tag{3.1}$$

where the a's and b's are given real numbers. As illustrated in Example 1.4, eqns (3.1) can be written as

$$Ax = b, \tag{3.2}$$

where $A = [a_{ij}]$ is the matrix of coefficients and

$$x = [x_1, x_2, \ldots, x_n]^\mathsf{T}, \qquad b = [b_1, b_2, \ldots, b_n]^\mathsf{T}. \tag{3.3}$$

It is tempting to try to extend the ideas of matrix algebra developed in Chapter 2 so as to include 'division'. We could then solve (3.2) by writing $x = b \div A$, or borrowing the notation from ordinary algebra,

$$x = A^{-1}b. \tag{3.4}$$

If we could find the matrix denoted by A^{-1} in (3.4), then this would give the desired solution vector x. However, it turns out that the best way to proceed is the reverse: we solve the equations, and then use this solution to find the matrix A^{-1}. The second part of this procedure will be delayed until Chapter 4. Notice incidentally that by applying the rule (2.41) to (3.2) these equations could equally well be written

$$x^\mathsf{T}A^\mathsf{T} = b^\mathsf{T},$$

where both x^T and b^T are *row* vectors.

Equations with complex coefficients can be converted into the real case, as indicated in Problem 3.7.

3.1 Two equations and unknowns

Example 3.1 To solve the equations

$$x_1 + x_2 = 3 \tag{3.5}$$

$$2x_1 - 3x_2 = -4 \tag{3.6}$$

we use the familiar method of *elimination*: subtract twice eqn (3.5) from eqn (3.6) to obtain

$$-5x_2 = -10,$$

so that $x_2 = 2$. Substitution of this value into (3.5) then gives $x_1 = 3 - 2 = 1$, so the equations have the unique solution $x_1 = 1$, $x_2 = 2$. In geometrical terms we can represent (3.5) and (3.6) as straight lines in the plane. The solution corresponds to the point of intersection of the lines (see Fig. 3.1).

Example 3.2 Suppose that (3.6) is replaced by

$$2x_1 + 2x_2 = -4. \tag{3.7}$$

Subtracting twice eqn (3.5) from (3.7) gives $0 = -10$. This means that there are no values of x_1 and x_2 which satisfy (3.5) and (3.7) simultaneously. The two equations are called *inconsistent*. Geometrically, the two lines corresponding to (3.5) and (3.7) are parallel, and so have no point of intersection (see Fig. 3.1).

Example 3.3 Suppose instead that (3.6) is replaced by

$$2x_1 + 2x_2 = 6. \tag{3.8}$$

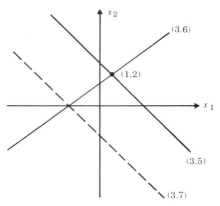

Fig. 3.1 The straight lines representing the equations (3.5), (3.6), (3.7).

On subtracting twice (3.5) from (3.8) we get $0 = 0$, which doesn't seem to get us very far. What it means is that eqns (3.5) and (3.8) do have a solution (they are *consistent*) but the second equation of the pair provides no information which is not given by the first. In other words, the solution of (3.5) and (3.8) is given by

$$x_1 = 3 - x_2, \tag{3.9}$$

where x_2 is arbitrary. Thus (3.9) represents an *infinite* number of solutions to the pair (3.5) and (3.8).

The three preceding examples illustrate what can happen in general for (3.1): either the equations have a unique solution, or they have an infinite number of solutions, or they are inconsistent (no solution). The problem of distinguishing between these cases is clearly vital, and will be tackled in Chapter 5. Our first concern is to describe systematic methods for finding the solution of (3.1) *assuming* it exists and is unique.

Exercise 3.1 Using elimination, determine the solution of the eqns (3.1) when $n = 2$. What are the conditions for (a) uniqueness of solution, (b) an infinite number of solutions, (c) inconsistency?

3.2 Gaussian elimination

A powerful method attributed to Gauss is a formalization of the elimination procedure of Example 3.1.

Example 3.4 To solve

$$x_1 - 3x_2 + 7x_3 = 2 \tag{3.10a}$$

$$2x_1 + 4x_2 - 3x_3 = -1 \tag{3.10b}$$

$$-3x_1 + 7x_2 + 2x_3 = 3 \tag{3.10c}$$

first eliminate x_1 from (3.10b) and (3.10c) by subtracting respectively 2 and -3 times (3.10a), to obtain

$$x_1 - 3x_2 + 7x_3 = 2 \tag{3.11a}$$

$$10x_2 - 17x_3 = -5 \tag{3.11b}$$

$$-2x_2 + 23x_3 = 9. \tag{3.11c}$$

Next eliminate x_2 from (3.11c) by multiplying (3.11b) by $-2/10$ and subtracting the result from (3.11c), so that the final set of equations becomes

$$x_1 - 3x_2 + 7x_3 = 2 \qquad (3.12a)$$

$$10x_2 - 17x_3 = -5 \qquad (3.12b)$$

$$\frac{98}{5}x_3 = 8. \qquad (3.12c)$$

From (3.12c) $x_3 = 20/49$, and substituting into (3.12b) gives

$$x_2 = -\frac{1}{2} + \frac{17}{10}\left(\frac{20}{49}\right) = \frac{19}{98}.$$

Finally, substituting into (3.12a)

$$x_1 = 2 + 3\left(\frac{19}{98}\right) - 7\left(\frac{20}{49}\right) = -\frac{27}{98},$$

so the solution is $(-27/98, 19/98, 20/49)$. The procedure for solving eqns (3.12) is called *back substitution*, since the unknowns are found in the order x_3, x_2, x_1.

It saves writing to record only the coefficients in the equations at each step of the process, and this can be done in matrix form. For (3.10) we have the *augmented matrix*

$$B = \begin{bmatrix} 1 & -3 & 7 & \vdots & 2 \\ 2 & 4 & -3 & \vdots & -1 \\ -3 & 7 & 2 & \vdots & 3 \end{bmatrix}, \qquad (3.13)$$

which consists of A together with the right-hand side of the equations as a fourth column, so in general for (3.1), B is the $n \times (n+1)$ matrix

$$B = [A, b]. \qquad (3.14)$$

In the solution of (3.10), at the first step which produces eqns (3.11) the first column of B is reduced so that it has all zeros below the 1,1 position, i.e.

$$B \rightarrow \begin{bmatrix} 1 & -3 & 7 & \vdots & 2 \\ 0 & 10 & -17 & \vdots & -5 \\ 0 & -2 & 23 & \vdots & 9 \end{bmatrix} \begin{matrix} \\ (R2) - 2(R1) \\ (R3) - (-3)(R1) \end{matrix}, \qquad (3.15)$$

where the notation $(R2) - 2(R1)$ means the second row of B in (3.13) has twice the first row subtracted from it, and so on. The convention adopted is that the row which is altered is written *first*. Similarly the next step, corresponding to production of (3.12) is

$$\rightarrow \begin{bmatrix} 1 & -3 & 7 & \vdots & 2 \\ 0 & 10 & -17 & \vdots & -5 \\ 0 & 0 & \frac{98}{5} & \vdots & 8 \end{bmatrix} \quad (R3) - (-2/10)(R2). \qquad (3.16)$$

Notice that the submatrix composed of the first three columns of the matrix in (3.16) is in *upper triangular* form—all the elements below the principal diagonal are zero (see Problem 2.8). The corresponding equations (3.12) are called a *triangular* set, and as we have seen are easily solved.

Exercise 3.2 Solve the equations

$$x_1 - x_2 + 3x_3 = 5$$
$$2x_1 - 4x_2 + 7x_3 = 7$$
$$4x_1 - 9x_2 + 2x_3 = -15$$

by gaussian elimination.

For $n > 3$, the method proceeds in exactly the same way. The first n columns of B are reduced one at a time, working from left to right, so as to obtain zeros below the principal diagonal. This is done by subtracting suitable multiples of rows. The resulting triangular system of equations is then solved by back substitution, starting with x_n.

In (3.15) the 1, 1 element (equal to 1) is used to reduce the first column; in (3.16) the 2, 2 element (equal to 10) is used to reduce the second column. These numbers are called the *pivots*, from the concept that the matrix changes or '*pivots*' around these numbers. Generally, the ith pivot is the number in the (i, i) position which is used to reduce the ith column.

Suppose that columns $1, 2, \ldots, (j-1)$ of B have been reduced, and denote the elements in column j of B at this stage by b_{ij}, $i = 1, 2, \ldots, n$. The element b_{jj} on the principal diagonal cannot be used as pivot if it is zero. In this case the jth row must be interchanged with some row *below* it, say the kth, having $b_{kj} \neq 0$, before the elimination can be continued. This simply corresponds to interchanging the jth and kth equations, so does not affect the solution. In fact, if b_{jj} is non-zero but small compared with the other elements, then use of b_{jj} as pivot can lead to large rounding errors.

One rule which is often used to overcome both of these difficulties is to choose as pivot the element b_{kj} having largest numerical value in column j, i.e.

$$|b_{kj}| = \max_i |b_{ij}|, \qquad j \leq i \leq n. \tag{3.17}$$

Rows j and k are then interchanged and the next elimination step is carried out. This procedure is known as *partial pivoting*, and is done at *every* step of the process. Notice that in (3.17), rows above the jth are not considered since this would upset the triangular form already obtained for the first $j - 1$ columns.

Example 3.5 Suppose at the third stage of a gaussian elimination with $n = 5$ the reduced matrix B is

$$\begin{bmatrix} 2 & 5 & 8 & \times & \times & \times \\ 0 & 3 & 1 & \times & \times & \times \\ 0 & 0 & 3 & \times & \times & \times \\ 0 & 0 & -7 & \times & \times & \times \\ 0 & 0 & 5 & \times & \times & \times \end{bmatrix}. \tag{3.18}$$

From (3.17) the element in the 4, 3 position (i.e. -7) would be used as the third pivot. The third and fourth rows in (3.18) would be interchanged, and the sub-diagonal elements in the third column would then be reduced to zero in the usual way.

The term 'partial' pivoting is used to contrast it with another scheme called 'complete pivoting'. In this the pivot is taken to be the element having largest numerical value in the lower right corner of the matrix—for example, in (3.18) this would be the array within dashed lines. This would involve both a row and a column interchange. In practice, however, partial pivoting is usually adequate.

If at stage j *all* the potential pivots b_{ij}, $j \leq i \leq n$, in column j are zero then the gaussian elimination procedure as described above stops. In fact this means that the equations do not have a unique solution. However, the method can be modified to determine whether the equations are consistent, and if so to calculate the (non-unique) solution. This will be discussed in detail in Chapter 5.

Exercise 3.3 Solve the following set of equations by gaussian elimination, using partial pivoting and exact arithmetic.

$$5x_1 + x_2 + 2x_3 = 29$$
$$3x_1 - x_2 + x_3 = 10$$
$$x_1 + 2x_2 + 4x_3 = 31.$$

Exercise 3.4 Carry out gaussian elimination on the following set of equations, and hence deduce that they are inconsistent.

$$x_1 - 2x_2 + x_3 - x_4 = -5$$
$$x_1 + 5x_2 - 7x_3 + 2x_4 = 2$$
$$3x_1 + x_2 - 5x_3 + 3x_4 = 1$$
$$2x_1 + 3x_2 - 6x_3 = 21.$$

When performing numerical calculations by hand it is useful to have checks against arithmetical errors. One simple check is to calculate the sums of the elements in each row of the original matrix B in (3.14), and regard these sums as forming an extra column. The elimination operations are then carried out in the usual way, including this new column. At each step the sum of the elements in each row of the reduced matrix B should equal the element in the same row of this extra column, if no slip has been made in the elimination procedure. A check on the back substitution process can be made by inserting the computed values of the variables into the original equations, which should be satisfied within the accuracy used.

Example 3.6 In (3.13) the row sums are $7, 2, 9$ respectively. We have, using the operations in (3.15) and (3.16),

$$\begin{bmatrix} 7 \\ 2 \\ 9 \end{bmatrix} \rightarrow \begin{bmatrix} 7 \\ -12 \\ 30 \end{bmatrix} \quad \begin{array}{l} (R2) - 2(R1) \\ (R3) - (-3)(R1) \end{array} \tag{3.19}$$

$$\rightarrow \begin{bmatrix} 7 \\ -12 \\ 27\tfrac{3}{5} \end{bmatrix} \quad (R3) - (-2/10)(R2). \tag{3.20}$$

The elements in (3.19) and (3.20) agree with the row sums in (3.15) and (3.16) respectively.

Exercise 3.5 Carry out the check on your working for Exercises 3.2 and 3.3.

Exercise 3.6 Use gaussian elimination to solve the equations

$$x_1 + x_2 + x_3 + x_4 = -1$$
$$2x_1 - x_2 + 3x_3 = 1$$
$$2x_2 + 3x_4 = -1$$
$$-x_1 + 2x_3 + x_4 = -2.$$

3.3   Triangular decomposition

We have seen that triangular systems of equations can be solved very easily. This leads to the idea of expressing A in (3.2) as a product

$$A = LU \tag{3.21}$$

where $U = [u_{ij}]$ is an $n \times n$ upper triangular matrix (defined in Problem 2.8), and $L = [l_{ij}]$ is an $n \times n$ *lower* triangular matrix (i.e. all elements of L above the principal diagonal are zero). It is convenient to take all $l_{ii} = 1$.

Example 3.7   Let A be the matrix of coefficients for the equations (3.10). Since $n = 3$ we have

$$LU = \begin{bmatrix} 1 & 0 & 0 \\ l_{21} & 1 & 0 \\ l_{31} & l_{32} & 1 \end{bmatrix} \begin{bmatrix} u_{11} & u_{12} & u_{13} \\ 0 & u_{22} & u_{23} \\ 0 & 0 & u_{33} \end{bmatrix} \tag{3.22}$$

$$= \begin{bmatrix} u_{11} & u_{12} & u_{13} \\ l_{21}u_{11} & (l_{21}u_{12} + u_{22}) & (l_{21}u_{13} + u_{23}) \\ l_{31}u_{11} & (l_{31}u_{12} + l_{32}u_{22}) & (l_{31}u_{13} + l_{32}u_{23} + u_{33}) \end{bmatrix}. \tag{3.23}$$

The nine unknown elements of L and U can be obtained by equating the nine elements of (3.23) to those of the 3×3 matrix A. This is done in the following order:

First row of A: $u_{11} = 1,$ $u_{12} = -3,$ $u_{13} = 7;$ (3.24)

First column of A: $l_{21}u_{11} = 2,$ $l_{31}u_{11} = -3,$ whence

$$l_{21} = 2, \quad l_{31} = -3; \tag{3.25}$$

Second row of A: $l_{21}u_{12} + u_{22} = 4,$ $l_{21}u_{13} + u_{23} = -3,$ whence

$$u_{22} = 10, \quad u_{23} = -17; \tag{3.26}$$

Second column of A: $l_{31}u_{12} + l_{32}u_{22} = 7,$ whence

$$l_{32} = -1/5; \tag{3.27}$$

Third row of A: $l_{31}u_{13} + l_{32}u_{23} + u_{33} = 2,$ whence

$$u_{33} = 98/5. \tag{3.28}$$

Notice the order in which the elements of U and L are obtained in (3.24)–(3.28): first row of U, first column of L, second row of U, second column of L, third row of U. Substituting for the elements in (3.22) gives

$$A = \begin{bmatrix} 1 & 0 & 0 \\ 2 & 1 & 0 \\ -3 & -\dfrac{1}{5} & 1 \end{bmatrix} \begin{bmatrix} 1 & -3 & 7 \\ 0 & 10 & -17 \\ 0 & 0 & \dfrac{98}{5} \end{bmatrix}. \tag{3.29}$$

$$\underset{L}{} \qquad \underset{U}{}$$

We now show how the decomposition is used to solve (3.2). Substituting (3.21) into (3.2) gives

$$LUx = b \tag{3.30}$$

and this is written as *two* triangular sets of equations:

$$Ly = b, \qquad Ux = y \tag{3.31}$$

where $y = [y_1, y_2, \ldots, y_n]^\mathrm{T}$. The first set is solved for y, and then the second set for x, which is the desired solution of the original equations.

Example 3.7 (continued) Using (3.29) the first set of equations in (3.31) is

$$\begin{aligned} y_1 & & = 2 \\ 2y_1 + & y_2 & = -1 \\ -3y_1 - & \tfrac{1}{5}y_2 + y_3 & = 3 \end{aligned}$$

(the b_i are as in (3.10)). The solution is therefore

$$y_1 = 2, \qquad y_2 = -1 - 2.2 = -5, \qquad y_3 = 3 + 3.2 + \frac{1}{5}(-5) = 8.$$

The second set in (3.31) is

$$\begin{aligned} x_1 - 3x_2 + 7x_3 &= 2 \\ 10x_2 - 17x_3 &= -5 \\ \frac{98}{5}x_3 &= 8, \end{aligned}$$

so $x_3 = 20/49$, $x_2 = (-5 + 17x_3)/10 = 19/98$, $x_1 = 2 + 3x_2 - 7x_3 = -27/98$, which agrees, of course, with the values found in Example 3.4.

Exercise 3.7 Determine L and U for the matrix

$$A = \begin{bmatrix} 2 & 3 & 4 \\ 4 & 10 & 9 \\ 6 & 17 & 20 \end{bmatrix}$$

and hence solve (3.2) in this case with $b = [23, 59, 101]^\mathrm{T}$.

We now show how the rather tedious arithmetic of (3.24)–(3.28) can be avoided. Consider the decomposition of A in (3.29), and compare this with the gaussian reduction of the same matrix A in (3.16). It will be seen that U is identical to the triangularized form of A. In fact this is no coincidence, and furthermore L can also be obtained from the gaussian elimination procedure, as follows. Continuing with the same example, consider the step (3.15): to obtain zero in the 2, 1 position we subtract 2 times (R1) from (R2), and for a zero in the 3, 1 position, -3 times (R1) is subtracted from (R3). It is convenient to denote by m_{ij} the *multiplier* which is used to obtain zero in the i, j position, corresponding to the operation $(Ri) - m_{ij}(Rj)$. Thus in our example $m_{21} = 2$, $m_{31} = -3$, and similarly from (3.16) we see that $m_{32} = -2/10 = -1/5$. If we now write down a lower triangular matrix having 1's on the principal diagonal, and m_{ij} as the i, j element, i.e.

$$\begin{bmatrix} 1 & 0 & 0 \\ 2 & 1 & 0 \\ -3 & -\dfrac{1}{5} & 1 \end{bmatrix},$$

we see that this is identical to L in the decomposition (3.29).

It can be shown that the above method for obtaining $A = LU$ holds in general: If A is reduced to triangular form by gaussian elimination (without row interchanges), then the resulting upper triangular matrix is U, and the matrix of multipliers constructed as described above is L. If partial pivoting is used, then we get triangular factors such that $LU = A'$, where A' is the matrix which is obtained from the original A by applying the row interchanges to it in the same order. Thus gaussian elimination and LU decomposition are equivalent procedures, and involve the same total computational effort.

Exercise 3.8 Repeat the decomposition for the matrix A in Exercise 3.7, using gaussian elimination.

Exercise 3.9 Solve the equations

$$\begin{aligned} 2x_1 + x_2 + 2x_3 &= 10 \\ 4x_1 + 4x_2 + 7x_3 &= 33 \\ 2x_1 + 5x_2 + 12x_3 &= 48 \end{aligned}$$

using LU decomposition. Hence find the solution if the right-hand sides of the equations change to 9, 22, 25 respectively.

An alternative way of performing the LU decomposition provides an interesting application of partitioning. Recall (Section 2.4) that the $r \times r$ leading principal submatrix A_r of A is the submatrix formed by rows and column numbers $1, 2, \ldots, r$ of A. Suppose that

$$A_r = L_r U_r, \qquad A_{r+1} = L_{r+1} U_{r+1} \qquad (3.32)$$

with L_r, L_{r+1} lower triangular and U_r, U_{r+1} upper triangular. Some manipulation shows that once L_r and U_r have been found then

$$L_{r+1} = \begin{bmatrix} L_r & 0 \\ c_r & 1 \end{bmatrix} \begin{matrix} r \\ 1 \end{matrix}, \qquad U_{r+1} = \begin{bmatrix} U_r & d_r \\ 0 & \alpha_r \end{bmatrix} \begin{matrix} r \\ 1 \end{matrix}, \qquad (3.33)$$

where the row r-vector c_r is given by

$$c_r U_r = [a_{r+1,1}, a_{r+1,2}, \ldots, a_{r+1,r}], \qquad (3.34)$$

the column r-vector d_r by

$$L_r d_r = [a_{1,r+1}, a_{2,r+1}, \ldots, a_{r,r+1}]^{\mathsf{T}}, \qquad (3.35)$$

and the scalar α_r by

$$\alpha_r = a_{r+1,r+1} - c_r d_r. \qquad (3.36)$$

The procedure is thus a repetitive one: L_1 and U_1 are both scalars so $L_1 = 1$, $U_1 = a_{11}$; (3.34) and (3.35) are solved for c_1 and d_1, and α_1 is found from (3.36), giving L_2 and U_2 in (3.33); the process is repeated with $r = 2$ to find L_3 and U_3, and so on, until $L_n = L$ and $U_n = U$ are determined.

Example 3.8 We repeat the decomposition of Example 3.7, using the matrix A of eqns (3.10). Since $A_1 = 1$ we have $L_1 = 1$, $U_1 = 1$. Next,

$$A_2 = \begin{bmatrix} 1 & -3 \\ 2 & 4 \end{bmatrix},$$

so with $r = 1$ in (3.34) and (3.35) these become

$$c_1 \cdot 1 = a_{21} = 2, \qquad 1 \cdot d_1 = a_{12} = -3$$

and in (3.36)

$$\alpha_1 = a_{22} - c_1 d_1 = 4 + 6 = 10.$$

Thus from (3.33) we obtain

$$L_2 = \begin{bmatrix} 1 & 0 \\ 2 & 1 \end{bmatrix}, \qquad U_2 = \begin{bmatrix} 1 & -3 \\ 0 & 10 \end{bmatrix}. \qquad (3.37)$$

The process is repeated with $r = 2$ in (3.34), to give

$$[c_{21}, c_{22}]U_2 = [a_{31}, a_{32}] = [-3, 7], \tag{3.38}$$

where U_2 is the matrix in (3.37). The equations in (3.38) are easily solved because U_2 is triangular:

$$c_{21} = -3, \qquad c_{22} = (7 + 3c_{21})/10 = -1/5.$$

Similarly, from (3.35)

$$L_2 \begin{bmatrix} d_{21} \\ d_{22} \end{bmatrix} = \begin{bmatrix} a_{13} \\ a_{23} \end{bmatrix} = \begin{bmatrix} 7 \\ -3 \end{bmatrix},$$

whence

$$d_{21} = 7, \qquad d_{22} = -17$$

and finally, from (3.36),

$$\alpha_2 = a_{33} - c_2 d_2 = 2 - \left[-3, -\frac{1}{5} \right]\begin{bmatrix} 7 \\ -17 \end{bmatrix} = 2 - 21 + \frac{17}{5} = \frac{98}{5}.$$

The desired L and U are then obtained by setting $r = 2$ in (3.33):

$$L_3 = \begin{bmatrix} L_2 & 0 \\ c_{21} & c_{22} & 1 \end{bmatrix}, \qquad U_2 = \begin{bmatrix} U_2 & \begin{matrix} d_{21} \\ d_{22} \end{matrix} \\ 0 & \alpha_2 \end{bmatrix}$$

and it can be seen that these agree with L and U found earlier in (3.29).

Exercise 3.10   Repeat the determination of L and U for the matrix A in Exercise 3.7 using the iterative process described above.

A word of warning is necessary here: not every matrix can be directly composed into factors L and U, as the following simple example demonstrates.

Exercise 3.9   For the equations

$$\begin{aligned} 4x_2 &= 8 \\ 4x_1 + 2x_2 &= 17, \end{aligned} \tag{3.39}$$

if we write

$$A = \begin{bmatrix} 0 & 4 \\ 4 & 2 \end{bmatrix} = \begin{bmatrix} 1 & 0 \\ l_1 & 1 \end{bmatrix}\begin{bmatrix} u_1 & u_2 \\ 0 & u_3 \end{bmatrix},$$

comparison of the 1, 1 and 2, 1 elements gives $u_1 = 0$, $l_1 u_1 = 4$, which cannot be satisfied. However, if the order of the equations in (3.39) is reversed then

$$A = \begin{bmatrix} 4 & 2 \\ 0 & 4 \end{bmatrix} = \underset{L}{\begin{bmatrix} 1 & 0 \\ 0 & 1 \end{bmatrix}} \underset{U}{\begin{bmatrix} 4 & 2 \\ 0 & 4 \end{bmatrix}}.$$

This example shows that it may be necessary to alter the order of the equations before the decomposition (3.21) can be found. This is to ensure that no u_{rr} (or α_r in the partitioned form) is zero. It can be shown that this rearrangement is always possible provided the original equations (3.2) have a unique solution. In fact it can also be shown that the reordering must be carried out so that the new leading principal submatrices A_r have non-zero determinants (see Problem 4.22). One important special case when A can be decomposed directly is when A is symmetric and positive definite (the latter is defined in Section 7.4) in which case L can be set equal to U^T (see Problem 3.6).

Exercise 3.11 Express the symmetric matrix

$$A = \begin{bmatrix} 4 & 1 & 3 \\ 1 & \dfrac{5}{2} & 0 \\ 3 & 0 & 15 \end{bmatrix}$$

in the form $U^T U$ where U is upper triangular.

3.4 Ill-conditioning

It should not be thought that all problems associated with solution of linear equations have now been overcome. The following example provides an introduction to the sort of numerical complications which can occur.

Example 3.10 Consider the equations

$$28x_1 + 25x_2 = 30 \tag{3.40}$$

$$19x_1 + 17x_2 = 20 \tag{3.41}$$

and suppose that by some means we have calculated (to two significant figures) a solution $x_1 = 18$, $x_2 = -19$. When these values are substituted into the left-hand sides of (3.40) and (3.41) they give 29 and 19 respectively, so it would seem reasonable to assume that this solution is a close approximation to the correct one. In fact the *exact* solution of (3.40) and (3.41) is $x_1 = 10$, $x_2 = -10$.

Suppose next that the right-hand side of (3.41) changes from 20 to 19, all else remaining unaltered. Then the exact solution changes to $x_1 = 35$, $x_2 = -38$, a disproportionately large variation.

Equations (3.40) and (3.41) are an example of an ill-conditioned set, when small changes in coefficients in the equations lead to much larger changes in the values of the x's. Such systems need special attention. One sign of ill-conditioning is the presence of a relatively small pivot. A simple (although not foolproof) test for detecting ill-conditioning is to determine by recalculation whether the solution alters drastically after small changes have been made to the coefficients in the original problem.

We shall consider another kind of solution method in Section 6.7.1, and problems of conditioning will be re-visited in Section 14.3.3. However, further discussion of practical difficulties and methods which have been developed to overcome them lies outside the scope of this book, and the reader is referred to Golub and Van Loan (1989) and Stewart (1973). Such details are of primary interest to numerical specialists, since in most cases where large systems of linear equations have to be solved there will usually be access to a computer and a well tried and tested library program.

Exercise 3.12 Using a pocket calculator, determine the solution of the equations

$$1.985x_1 - 1.358x_2 = 2.212$$
$$0.953x_1 - 0.652x_2 = b_2$$

(a) when $b_2 = 1.062$, (b) when $b_2 = 1.063$, giving your answers correct to three decimal places.

Problems

3.1 Using a pocket calculator, determine the solution of

$$3.41x + 1.23x_2 - 1.09x_3 = 4.72$$
$$2.71x_1 + 2.14x_2 + 1.29x_3 = 3.10$$
$$1.89x_1 - 1.91x_2 - 1.89x_3 = 2.91$$

correct to two decimal places, using gaussian elimination with partial pivoting, and applying checks.

3.2 A square matrix is called *tridiagonal* if its only non-zero elements lie on the principal diagonal and on the diagonals immediately above and below this, e.g.

$$A = \begin{bmatrix} a_1 & b_1 & 0 & 0 \\ c_1 & a_2 & b_2 & 0 \\ 0 & c_2 & a_3 & b_3 \\ 0 & 0 & c_3 & a_4 \end{bmatrix}. \tag{3.42}$$

Express A as the product LU in the form

$$A = \begin{bmatrix} 1 & 0 & 0 & 0 \\ l_1 & 1 & 0 & 0 \\ 0 & l_2 & 1 & 0 \\ 0 & 0 & l_3 & 1 \end{bmatrix} \begin{bmatrix} u_1 & v_1 & 0 & 0 \\ 0 & u_2 & v_2 & 0 \\ 0 & 0 & u_3 & v_3 \\ 0 & 0 & 0 & u_4 \end{bmatrix}, \tag{3.43}$$

where the u's are all non-zero, and determine the conditions on the a's, b's, c's for (3.43) to hold. This can be extended to $n \times n$ tridiagonal matrices with L, U *bidiagonal*, as in (3.43), and gives a simple method for solving (3.2) in this case.

3.3 Using (3.43), solve the equations

$$\begin{aligned} x_1 + 2x_2 &= 8 \\ 2x_1 - x_2 + x_3 &= -1 \\ 3x_2 - 3x_3 - x_4 &= 10 \\ 7x_3 + 4x_4 &= 6 \end{aligned} \quad .$$

3.4 The equation of a circle in a cartesian coordinates is $x_1^2 + x_2^2 + 2gx_1 + 2fx_2 + c = 0$. Determine the equation of the circle which passes through the points $(7, 5)$, $(6, -2)$, $(-1, -1)$, by solving the appropriate equations using gaussian elimination.

3.5 Use gaussian elimination to find the matrix X such that $AX = B$, where

$$A = \begin{bmatrix} 4 & 2 & -1 \\ 5 & 3 & -1 \\ 3 & -1 & 4 \end{bmatrix}, \qquad B = \begin{bmatrix} 1 & 3 \\ -1 & 1 \\ 6 & 4 \end{bmatrix}.$$

3.6 Show that a real symmetric 2×2 matrix A can be expressed as $U^T U$, with U a real upper triangular matrix having non-zero diagonal elements, only if $a_{11} > 0$, $a_{11}a_{22} > a_{12}^2$. (See Problem 7.2 for a generalization of this result.)

3.7 Systems of equations with complex coefficients can be converted into the real case as follows:

Let $A = A_1 - iA_2$, $b = b_1 + ib_2$, $x = x_1 + ix_2$ in (3.2) with the matrices A_1, A_2, and the column vectors b_1, b_2, x_1, x_2 all real. Hence show that (3.2) can be written in the real form

$$D\begin{bmatrix} x_1 \\ x_2 \end{bmatrix} = \begin{bmatrix} b_1 \\ b_2 \end{bmatrix},$$

where D is the $2n \times 2n$ real matrix defined in (2.76).

3.8 Using gaussian elimination, determine the condition to be satisfied by the parameter a so that the equations

$$x_1 + 2x_2 + 3x_3 = 9$$
$$3x_1 + 7x_2 + 6x_3 = 13$$
$$x_1 + ax_2 + 8x_3 = 30$$

possess a unique solution.

3.9 Use gaussian elimination to find the matrix X which satisfies the equation

$$\begin{bmatrix} 1 & 2 \\ -3 & 4 \end{bmatrix} X - X \begin{bmatrix} 0 & -2 \\ 1 & 5 \end{bmatrix} = \begin{bmatrix} -1 & 6 \\ -13 & -8 \end{bmatrix}.$$

References

Golub and Van Loan (1989), Stewart (1973).

4 Determinant and inverse

The reader who worked through Exercise 3.1 will have found that the solution of

$$a_{11}x_1 + a_{12}x_2 = b_1$$
$$a_{21}x_1 + a_{22}x_2 = b_2$$
(4.1)

is

$$x_1 = (a_{22}b_1 - a_{12}b_2)/d, \qquad x_2 = (-a_{21}b_1 + a_{11}b_2)/d, \quad (4.2)$$

where

$$d = a_{11}a_{22} - a_{12}a_{21}. \quad (4.3)$$

The solution (4.2) is valid only if $d \neq 0$, so d *determines* whether the equations (4.1) have a unique solution. For this reason d is called the *determinant* of the matrix of coefficients

$$A = \begin{bmatrix} a_{11} & a_{12} \\ a_{21} & a_{22} \end{bmatrix} \quad (4.4)$$

and is written $\det A$, $|A|$,

$$\det \begin{bmatrix} a_{11} & a_{12} \\ a_{21} & a_{22} \end{bmatrix} \quad \text{or} \quad \begin{vmatrix} a_{11} & a_{12} \\ a_{21} & a_{22} \end{vmatrix}.$$

We thus have the *definition*

$$\det A = a_{11}a_{22} - a_{12}a_{21} \quad (4.5)$$

for the general 2×2 matrix A in (4.4). In fact, determinants were studied long before the introduction of matrices but are now much less important.

The solution (4.2) can be written as

$$\begin{bmatrix} x_1 \\ x_2 \end{bmatrix} = \frac{1}{d} \begin{bmatrix} a_{22} & -a_{12} \\ -a_{21} & a_{11} \end{bmatrix} \begin{bmatrix} b_1 \\ b_2 \end{bmatrix}, \quad (4.6)$$

so by comparison with (3.4), i.e. $x = A^{-1}b$, we can identify in this case

$$A^{-1} = \frac{1}{d} \begin{bmatrix} a_{22} & -a_{12} \\ -a_{21} & a_{11} \end{bmatrix} \qquad (4.7)$$

as the 'inverse' of A. It is left as an easy exercise to verify that for the matrices in (4.4) and (4.7)

$$AA^{-1} = A^{-1}A = I_2, \qquad (4.8)$$

which adds further justification for the notation A^{-1}.

The aim of this chapter is to develop properties of the determinant and inverse for general square matrices.

Exercise 4.1 For the matrix $A^{(1)}$ in (2.33) associated with the complex number z_1, show (a) det $A^{(1)} = |z_1|^2$, (b) $(A^{(1)})^{-1} \sim z_1^{-1}$.

4.1 Determinant

4.1.1 3 × 3 case

It is tedious but straightforward to solve eqns (3.1) with $n = 3$, and it turns out, as in (4.2), that the expressions for x_1, x_2, x_3 have a common denominator which in this case is

$$d = a_{11}a_{22}a_{33} - a_{11}a_{23}a_{32} + a_{12}a_{23}a_{31} - a_{12}a_{21}a_{33}$$
$$+ a_{13}a_{21}a_{32} - a_{13}a_{22}a_{31}. \qquad (4.9)$$

Again, the equations have a unique solution if and only if (4.9) is non-zero. It is convenient to factorize (4.9) as follows:

$$d = a_{11}(a_{22}a_{33} - a_{23}a_{32}) - a_{12}(a_{21}a_{33} - a_{23}a_{31}) + a_{13}(a_{21}a_{32} - a_{22}a_{31}) \qquad (4.10)$$

$$= a_{11} \begin{vmatrix} a_{22} & a_{23} \\ a_{32} & a_{33} \end{vmatrix} - a_{12} \begin{vmatrix} a_{21} & a_{23} \\ a_{31} & a_{33} \end{vmatrix} + a_{13} \begin{vmatrix} a_{21} & a_{22} \\ a_{31} & a_{32} \end{vmatrix}, \qquad (4.11)$$

the transition from (4.10) to (4.11) being accomplished using (4.5). The expression (4.11) can be regarded as the *definition* of the determinant of a general 3 × 3 matrix $A = [a_{ij}]$.

The notation introduced at the beginning of this chapter is used in general: for a square array the matrix is denoted by brackets and the determinant by vertical lines.

Exercise 4.2 Using (4.11) calculate det A when

$$A = \begin{bmatrix} 2 & -2 & 5 \\ 1 & 7 & -2 \\ 4 & -3 & 6 \end{bmatrix}.$$

Our definition of a 3×3 determinant in (4.11) is in terms of 2×2 determinants. To develop the definition of an $n \times n$ determinant depends upon appreciating how (4.11) is built up. The first 2×2 determinant in (4.11) is the determinant of the 2×2 submatrix obtained from A by deleting row 1 and column 1:

$$
\begin{matrix}
\textcircled{a_{11}} & a_{12} & a_{13} \\
a_{21} & a_{22} & a_{23} \\
a_{31} & a_{32} & a_{33}.
\end{matrix}
\tag{4.12}
$$

To obtain the second term in (4.11) row 1 and column 2 are deleted:

$$
\begin{matrix}
a_{11} & \textcircled{a_{12}} & a_{13} \\
a_{21} & a_{22} & a_{23} \\
a_{31} & a_{32} & a_{33};
\end{matrix}
\tag{4.13}
$$

and similarly for the third term:

$$
\begin{matrix}
a_{11} & a_{12} & \textcircled{a_{13}} \\
a_{21} & a_{22} & a_{23} \\
a_{31} & a_{32} & a_{33}.
\end{matrix}
\tag{4.14}
$$

It is convenient to define the *minor M_{ij}* of a_{ij} as the determinant of the submatrix obtained from A by deleting row i and column j. We can then write (4.12), (4.13), (4.14) respectively as

$$M_{11} = \begin{vmatrix} a_{22} & a_{23} \\ a_{32} & a_{33} \end{vmatrix}, \qquad M_{12} = \begin{vmatrix} a_{21} & a_{23} \\ a_{31} & a_{33} \end{vmatrix}, \qquad M_{13} = \begin{vmatrix} a_{21} & a_{22} \\ a_{31} & a_{32} \end{vmatrix} \tag{4.15}$$

and (4.11) then becomes

$$\det A = a_{11}M_{11} - a_{12}M_{12} + a_{13}M_{13}. \tag{4.16}$$

The negative term in (4.16) can be avoided by defining the *cofactor A_{ij}* of a_{ij} as

$$A_{ij} = (-1)^{i+j}M_{ij}, \tag{4.17}$$

so $A_{11} = (-1)^2 M_{11}$, $A_{12} = (-1)^3 M_{12}$, $A_{13} = (-1)^4 M_{13}$, and (4.16) becomes

$$\det A = a_{11}A_{11} + a_{12}A_{12} + a_{13}A_{13}. \qquad (4.18)$$

The formula (4.18) is called the *expansion* of $\det A$ by the first row, since it expresses $\det A$ as a term-by-term product of the elements in the first row with their cofactors.

Example 4.1   Consider the matrix of coefficients of eqns (3.10), namely

$$A = \begin{bmatrix} 1 & -3 & 7 \\ 2 & 4 & -3 \\ -3 & 7 & 2 \end{bmatrix}. \qquad (4.19)$$

Then (4.18) gives

$$\det A = 1(-1)^2 \begin{vmatrix} 4 & -3 \\ 7 & 2 \end{vmatrix} + (-3)(-1)^3 \begin{vmatrix} 2 & -3 \\ -3 & 2 \end{vmatrix} + 7(-1)^4 \begin{vmatrix} 2 & 4 \\ -3 & 7 \end{vmatrix}$$

$$= (8 + 21) + 3(4 - 9) + 7(14 + 12)$$

$$= 196.$$

An unexpected fact is that $\det A$ can be expanded by *any* row (or *any* column): form the term-by-term product of the elements in any row (or column) with their cofactors. For example, expanding by the third row produces

$$\det A = a_{31}A_{31} + a_{32}A_{32} + a_{33}A_{33}$$

and the reader can easily verify that this gives the same expression as (4.10).

Example 4.1   *(continued)*   Expanding $\det A$ by the second column gives

$$\det A = a_{12}A_{12} + a_{22}A_{22} + a_{32}A_{32} \qquad (4.20)$$

$$= (-3)(-1)^3 \begin{vmatrix} 2 & -3 \\ -3 & 2 \end{vmatrix} + 4(-1)^4 \begin{vmatrix} 1 & 7 \\ -3 & 2 \end{vmatrix} + 7(-1)^5 \begin{vmatrix} 1 & 7 \\ 2 & -3 \end{vmatrix}$$

$$= 3(4 - 9) + 4(2 + 21) - 7(-3 - 14)$$

$$= 196, \text{ as before.}$$

Exercise 4.3   Evaluate (4.20) for a general 3×3 matrix and confirm that it agrees with (4.10).

A second unexpected property is that, if the elements in any row are multiplied term-by-term with the cofactors of a *different* row, the result is zero. For example, the first row taken with the cofactors of the third row gives

$$a_{11}A_{31} + a_{12}A_{32} + a_{13}A_{33} \qquad (4.21)$$

$$= a_{11}(-1)^4 \begin{vmatrix} a_{12} & a_{13} \\ a_{22} & a_{23} \end{vmatrix} + a_{12}(-1)^5 \begin{vmatrix} a_{11} & a_{13} \\ a_{21} & a_{23} \end{vmatrix} + a_{13}(-1)^6 \begin{vmatrix} a_{11} & a_{12} \\ a_{21} & a_{22} \end{vmatrix}$$

$$= a_{11}(a_{12}a_{23} - a_{13}a_{22}) - a_{12}(a_{11}a_{23} - a_{13}a_{21}) + a_{13}(a_{11}a_{22} - a_{12}a_{21})$$

$$= 0.$$

The same fact holds for columns: for example, taking the second column with the cofactors of the first column gives

$$a_{12}A_{11} + a_{22}A_{21} + a_{32}A_{31} = 0. \qquad (4.22)$$

Exercise 4.4 Verify (4.22) by evaluating in full.

4.1.2 General properties

The preceding results can be generalized for any $n \times n$ matrix A: we define the nth-order determinant by

$$\det A = a_{i1}A_{i1} + a_{i2}A_{i2} + \cdots + a_{in}A_{in} \quad \text{(expansion by ith row)}$$
$$(4.23)$$

$$= a_{1j}A_{1j} + a_{2j}A_{2j} + \cdots + a_{nj}A_{nj} \quad \text{(expansion by jth column)},$$
$$(4.24)$$

where A_{ij}, the cofactor of a_{ij}, is defined by (4.17). The minor M_{ij} is also defined as previously, being the determinant of the submatrix obtained by deleting row i, column j of A, but now having dimensions $(n - 1) \times (n - 1)$. Thus, for example, a 4×4 determinant is defined in terms of four 3×3 determinants, and so on. It is helpful to note that the sign $(-1)^{i+j}$ associated with A_{ij} can be taken from the pattern in Table 4.1, for example $A_{33} = +M_{33}$.

Exercise 4.5 Evaluate the following by expanding by the first row

$$\begin{vmatrix} 1 & 0 & 2 & 0 \\ -1 & 4 & 3 & 6 \\ 0 & -2 & 5 & -3 \\ 3 & 1 & 1 & 0 \end{vmatrix}.$$

Exercise 4.6 Prove that if all the elements in a row (or column) are zero then $\det A = 0$.

Table 4.1

		j				
		1	2	3	4	$\cdots$
	1	+	−	+	−	$\cdots$
	2	−	+	−	+	$\cdots$
i	3	+	−	+	−	$\cdots$
	$\vdots$	$\vdots$				

The result on multiplying rows with cofactors of different rows still holds:

$$a_{i1}A_{k1} + a_{i2}A_{k2} + \cdots + a_{in}A_{kn} = 0, \qquad k \neq i, \qquad (4.25)$$

and similarly for columns:

$$a_{1j}A_{1k} + a_{2j}A_{2k} + \cdots + a_{nj}A_{nk} = 0, \qquad k \neq j. \qquad (4.26)$$

It must be stressed that, except for $n = 2$ or 3, or when many of the elements are zero, the rules (4.23) and (4.24) are in general too unwieldy for practical evaluation of determinants. This topic will be covered separately in Section 4.2. However, (4.23) and (4.24) are needed when the elements of A are algebraic quantities rather than numbers, and this is important for the determination of the characteristic equation, studied in Chapter 6. The formulae (4.23)–(4.26) are also valuable for theoretical purposes, and can be used to derive the following properties of $n \times n$ determinants.

Property PD1

$$\det A^{\mathsf{T}} = \det A, \qquad \det A^* = \overline{(\det A)}. \qquad (4.27)$$

Property PD2

$$\det(kA) = k^n \det A, \qquad (4.28)$$

where k is a scalar (recall that kA was defined in Section 2.2.2).

Instead of always saying 'rows or columns' it is useful to use the word 'line' to refer to both, with the understanding that when we mention two lines they are to be parallel.

Property PD3   If any two lines of A are identical then $\det A = 0$.

Example 4.2 The preceding three properties can be demonstrated for the 2×2 matrix in (4.4). First we have

$$A^{\mathsf{T}} = \begin{bmatrix} a_{11} & a_{21} \\ a_{12} & a_{22} \end{bmatrix}, \qquad A^* = \begin{bmatrix} \bar{a}_{11} & \bar{a}_{21} \\ \bar{a}_{12} & \bar{a}_{22} \end{bmatrix};$$

so from (4.5)

$$\det A^{\mathsf{T}} = a_{11}a_{22} - a_{21}a_{12} = \det A$$
$$\det(A^*) = \bar{a}_{11}\bar{a}_{22} - \bar{a}_{21}\bar{a}_{12} = \overline{(a_{11}a_{22} - a_{21}a_{12})} = \overline{(\det A)}.$$

Next,

$$kA = \begin{bmatrix} ka_{11} & ka_{12} \\ ka_{21} & ka_{22} \end{bmatrix};$$

so from (4.5)

$$\det A = (ka_{11})(ka_{22}) - (ka_{12})(ka_{21})$$
$$= k^2 \det A,$$

agreeing with (4.28). Finally, if the second column of A is identical to the first, i.e. $a_{12} = a_{11}$, $a_{22} = a_{21}$, then again from (4.5)

$$\det A = a_{11}a_{22} - a_{11}a_{22} = 0,$$

showing that PD3 holds.

   The properties can also be used to prove facts about determinants without consideration of elements, as the following example illustrates.

Example 4.3 If A is skew symmetric and n is an odd integer, then by definition $A^{\mathsf{T}} = -A$ (see Section 2.3.2) so

$$\det A^{\mathsf{T}} = \det(-A)$$
$$= (-1)^n \det A, \quad \text{by (4.28)}.$$

Thus applying (4.27) we deduce that

$$\det A = (-1)^n \det A$$
$$= -\det A,$$

since n is odd, whence $\det A = 0$ for any such matrix A.

Exercise 4.7 Prove that the determinant of any hermitian matrix is a purely real number.

Property PD4 If A is a diagonal matrix then

$$\det A = a_{11}a_{22}a_{33} \cdots a_{nn}. \tag{4.29}$$

Property PD5   If A is a triangular matrix then

$$\det A = a_{11}a_{22}a_{33}\cdots a_{nn}. \tag{4.30}$$

Example 4.4   Using PD4 and PD5, we have

$$\begin{vmatrix} 2 & 0 & 0 \\ 0 & 4 & 0 \\ 0 & 0 & 7 \end{vmatrix} = 56, \qquad \begin{vmatrix} 2 & 3 & 5 \\ 0 & 4 & 1 \\ 0 & 0 & 7 \end{vmatrix} = 56.$$

The simple result (4.30) for evaluation of triangular determinants is interesting because it suggests a link with the gaussian elimination method of Section 3.2, whereby a square matrix can be reduced to triangular form. This relationship will be developed in Section 4.2, but first we need some further important properties.

Property PD6

(a) If any two lines of A are interchanged the value of $\det A$ is multiplied by -1.

(b) If any line of A is multiplied by a non-zero scalar k then $\det A$ is also multiplied by k.

(c) The value of $\det A$ is unchanged if an arbitrary multiple of any line is added to any other line.

Example 4.5   Let A be the 3×3 matrix in (4.19), which has $\det A = 196$. Then interchanging the first and third rows, denoted by $(R1) \leftrightarrow (R3)$, gives

$$\begin{vmatrix} -3 & 7 & 2 \\ 2 & 4 & -3 \\ 1 & -3 & 7 \end{vmatrix} = -196, \quad \text{by PD6(a)}.$$

Similarly, by PD6(c), $(R2) - 2(R1)$ and $(R3) + 3(R1)$ applied to (4.19) produce

$$\begin{vmatrix} 1 & -3 & 7 \\ 0 & 10 & -17 \\ 0 & -2 & 23 \end{vmatrix} = 196. \tag{4.31}$$

Exercise 4.8   Prove that if one line of A is a multiple of another line, then $\det A = 0$.

The properties PD6 are also useful when the elements of a determinant are given as algebraic rather than numerical quantities, as the following two examples illustrate.

Example 4.6

$$
\begin{vmatrix} 1 & 1 & 1 \\ a & b & c \\ a^2 & b^2 & c^2 \end{vmatrix} = \begin{vmatrix} 1 & 0 & 0 \\ a & (b-a) & (c-a) \\ a^2 & (b^2-a^2) & (c^2-a^2) \end{vmatrix} \quad \begin{array}{l} (C2)-(C1) \\ (C3)-(C1) \end{array}
$$

$$
= \begin{vmatrix} (b-a) & (c-a) \\ (b^2-a^2) & (c^2-a^2) \end{vmatrix}, \qquad \text{expanding by first row,}
$$

$$
= (b-a)(c-a) \begin{vmatrix} 1 & 1 \\ b+a & c+a \end{vmatrix} \tag{4.32}
$$

$$
= (b-a)(c-a)(c+a-b-a)
$$

$$
= (b-a)(c-a)(c-b).
$$

Note that (4.32) is obtained from the preceding determinant by applying PD6(b) in reverse, so that the factors $(b-a)$ and $(c-a)$ can be removed from the first and second columns respectively. (Following Section 3.2, the notation $(C2)-(C1)$ means column 2 has column 1 subtracted from it, etc.)

Example 4.7

$$
\begin{vmatrix} a & b+c & 1 \\ b & c+a & 1 \\ c & a+b & 1 \end{vmatrix} = \begin{vmatrix} a & a+b+c & 1 \\ b & a+b+c & 1 \\ c & a+b+c & 1 \end{vmatrix} \quad (C2)+(C1)
$$

$$
= (a+b+c) \begin{vmatrix} a & 1 & 1 \\ b & 1 & 1 \\ c & 1 & 1 \end{vmatrix}, \qquad \text{by PD6(b),}
$$

$$
= 0, \quad \text{by PD3.}
$$

Exercise 4.9 Express the 4×4 determinant in Exercise 4.5 in terms of a single 3×3 determinant by reducing the first row to $1, 0, 0, 0$. Hence evaluate the determinant.

Exercise 4.10 Without evaluating directly any of the determinants, prove

(a) $\begin{vmatrix} a-b & b-c & c-a \\ b-c & c-a & a-b \\ c-a & a-b & b-c \end{vmatrix} = 0;$

(b) $\begin{vmatrix} (1+a) & b & c \\ a & (1+b) & c \\ a & b & (1+c) \end{vmatrix} = 1+a+b+c;$

(c) $\begin{vmatrix} bcd & a & a^2 & a^3 \\ acd & b & b^2 & b^3 \\ abd & c & c^2 & c^3 \\ abc & d & d^2 & d^3 \end{vmatrix} = \begin{vmatrix} 1 & a^2 & a^3 & a^4 \\ 1 & b^2 & b^3 & b^4 \\ 1 & c^2 & c^3 & c^4 \\ 1 & d^2 & d^3 & d^4 \end{vmatrix}.$

Property PD7 If A and B are two $n \times n$ matrices then

$$\det(AB) = (\det A)(\det B). \tag{4.33}$$

The result in (4.33) is by no means obvious: it states that if the product AB is evaluated then $\det(AB)$ is equal to the product of the determinants of A and of B. Note, however, that in general $\det(A + B) \neq \det A + \det B$ (try some simple examples).

Example 4.8 If A is decomposed into triangular factors as in (3.21), then

$$\begin{aligned}
\det A &= \det(LU) \\
&= (\det L)(\det U), \quad \text{by PD7,} \\
&= (l_{11}l_{22}\cdots l_{nn})(u_{11}u_{22}\cdots u_{nn}), \quad \text{by PD5,} \\
&= u_{11}u_{22}\cdots u_{nn}
\end{aligned}$$

since all $l_{ii} = 1$.

Exercise 4.11 If A and B are two arbitrary $n \times n$ matrices, prove (a) $\det(AB) = \det(BA)$, (b) $\det(A^k) = (\det A)^k$ for any positive integer k.

To close this section it is worth noting that the term 'minor' is also used to denote the determinant of *any* square submatrix of A, not just one obtained by deleting only a single row and column. In particular the determinants of the (leading) principal submatrices of A (defined in Section 2.4) are called (*leading*) *principal minors* of A.

We have deliberately not given proofs of the properties of determinants presented in this section. Such proofs are relatively unimportant, except to mathematicians, and details can be found in very many books (e.g. Mirsky (1963)). What *is* important for readers of this book is an appreciation of the properties and the ways in which they can be applied.

4.1.3 Some applications

It is sometimes fashionable to play down the usefulness of determinants. However, they still play a role in many problems of practical interest not always directly connected with sets of linear equations.

Example 4.9 In the algebra of vectors in three dimensions, the *vector product* of two vectors **a** and **b** is a vector perpendicular to both of them defined by

$$\mathbf{a} \times \mathbf{b} = \begin{vmatrix} \mathbf{i} & \mathbf{j} & \mathbf{k} \\ a_1 & a_2 & a_3 \\ b_1 & b_2 & b_3 \end{vmatrix}, \qquad (4.34)$$

where the determinant is to be expanded by the first row (expansion in any other way is not meaningful here). In (4.34) $\mathbf{i}$, $\mathbf{j}$ and $\mathbf{k}$ are the unit vectors along the rectangular cartesian coordinate axes and a_1, a_2, a_3, b_1, b_2, b_3 are the components of $\mathbf{a}$ and $\mathbf{b}$ respectively in these directions. The *modulus* (or length) of any vector $\mathbf{x} = x_1\mathbf{i} + x_2\mathbf{j} + x_3\mathbf{k}$ is

$$|\mathbf{x}| = (x_1^2 + x_2^2 + x_3^2)^{1/2}. \qquad (4.35)$$

It can be shown that the area of a triangle having $\mathbf{a}$ and $\mathbf{b}$ as two sides is $\frac{1}{2}|\mathbf{a} \times \mathbf{b}|$.

Exercise 4.12 If $\mathbf{a} = 2\mathbf{i} + 3\mathbf{j} - \mathbf{k}$, $\mathbf{b} = \mathbf{i} - \mathbf{j} + \mathbf{k}$, determine $\mathbf{a} \times \mathbf{b}$. Hence evaluate the area of the triangle having $\mathbf{a}$ and $\mathbf{b}$ as adjacent sides.

Exercise 4.13 When $\mathbf{x}$ and $\mathbf{y}$ are vectors in three dimensions the *scalar product* in (2.44) is written

$$\mathbf{x} \cdot \mathbf{y} = x_1 y_1 + x_2 y_2 + x_3 y_3.$$

It can be shown that the volume V of a tetrahedron having the vectors $\mathbf{a}$, $\mathbf{b}$, $\mathbf{c}$ as concurrent edges is given in terms of the *scalar triple product* $(\mathbf{a} \times \mathbf{b}) \cdot \mathbf{c}$, i.e.

$$V = \frac{1}{6}(\mathbf{a} \times \mathbf{b}) \cdot \mathbf{c}.$$

Use (4.34) to prove that

$$V = \frac{1}{6}\begin{vmatrix} a_1 & a_2 & a_3 \\ b_1 & b_2 & b_3 \\ c_1 & c_2 & c_3 \end{vmatrix}.$$

Example 4.10 In control theory a system is said to be *controllable* if it is possible to manipulate the control variables in such a way that the system starts out from any initial state and finishes up in any desired state—for example, transferring a spacecraft from an orbit round the Earth to a specified orbit round the Moon, or to a 'soft landing' on the Moon, by suitably controlling the rocket motors. A large class of control systems can be described by a set of linear differential equations

$$\frac{\mathrm{d}x}{\mathrm{d}t} = Ax + bu, \qquad (4.36)$$

where $x = [x_1, x_2, \ldots, x_n]^T$ is the vector of variables describing the state of the system, $dx/dt = [dx_1/dt, \ldots, dx_n/dt]^T$, A is a given constant $n \times n$ matrix, b is a given constant column n-vector, and u is the scalar control variable which can be manipulated.

It can be shown that (4.36) is controllable if and only if $\det \mathscr{C} \neq 0$, where $\mathscr{C}$ is the *controllability* matrix and has columns b, Ab, $A^2b, \ldots, A^{n-1}b$, i.e.

$$\mathscr{C} = [b, Ab, A^2b, \ldots, A^{n-1}b]. \tag{4.37}$$

Exercise 4.14 It can be shown that a certain circuit, represented in Fig. 4.1, is described by (4.36) with

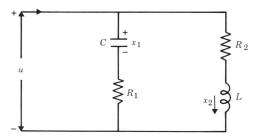

Fig. 4.1   Electric circuit for Exercise 4.14.

$$A = \begin{bmatrix} -\dfrac{1}{R_1 C} & 0 \\ 0 & -\dfrac{R_2}{L} \end{bmatrix}, \qquad b = \begin{bmatrix} \dfrac{1}{R_1 C} \\ \dfrac{1}{L} \end{bmatrix}.$$

Obtain the condition which R_1, R_2, L and C must satisfy for the circuit to be controllable (i.e. it must be possible to change the voltage x_1 across the capacitor and the current x_2 through the inductor from any initial values to any other values merely by suitably altering the input voltage u).

Some further applications of determinants are included in the problems at the end of this chapter.

4.2   Evaluation of determinants

As remarked in Section 4.1.2, the definition (4.23) does not provide a feasible way of evaluating large determinants with numerical elements. The reason is easily seen: the expansion of a

3×3 determinant involves three 2×2 determinants, so involves $3 \times 2 = 3!$ products; similarly a 4×4 determinant involves $4 \times 3 \times 2 = 4!$ products, etc., and generally an $n \times n$ determinant contains $n!$ products if expanded by repeated use of (4.23) or (4.24). The number $n!$ increases very rapidly as n increases, for example $18! \simeq 6.4 \times 10^{15}$. Assuming a computer multiplication time of 10^{-7} s, an 18×18 determinant evaluated in this way would take approximately 20 years! Clearly an alternative method must be used.

The easiest way to proceed is to reduce a given $n \times n$ matrix A to triangular form using gaussian elimination, as described in Section 3.2. The operations which are performed on the *rows* of A to achieve the triangularization can now be recognized as precisely those described in properties PD6(a) and (c). Operations of type (c) leave $\det A$ unaltered, but those of type (a) introduce a factor -1. The final triangular determinant is equal, by PD5, to the product of the elements on the principal diagonal (i.e. the product of the pivots) so $\det A$ is equal to this product multiplied by $(-1)^k$, where k is the number of row interchanges (these are introduced when partial pivoting is used). It can be shown that the number of multiplications involved in gaussian elimination is of the order of $\frac{1}{3}n^3$ for an $n \times n$ determinant, so computer evaluation when $n = 18$ would take only about 2×10^{-4} s.

Example 4.11 Let A be the matrix in (4.19). To reduce the first column of $\det A$ we apply (R2) $-$ 2(R1), (R3) $-$ (-3)(R1) so that (see (4.31))

$$\det A = \begin{vmatrix} 1 & -3 & 7 \\ 0 & 10 & -17 \\ 0 & -2 & 23 \end{vmatrix}$$

$$= \begin{vmatrix} 1 & -3 & 7 \\ 0 & 10 & -17 \\ 0 & 0 & \dfrac{196}{10} \end{vmatrix} \quad (R3) - (-2/10)(R2)$$

$$= 1 \times 10 \times \frac{196}{10} = 196,$$

which agrees with the value found in Example 4.1.

In fact the matrix A in (4.19) is the matrix of eqns (3.10). The elimination operations used to reduce $\det A$ above are exactly the same as those used to reduce the augmented matrix B in (3.15) and (3.16).

It can now be appreciated that when gaussian elimination is performed on a system of equations $Ax = b$, the determinant of A is obtained as a by-product of the calculations, provided the number of row interchanges is recorded.

Exercise 4.15   From your working for Exercises 3.2 and 3.3 write down the values of the determinants of the corresponding matrices A.

Exercise 4.16   Using gaussian elimination evaluate

$$\text{(a)} \quad \begin{vmatrix} 1 & 1 & 1 \\ 35 & 37 & 34 \\ 23 & 26 & 25 \end{vmatrix}, \quad \text{(b)} \quad \begin{vmatrix} 1 & 2 & 3 & 4 \\ -1 & 1 & 2 & 3 \\ 1 & -1 & 1 & 2 \\ -1 & 1 & -1 & 5 \end{vmatrix}.$$

It will be recalled that the gaussian elimination method breaks down if at any stage it is impossible to find a non-zero pivot. In this case $\det A = 0$, as the following argument shows.

Example 4.12   Suppose that at the third stage of reducing a 4×4 determinant we obtain

$$\begin{vmatrix} k_1 & k_2 & k_4 & k_6 \\ 0 & k_3 & k_5 & k_7 \\ 0 & 0 & 0 & k_8 \\ 0 & 0 & 0 & k_9 \end{vmatrix}. \tag{4.38}$$

It is impossible to obtain a non-zero pivot in the 3, 3 position by row interchanges. Expanding (4.38) by the first column twice in succession gives

$$k_1 k_3 \begin{vmatrix} 0 & k_8 \\ 0 & k_9 \end{vmatrix},$$

which is zero because of the zero column (see Exercise 4.6).

A square matrix A for which $\det A = 0$ is called *singular*, otherwise if $\det A \neq 0$ then A is *non-singular*.

Exercise 4.17   Prove, using gaussian elimination, that

$$\det \begin{bmatrix} 1 & -2 & 1 & -1 \\ 1 & 5 & -7 & 2 \\ 3 & 1 & -5 & 3 \\ 2 & 3 & -6 & 0 \end{bmatrix} = 0.$$

It should be noted that the matrix in Exercise 4.17 is that of the equations in Exercise 3.4 which were found to be inconsistent. This again illustrates the point that eqns (3.2), i.e. $Ax = b$, have a unique solution only if $\det A \neq 0$ (a formal treatment of this important result will be given in Section 5.4.3).

Exercise 4.18 If X and Y are two arbitrary $n \times n$ non-singular matrices prove that each of X^TX, XX^T, XY, YX, and X^k ($k = 2, 3, 4, \ldots$) is also non-singular.

Exercise 4.19 Let A be a general $n \times n$ 'reversed lower triangular' matrix, having all elements above the secondary diagonal equal to zero, e.g. when $n = 3$

$$A = \begin{bmatrix} 0 & 0 & a_{13} \\ 0 & a_{22} & a_{23} \\ a_{31} & a_{32} & a_{33} \end{bmatrix}.$$

Evaluate $\det A$ by transforming it into usual lower triangular form. In particular, what is $\det K_n$, where K_n is the matrix defined in Problem 2.4?

4.3 Inverse

4.3.1 Definition and properties

Following (4.8), the *inverse B* of an $n \times n$ matrix A is defined to be a matrix satisfying

$$AB = BA = I_n. \tag{4.39}$$

Suppose there exist *two* matrices B and B_1 which both satisfy (4.39). Then $AB_1 = I$, which on premultiplication by B gives $BAB_1 = B$, so $IB_1 = B$ in view of (4.39), whence $B_1 = B$. Hence, if an inverse does exist then it is *unique*. We have already encountered the notation A^{-1} for the inverse of A. Next, application of (4.33) to (4.39) gives

$$\det(AB) = (\det A)(\det B) = \det I_n = 1, \tag{4.40}$$

the latter part of (4.40) following by use of (4.29). Now since B is an $n \times n$ matrix with finite elements, $\det B$ must also be a finite number, so if $\det A = 0$ then (4.40) implies that B does not exist. In other words, A^{-1} exists only if A is non-singular, in which case (4.39) can be written

$$AA^{-1} = A^{-1}A = I_n \tag{4.41}$$

and (4.40) implies

$$\det(A^{-1}) = \frac{1}{\det A}.$$ (4.42)

The converse also applies, namely that if A is non-singular then A^{-1} exists. In fact this can be demonstrated by giving an explicit formula for A^{-1}. For simplicity we detail the case $n = 3$. Consider the product AC where

$$C = \begin{bmatrix} A_{11} & A_{21} & A_{31} \\ A_{12} & A_{22} & A_{32} \\ A_{13} & A_{23} & A_{33} \end{bmatrix}$$ (4.43)

and the A_{ij} are the cofactors defined in Section 4.1. Using (4.23) and (4.25) it is easily verified that

$$AC = \begin{bmatrix} |A| & 0 & 0 \\ 0 & |A| & 0 \\ 0 & 0 & |A| \end{bmatrix}.$$ (4.44)

For example, the $1, 1$ element of AC is the term in (4.18); the $1, 3$ element of AC is the term in (4.21). Thus

$$AC = (|A|)I_3$$ (4.45)

and since $|A|$ is assumed non-zero we can divide in (4.45) to obtain

$$A\left(\frac{C}{|A|}\right) = I_3.$$

Similarly, it can be verified that $(C/|A|)A = I_3$, using (4.24) and (4.26). Thus the matrix $C/|A|$ is the inverse of A since it satisfies (4.39). The matrix C is called the *adjoint* of A, usually denoted by adj A. It is the *transpose* of the matrix of cofactors (note that in (4.43) the cofactors of the rows of A form the columns of C). The argument generalizes immediately for any value of n, so we have shown that if A is non-singular its inverse is

$$A^{-1} = \left(\frac{1}{\det A}\right) \text{adj } A.$$ (4.46)

Example 4.13 We use (4.46) to calculate the inverse of

$$A = \begin{bmatrix} 2 & -1 & 1 \\ 1 & 0 & 1 \\ 3 & -1 & 4 \end{bmatrix}. \tag{4.47}$$

Using the definition (4.17) the cofactors of the first row are

$$A_{11} = + \begin{vmatrix} 0 & 1 \\ -1 & 4 \end{vmatrix} = 1, \qquad A_{12} = - \begin{vmatrix} 1 & 1 \\ 3 & 4 \end{vmatrix} = -1,$$

$$A_{13} = + \begin{vmatrix} 1 & 0 \\ 3 & -1 \end{vmatrix} = -1,$$

which gives the first column of adj A, and expanding by the first row

$$\det A = 2 \cdot 1 + (-1)(-1) + 1 \cdot (-1) = 2.$$

The other cofactors are evaluated similarly, and substitution into (4.46) gives

$$A^{-1} = \frac{1}{2} \begin{bmatrix} 1 & 3 & -1 \\ -1 & 5 & -1 \\ -1 & -1 & 1 \end{bmatrix}. \tag{4.48}$$

In general (4.46) is useless as a numerical method for finding A^{-1} since it involves calculation of n^2 determinants, each having order $n - 1$. We shall discuss practical evaluation of A^{-1} in Section 4.4. The special case of (4.46) when $n = 2$ was given in (4.7) and is worth remembering.

Exercise 4.20 Verify by direct multiplication that if D is the diagonal matrix in (2.36) then

$$D^{-1} = \text{diag}[1/d_{11}, 1/d_{22}, \ldots, 1/d_{nn}], \tag{4.49}$$

provided all $d_{ii} \neq 0$.

Exercise 4.21 Use (4.46) to calculate the inverse of

$$A = \begin{bmatrix} 2 & 1 & -5 \\ 3 & 2 & 4 \\ 1 & 0 & 3 \end{bmatrix}. \tag{4.50}$$

Going back to the discussion at the beginning of Chapter 3, we can now settle the question of 'division': provided $|A| \neq 0$ the unique solution of the equations $Ax = b$ is given by $x = A^{-1}b$.

Example 4.14 If A is the matrix in (4.47) and $b = [2, 4, -1]^T$, the solution of $Ax = b$ is

$$x = A^{-1} \begin{bmatrix} 2 \\ 4 \\ -1 \end{bmatrix} = \frac{1}{2} \begin{bmatrix} 15 \\ 19 \\ -7 \end{bmatrix},$$

using the expression for A^{-1} in (4.48).

Exercise 4.22 Using the result of Exercise 4.21 solve the equations

$$2x_1 + x_2 - 5x_3 = 1$$
$$3x_1 + 2x_2 + 4x_3 = 0$$
$$x_1 \qquad + 3x_3 = 2.$$

We can also now analyse a difficulty encountered in Section 2.2.3. We saw that, if $AC = AD$, it does not necessarily follow that $C = D$. However, if A is non-singular we have

$$A^{-1}AC = A^{-1}AD$$

which gives

$$IC = ID,$$

whence $C = D$. Thus the concept of inverse allows us to manipulate equations involving matrices in a manner analogous to ordinary algebra.

As a further example, which is of considerable importance, let A and B be two $n \times n$ non-singular matrices and let $X = (AB)^{-1}$. This implies from (4.39) that $XAB = I$, so we can construct the following sequence of identities:

$$XABB^{-1} = IB^{-1} = B^{-1},$$
$$XAI = B^{-1}, \qquad XAA^{-1} = B^{-1}A^{-1}, \qquad XI = B^{-1}A^{-1},$$

showing that $X = B^{-1}A^{-1}$, i.e.

$$(AB)^{-1} = B^{-1}A^{-1}. \tag{4.51}$$

Note again the reversal of order, as encountered previously in the result for $(AB)^T$ in (2.41).

We can similarly find the inverse of the transpose of a matrix. Writing $X = (A^T)^{-1}$ we have

$$A^T X = I \tag{4.52}$$

and taking the transpose of both sides of (4.52) gives, using (2.38) and (2.41),

$$X^T A = I^T = I. \tag{4.53}$$

Equation (4.53) implies that $X^{\mathsf{T}} = A^{-1}$, and transposing this gives $(X^{\mathsf{T}})^{\mathsf{T}} = X = (A^{-1})^{\mathsf{T}}$, so we have proved

$$(A^{\mathsf{T}})^{-1} = (A^{-1})^{\mathsf{T}}. \tag{4.54}$$

The identity (4.54) states that the order of the operations of transposing and inverting a matrix can be interchanged.

Exercise 4.23 If A, B, C are three $n \times n$ non-singular matrices, prove that $(ABC)^{-1} = C^{-1}B^{-1}A^{-1}$.

Exercise 4.24 Prove that $(A^2)^{-1} = (A^{-1})^2$, and similarly show that $(A^k)^{-1} = (A^{-1})^k$ for any positive integer k (thus we can write $(A^k)^{-1}$ as A^{-k}).

Exercise 4.25 If B is non-singular and $AB^{-1} = B^{-1}A$, what condition must A and B satisfy?

Exercise 4.26 Use (2.76) to prove that if A and B are two $n \times n$ non-singular matrices then $(A \otimes B)^{-1} = A^{-1} \otimes B^{-1}$.

Notice that in comparison with (4.51) the Kronecker product formula does not involve reversal of order (as for the transpose formula, (2.68)).

Exercise 4.27 Prove that if A is non-singular and symmetric then A^{-1} is also symmetric.

Exercise 4.28 If A is a non-singular matrix with complex elements, prove that $(A^*)^{-1} = (A^{-1})^*$.

Exercise 4.29 If A is the matrix in (4.50), find the matrix X such that $A^{\mathsf{T}}XA = I$.

4.3.2 Partitioned form

It is sometimes useful to obtain the inverse of a matrix in partitioned form. For example, suppose

$$F = \begin{bmatrix} A & B \\ C & D \end{bmatrix} \begin{matrix} n \\ k \end{matrix} \quad \begin{matrix} n & k \end{matrix}$$

and let

$$F^{-1} = \begin{bmatrix} W & X \\ Y & Z \end{bmatrix} \begin{matrix} n \\ k \end{matrix} \quad \begin{matrix} n & k \end{matrix}.$$

By definition $FF^{-1} = I_{n+k}$, so

$$\begin{bmatrix} A & B \\ C & D \end{bmatrix} \begin{bmatrix} W & X \\ Y & Z \end{bmatrix} = \begin{bmatrix} I_n & 0 \\ 0 & I_k \end{bmatrix} \tag{4.55}$$

and applying the rules of partitioned multiplication (Section 2.4) to (4.55) produces

$$AW + BY = I_n \tag{4.56}$$

$$AX + BZ = 0 \tag{4.57}$$

$$CW + DY = 0 \tag{4.58}$$

$$CX + DZ = I_k. \tag{4.59}$$

Suppose that F has been partitioned so that A is non-singular. Premultiplication of (4.56) by A^{-1} gives

$$W = A^{-1} - A^{-1}BY \tag{4.60}$$

and substitution into (4.58) gives

$$CA^{-1} - CA^{-1}BY + DY = 0,$$

from which

$$Y = -(D - CA^{-1}B)^{-1}CA^{-1} \tag{4.61}$$

provided $G = D - CA^{-1}B$ is non-singular.

Similarly from (4.57) and (4.59)

$$Z = G^{-1}, \qquad X = -A^{-1}BG^{-1}. \tag{4.62}$$

An advantage of using (4.60), (4.61) and (4.62) to calculate the inverse of the $(n + k) \times (n + k)$ matrix F is that the matrices to be inverted (A and G) have orders only n and k.

Exercise 4.30 Obtain the inverse of A in (4.47) using the partitioned form

$$\begin{bmatrix} 2 & -1 & 1 \\ 1 & 0 & 1 \\ \hline 3 & -1 & 4 \end{bmatrix}.$$

4.4 Calculation of inverse

As stated at the beginning of Chapter 3, A^{-1} is best evaluated by computing the solution of associated sets of linear equations. If

$X = A^{-1}$ and the columns of X are denoted by $X_1, X_2, \ldots, X_n$ then $AX = I$ implies

$$AX_i = e_i^T, \qquad i = 1, 2, \ldots, n, \qquad (4.63)$$

where e_i^T denotes the ith column of I_n (see Exercise 2.19). The equations (4.63) show that the ith column X_i of A^{-1} is the solution of the set of equations (3.2) with right-hand side equal to the ith column of I_n. This solution can be determined by gaussian elimination and back substitution.

Example 4.15 Let us determine the second column of the inverse of A in (4.19), which is the matrix of eqns (3.10). Replace the right-hand sides of eqns (3.10) by $0, 1, 0$ so the augmented matrix (3.14) is $B = [A, e_2^T]$, i.e.

$$B = \begin{bmatrix} 1 & -3 & 7 & \vdots & 0 \\ 2 & 4 & -3 & \vdots & 1 \\ -3 & 7 & 2 & \vdots & 0 \end{bmatrix} \rightarrow \begin{bmatrix} 1 & -3 & 7 & 0 \\ 0 & 10 & -17 & 1 \\ 0 & 0 & \dfrac{196}{10} & \dfrac{2}{10} \end{bmatrix}, \qquad (4.64)$$

using the same operations as in (3.15) and (3.16). The equations corresponding to (4.64) are

$$\begin{aligned} x_1 - 3x_2 + 7x_3 &= 0 \\ 10x_2 - 17x_3 &= 1 \\ 196x_3 &= 2 \end{aligned} \qquad (4.65)$$

and back substitution gives $x_3 = 2/196$, $x_2 = 23/196$, $x_1 = 55/196$, so the second column of A^{-1} is $(1/196)[55, 23, 2]^T$.

The transformations on each of the columns e_i^T can be carried out simultaneously by using an augmented matrix

$$B = [A \; \vdots \; I_n]. \qquad (4.66)$$

However, each column of A^{-1} must be found separately by back substitution on a set of equations like (4.65). The following modified form of gaussian elimination (called *Gauss–Jordan elimination*) is more convenient when carrying out calculations by hand.

The basic idea is to take the reduced triangular form of the augmented matrix and apply *further* row operations until the array of the first n columns is *diagonal*, with unit pivots. No back substitution is then required. The procedure is first illustrated for B in (4.64).

Example 4.15 (continued) Starting with the triangular form in (4.64), first reduce the pivots to unity by

$$\left(\frac{1}{10}\right)(R2), \qquad \left(\frac{10}{196}\right)(R3) \tag{4.67}$$

giving

$$\begin{bmatrix} 1 & -3 & 7 & \vdots & 0 \\ 0 & 1 & -\dfrac{17}{10} & \vdots & \dfrac{1}{10} \\ 0 & 0 & 1 & \vdots & \dfrac{2}{196} \end{bmatrix}. \tag{4.68}$$

Then reduce the upper part of the first three columns of (4.68), working from left to right:

$$\rightarrow \begin{bmatrix} 1 & 0 & \dfrac{19}{10} & \vdots & \dfrac{3}{10} \\ 0 & 1 & -\dfrac{17}{10} & \vdots & \dfrac{1}{10} \\ 0 & 0 & 1 & \vdots & \dfrac{2}{196} \end{bmatrix} \quad (R1) + 3(R2) \tag{4.69}$$

$$\rightarrow \begin{bmatrix} 1 & 0 & 0 & \vdots & \dfrac{55}{196} \\ 0 & 1 & 0 & \vdots & \dfrac{23}{196} \\ 0 & 0 & 1 & \vdots & \dfrac{2}{196} \end{bmatrix} \quad \begin{array}{l} (R1) - \left(\dfrac{19}{10}\right)(R3) \\[2mm] (R2) + \left(\dfrac{17}{10}\right)(R3). \end{array} \tag{4.70}$$

This is equivalent to converting the equations (4.65) into

$$x_1 = \frac{55}{196}$$

$$x_2 = \frac{23}{196}$$

$$x_3 = \frac{2}{196},$$

so the last column in (4.70) is precisely the desired second column of A^{-1}. *All* the columns of A^{-1} can be obtained simultaneously by commencing with the augmented matrix (4.66). Continuing with Example 4.15:

$$B = \begin{bmatrix} 1 & -3 & 7 & \vdots & 1 & 0 & 0 \\ 2 & 4 & -3 & \vdots & 0 & 1 & 0 \\ -3 & 7 & 2 & \vdots & 0 & 0 & 1 \end{bmatrix}$$

$$\rightarrow \begin{bmatrix} 1 & -3 & 7 & \vdots & 1 & 0 & 0 \\ 0 & 10 & -17 & \vdots & -2 & 1 & 0 \\ 0 & 0 & \dfrac{98}{5} & \vdots & \dfrac{26}{10} & \dfrac{2}{10} & 1 \end{bmatrix} \quad \begin{array}{l} (R2) - 2(R1) \\ (R3) + 3(R1) \\ \\ (R3) + \dfrac{2}{10}(R2), \end{array} \qquad (4.71)$$

these steps being just the gaussian elimination used to obtain (4.64). The operations in (4.67), (4.69), and (4.70) are then applied to (4.71), giving the following (only the effect on the last three columns is presented, the reduction of the first three columns is as before, in (4.68), (4.69), and (4.70)):

$$\begin{array}{c} \dfrac{1}{10}(R2) \\ \xrightarrow{\hspace{1cm}} \\ \dfrac{10}{196}(R3) \end{array} \begin{bmatrix} 1 & 0 & 0 \\ \dfrac{-2}{10} & \dfrac{1}{10} & 0 \\ \dfrac{26}{196} & \dfrac{2}{196} & \dfrac{10}{196} \end{bmatrix} \xrightarrow{(R1)+3(R2)} \begin{bmatrix} \dfrac{4}{10} & \dfrac{3}{10} & 0 \\ \dfrac{-2}{10} & \dfrac{1}{10} & 0 \\ \dfrac{26}{196} & \dfrac{2}{196} & \dfrac{10}{196} \end{bmatrix}$$

$$\begin{array}{c} (R1) - \left(\dfrac{19}{10}\right)(R3) \\ \xrightarrow{\hspace{2cm}} \\ (R2) + \left(\dfrac{17}{10}\right)(R3) \end{array} \begin{bmatrix} \dfrac{29}{196} & \dfrac{55}{196} & \dfrac{-19}{196} \\ \dfrac{5}{196} & \dfrac{23}{196} & \dfrac{17}{196} \\ \dfrac{26}{196} & \dfrac{2}{196} & \dfrac{10}{196} \end{bmatrix}.$$

Thus the inverse of A in (4.19) is

$$A^{-1} = \frac{1}{196} \begin{bmatrix} 29 & 55 & -19 \\ 5 & 23 & 17 \\ 26 & 2 & 10 \end{bmatrix}. \qquad (4.72)$$

Summarizing, to compute the inverse of an $n \times n$ non-singular matrix A:

(a) form the $n \times 2n$ matrix B in (4.66);

(b) reduce the first n columns of B to upper triangular form by gaussian elimination;

(c) divide each row of the reduced matrix by the pivot for that row;

(d) reduce the upper part of the first n columns to zero, starting with column 2 and working from left to right;

(e) the last n columns give A^{-1}.

Symbolically we can write

$$[A \mathbin{\vdots} I_n] \to [I_n \mathbin{\vdots} A^{-1}].$$

Notice that *any* set of equations $Ax = b$ could be solved by Gauss–Jordan elimination instead of ordinary gaussian elimination. However, the latter method is generally to be preferred in this case, since it involves a smaller total computational effort.

Exercise 4.31 Calculate the inverse of the matrices in Exercise 3.2 and in (4.50) using the Gauss–Jordan method.

Exercise 4.32 Calculate the inverse of the 4×4 coefficient matrix in Exercise 3.6 using the Gauss–Jordan procedure.

Notice that the Gauss–Jordan method is particularly simple when applied to an $n \times n$ triangular matrix (upper or lower), since only half the procedure is needed. For example, consider a 3×3 upper triangular non-singular matrix U. Applying the preceding rules (c) and (d) we obtain

$$[U \mathbin{\vdots} I] \to \begin{bmatrix} 1 & \dfrac{u_2}{u_1} & \dfrac{u_3}{u_1} & \vdots & \dfrac{1}{u_1} & 0 & 0 \\[2ex] 0 & 1 & \dfrac{u_5}{u_4} & \vdots & 0 & \dfrac{1}{u_4} & 0 \\[2ex] 0 & 0 & 1 & \vdots & 0 & 0 & \dfrac{1}{u_6} \end{bmatrix} \begin{matrix} \dfrac{1}{u_1}(\mathrm{R1}) \\[2ex] \dfrac{1}{u_4}(\mathrm{R2}) \\[2ex] \dfrac{1}{u_6}(\mathrm{R3}) \end{matrix}$$

$$\to \begin{bmatrix} 1 & 0 & \alpha & \vdots & \dfrac{1}{u_1} & -\dfrac{u_2}{u_1 u_4} & 0 \\[2ex] 0 & 1 & \dfrac{u_5}{u_4} & \vdots & 0 & \dfrac{1}{u_4} & 0 \\[2ex] 0 & 0 & 1 & \vdots & 0 & 0 & \dfrac{1}{u_6} \end{bmatrix} \begin{matrix} (\mathrm{R1}) - \dfrac{u_2}{u_1}(\mathrm{R2}), \\[2ex] {} \\ {} \end{matrix}$$

where $\alpha = (u_3/u_1) - (u_2/u_1)(u_5/u_4)$, and the operations $(\mathrm{R1}) - \alpha(\mathrm{R3})$ and $(\mathrm{R2}) - (u_5/u_4)(\mathrm{R3})$ then give U^{-1}. From the nature of the operations it is obvious that U^{-1} will also be upper triangular, and this applies for arbitrary n. Similarly, the inverse of an arbitrary non-singular lower triangular matrix is lower triangular. In either case it is straightforward to derive explicit algebraic formulae for the elements of the inverse for arbitrary n, as is shown below.

Exercise 4.33 Complete the above derivation of U^{-1}. Similarly, obtain the inverse of the 3×3 lower triangular matrix in (3.22).

Exercise 4.34 If the non-singular upper triangular matrix $U = [u_{ij}]$ has the upper triangular inverse $U^{-1} = [v_{ij}]$ verify by considering the identify $U^{-1}U = I$ that

$$v_{ij} = -\left[\sum_{r=i}^{j-1} v_{ir} u_{rj}\right]\Big/ u_{jj}, \qquad i < j \tag{4.73}$$

with $v_{ii} = 1/u_{ii}$.

It is interesting to link the Gauss–Jordan method with the triangular decomposition $A = LU$ described in Section 3.3. The equation $AX = I$ is equivalent to $LUX = I$, and the first (gaussian) part of the Gauss–Jordan procedure produces $UX = L^{-1}$, so the reduced array at this stage is $[U \vdots L^{-1}]$. Since $A^{-1} = (LU)^{-1} = U^{-1}L^{-1}$, on inverting U (as described above, or by using (4.73)) we can therefore express A^{-1} as a product of triangular factors.

Example 4.16 Consider again the matrix A in (4.19). The LU decomposition of A was given in (3.29). The gaussian elimination applied to $[A \vdots I]$ is displayed in (4.71), and it can be seen that the first three columns in (4.71) are identical to the matrix U in (3.29). The last three columns in (4.71) therefore represent the inverse L^{-1} of the matrix L in (3.29).

To summarize the procedure in general:
the first triangularization produces

$$[A \vdots I] \rightarrow [U \vdots L^{-1}],$$

the second application of gaussian elimination gives

$$[U \vdots I] \rightarrow [I \vdots U^{-1}],$$

and finally $A = U^{-1}L^{-1}$.

Exercise 4.35 Calculate U^{-1} for the matrix A in Example 4.16 using both the elimination procedure, and the explicit formula (4.73). Hence express A^{-1} as a product of triangular factors. Check, by evaluating the product, that your expression agrees with (4.72).

Repeat the procedure to express the inverse of A in (4.50) as a product of triangular factors. Check your answer by comparing the product with the inverse obtained in Exercise 4.21.

Once again, the book by Golub and Van Loan (1989) should be consulted for numerical details of matrix inversion.

4.5   Cramer's rule

Consider again the n equations in n unknowns, $Ax = b$, when they have the unique solution $x = A^{-1}b$. Substituting the expression (4.46) for A^{-1} gives

$$x = \left(\frac{1}{\det A}\right)(\text{adj } A)b. \tag{4.74}$$

For simplicity take $n = 3$, and use the expression for adj A given in (4.43). Then (4.74) becomes

$$x_1 = \left(\frac{1}{\det A}\right)(A_{11}b_1 + A_{21}b_2 + A_{31}b_3) \tag{4.75}$$

with similar expressions for x_2 and x_3. The interesting feature to notice about (4.75) is that the numerator is precisely the determinant

$$\begin{vmatrix} b_1 & a_{12} & a_{13} \\ b_2 & a_{22} & a_{23} \\ b_3 & a_{32} & a_{33} \end{vmatrix} \tag{4.76}$$

expanded by the first column, since the cofactors of this first column are A_{11}, A_{21}, A_{31}, the same as the cofactors of the first column of $\det A$. The generalization of (4.75) and (4.76) is similarly obtained as

$$x_i = \left(\frac{1}{\det A}\right) \times \left(\begin{array}{l}\text{determinant of the matrix obtained} \\ \text{from } A \text{ by replacing the } i\text{th column by } b\end{array}\right) \tag{4.77}$$

for $i = 1, 2, \ldots, n$. This result is known as *Cramer's rule*. It is occasionally useful for small values of n when not all of the coefficients in the equations are known numerically, but is worthless as a numerical method since it involves calculating $n + 1$ determinants, each having order n.

Example 4.17 For eqns (3.10) in Example 3.4, Cramer's rule gives

$$x_1 = \frac{1}{196} \begin{vmatrix} 2 & -3 & 7 \\ -1 & 4 & -3 \\ 3 & 7 & 2 \end{vmatrix} = -\frac{27}{98}$$

$$x_2 = \frac{1}{196} \begin{vmatrix} 1 & 2 & 7 \\ 2 & -1 & -3 \\ -3 & 3 & 2 \end{vmatrix} = \frac{19}{98}$$

$$x_3 = \frac{1}{196} \begin{vmatrix} 1 & -3 & 2 \\ 2 & 4 & -1 \\ -3 & 7 & 3 \end{vmatrix} = \frac{20}{49}$$

(det A was calculated to be 196 in Example 4.1).

Exercise 4.36 Solve the equations in Exercise 3.2 by Cramer's rule (det A has been evaluated in Exercise 4.15).

Problems

4.1 It can be shown that the quadratic equations $a_0\lambda^2 + a_1\lambda + a_2 = 0$ and $b_0\lambda^2 + b_1\lambda + b_2 = 0$ ($a_0 \neq 0$, $b_0 \neq 0$) have a common root if and only if

$$\begin{vmatrix} a_0 & a_1 & a_2 & 0 \\ 0 & a_0 & a_1 & a_2 \\ b_0 & b_1 & b_2 & 0 \\ 0 & b_0 & b_1 & b_2 \end{vmatrix} = 0$$

(this determinant is a special case of the *resultant* of two polynomials, and will be investigated in some detail in Section 11.2.2). Use this result to show that the equations

$$ax^2 + x + (1 - a) = 0, \qquad (1 - b)x^2 + x + b = 0$$

have a common root for *any* values of a and b.

4.2 An equation with real coefficients

$$a_0\lambda^3 + a_1\lambda^2 + a_2\lambda + a_3 = 0$$

($a_0 \neq 0$) has all its roots with negative real parts if and only if the following conditions are satisfied:

$$a_1 > 0, \qquad \begin{vmatrix} a_1 & a_3 \\ a_0 & a_2 \end{vmatrix} > 0, \qquad \begin{vmatrix} a_1 & a_3 & 0 \\ a_0 & a_2 & 0 \\ 0 & a_1 & a_3 \end{vmatrix} > 0.$$

These conditions are important in the study of stability of linear continuous-time systems (like that in Example 1.7), and can be generalized for nth-degree equations (see Section 11.4.1).

Determine for what values of k the equation

$$(3 - k)\lambda^3 + 2\lambda^2 + (5 - 2k)\lambda + 2 = 0$$

has all its roots with negative real parts.

4.3   The equations

$$i_1 R_6 + (i_1 - i_2)R_3 + (i_1 - i_3)R_4 = e$$
$$(i_1 - i_3)R_4 - i_3 R_1 - (i_3 - i_2)R_5 = 0$$
$$(i_1 - i_2)R_3 + (i_3 - i_2)R_5 - i_2 R_2 = 0$$

describe the Wheatstone bridge shown in Fig. 4.2. Obtain expressions for i_2 and i_3 as ratios of determinants using Cramer's rule. Hence show that if there is no current through the galvanometer (i.e. $i_2 = i_3$) then $R_1 R_3 = R_2 R_4$.

4.4   In Example 1.7 let a force $u(t)$ be applied to the right-hand mass as shown in Fig. 1.4. This leads to a term $[0, 0, 1/m_1, 0]^T u$ added to the right-hand side of eqns (1.10). If $m_1 = m_2 = 1$, $d_1 = d_2 = 1$, and

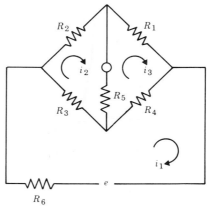

Fig. 4.2   Wheatstone bridge for Problem 4.3.

$k_2 = \frac{1}{4}$, use (4.37) to show that the condition for the system to be controllable is $k_1 \neq \frac{1}{2}$.

4.5 If

$$M = \begin{bmatrix} \overset{n}{A} & \overset{m}{0} \\ C & D \end{bmatrix} \begin{matrix} n \\ m \end{matrix}, \qquad (4.78)$$

then M is called *lower block triangular*.

(a) If A is a lower triangular matrix, prove that

$$\det M = (a_{11}a_{22}\cdots a_{nn})\det D.$$

(b) Hence deduce that if A is an *arbitrary* $n \times n$ matrix then $\det M = \det A \det D$.

4.6 If in (4.78) A and D are both non-singular, obtain M^{-1} in the same partitioned form.

4.7 Let

$$U = \begin{bmatrix} \overset{n}{W} & \overset{m}{X} \\ Y & Z \end{bmatrix} \begin{matrix} n \\ m \end{matrix}, \qquad V = \begin{bmatrix} W^{-1} & 0 \\ -YW^{-1} & I_m \end{bmatrix},$$

where W is non-singular. By applying PD7 to the product VU and using the result of Problem 4.5(b), prove that

$$\det U = \det W \det(Z - YW^{-1}X). \qquad (4.79)$$

The matrix $Z - YW^{-1}X = (U/W)$ is called the *Schur complement* of W in U.

Similarly, it can be shown that if Z is non-singular then $\det U = \det Z \det(W - XZ^{-1}Y)$.

4.8 Consider eqns (4.36) describing a linear control system. In practice it may be possible to measure only a scalar *output*

$$y = cx = c_1x_1 + c_2x_2 + \cdots + c_nx_n.$$

If $\bar{x}(s) = \mathcal{L}\{x(t)\}$ denotes the Laplace transform of $x(t)$, then the transform of dx/dt is $s\bar{x}$, assuming $x(0) = 0$.

Taking the Laplace transform of (4.36) thus gives

$$s\bar{x} = A\bar{x} + b\bar{u}$$

and on rearrangement we have

$$(sI_n - A)\bar{x} = b\bar{u}.$$

Hence

$$\bar{y} = c\bar{x}$$
$$= c(sI_n - A)^{-1}b\bar{u}$$
$$= g(s)\bar{u},$$

where the rational function $g(s) = c(sI_n - A)^{-1}b$ is called the *transfer function* of the system, since it relates the Laplace transform of the output y to that of the input u.

Use (4.79) to prove that

$$g(s) = \left.\begin{vmatrix} (sI - A) & b \\ -c & 0 \end{vmatrix}\right/ |sI - A| .$$

4.9   Prove that if A is non-singular then $\det(\text{adj } A) = (\det A)^{n-1}$ (hint: use PD2 and PD7). Show also that $\text{adj}(A^{\mathsf{T}}) = (\text{adj } A)^{\mathsf{T}}$.

4.10   A real $n \times n$ matrix U is called *orthogonal* if $U^{\mathsf{T}}U = I_n$. Prove

   (a)   The 2×2 matrix in (1.7) is orthogonal.
   (b) $U^{-1} = U^{\mathsf{T}}$.
   (c) $\det U = \pm 1$.
   (d) U^{T} and U^{-1} are orthogonal.
   (e)   The product of any two $n \times n$ orthogonal matrices is also orthogonal.
   (f) $\dfrac{1}{\sqrt{2}}\begin{bmatrix} U & U \\ -U & U \end{bmatrix}$ is a $2n \times 2n$ orthogonal matrix.
   (g)   The diagonal elements u_{ii} of U satisfy $|u_{ii}| \leq 1$.

4.11   A *complex* $n \times n$ matrix U is called *unitary* if $U^*U = I_n$.

   (a)   Prove that $|\det U| = 1$.
   (b)   If S is an arbitrary $n \times n$ skew hermitian matrix it can be shown (see Problem 6.2) that $I_n + S$ is non-singular. Prove that $(I_n - S)(I_n + S)^{-1}$ is unitary.
   (c)   If U is such that $I_n + U$ is non-singular, prove that $(I_n - U)(I_n + U)^{-1}$ is skew hermitian.

4.12   If A and B are two $n \times n$ real orthogonal matrices, prove that $A + B = A(A^{\mathsf{T}} + B^{\mathsf{T}})B$. Hence prove that if $\det A + \det B = 0$, then $A + B$ is singular.

4.13   If P is an $n \times n$ non-singular matrix and $B = P^{-1}AP$, prove that

$B^k = P^{-1}A^kP$ for any positive integer k. If

$$A = \begin{bmatrix} 2 & 1 \\ 1 & 2 \end{bmatrix}, \qquad P = \begin{bmatrix} 1 & 1 \\ -1 & 1 \end{bmatrix}$$

verify that B is diagonal. Write down B^5 and hence calculate A^5.

4.14 If X is an $n \times n$ matrix satisfying the equation $X^2 = X$, it is called *idempotent*. Deduce that either $X = I_n$ or X is singular.

4.15 If A is an $n \times n$ matrix such that $A^k = 0$, $A^{k-1} \neq 0$ for some positive integer k, it is called *nilpotent of index k* (an example was given in Problem 2.8). Prove (a) A is singular; (b) $(I - A)^{-1} = I + A + A^2 + \cdots + A^{k-1}$.

4.16 The $n \times n$ *Vandermonde* matrix V_n is defined by

$$V_n = \begin{bmatrix} 1 & 1 & \cdots & 1 \\ \lambda_1 & \lambda_2 & \cdots & \lambda_n \\ \lambda_1^2 & \lambda_2^2 & \cdots & \lambda_n^2 \\ \vdots & \vdots & & \vdots \\ \lambda_1^{n-1} & \lambda_2^{n-1} & \cdots & \lambda_n^{n-1} \end{bmatrix} \tag{4.80}$$

Show that $\det V_2 = (\lambda_2 - \lambda_1)$, and by reducing the last column of $\det V_3$ to $[1, 0, 0]^T$, that

$$\det V_3 = (\lambda_3 - \lambda_2)(\lambda_3 - \lambda_1)\det V_2,$$

Similarly, prove that

$$\det V_n = (\lambda_n - \lambda_{n-1})(\lambda_n - \lambda_{n-2}) \cdots (\lambda_n - \lambda_1)\det V_{n-1}$$

and hence deduce that

$$\det V_n = \prod_{n \geq j > i \geq 1} (\lambda_j - \lambda_i). \tag{4.81}$$

Thus the Vandermonde matrix is non-singular if and only if all the λ's are different from each other.

4.17 It is required to determine a polynomial

$$y = a_0 + a_1 x + a_2 x^2 + \cdots + a_{n-1}x^{n-1}$$

so that it passes through n given points (x_1, y_1), (x_2, y_2), ..., (x_n, y_n). Show that the coefficients are determined by $aV = y$, where $a = [a_0, a_1, \ldots, a_{n-1}]$, $y = [y_1, \ldots, y_n]$ and V is a Vandermonde matrix in the form (4.80). (Notice that the solution for a is unique if all the x's are different, because of (4.81).)

4.18 If $A = \text{diag}[\lambda_1, \lambda_2, \ldots, \lambda_n]$ and $b = [b_1, b_2, \ldots, b_n]^T$ show that the controllability matrix in (4.37) is equal to

$$(\text{diag}[b_1, b_2, \ldots, b_n])V_n^T,$$

where V_n is defined in (4.80). Hence deduce that in this case (4.36) is controllable if and only if all the λ's are different from each other and all $b_i \neq 0$.

4.19 If A is non-singular and dA/dt and $d(A^{-1})/dt$ both exist, use (2.70) to prove that

$$\frac{d(A^{-1})}{dt} = -A^{-1}\frac{dA}{dt}A^{-1}$$

(this corresponds to $d(a^{-1})/dt = -a^{-2}\,da/dt$ for a scalar function $a(t)$). Hence obtain a formula for $d(BA^{-1})/dt$.

4.20 If the elements of a 2×2 matrix A are differentiable functions of t, prove that

$$\frac{d}{dt}(\det A) = \begin{vmatrix} \dot{a}_{11} & \dot{a}_{12} \\ a_{21} & a_{22} \end{vmatrix} + \begin{vmatrix} a_{11} & a_{12} \\ \dot{a}_{21} & \dot{a}_{22} \end{vmatrix}.$$

This can be extended to an $n \times n$ matrix:

$$\frac{d}{dt}(\det A) = \sum_{i=1}^{n} \det A_i,$$

where A_i is the matrix obtained from A by differentiating the ith row only.

4.21 Consider the matrix D defined in (2.77), associated with the complex matrix $A = A_1 + iA_2$. Deduce, using the result (b) of Problem 2.14, that $D^{-1} \sim A^{-1}$.

4.22 Consider the LU decomposition of Section 3.3, as described in (3.32) and (3.33). Prove that $|A_2| = a_{11}\alpha_1$, $|A_3| = a_{11}\alpha_1\alpha_2, \ldots$, $|A| = a_{11}\alpha_1\alpha_2 \cdots \alpha_{n-1}$. Hence deduce that for L and U to exist and be unique, all the leading principal submatrices A_r of A must be non-singular.

4.23 Consider the 4×4 triadiagonal matrix A in (3.42), and the product $A = LU$ in (3.43) with L and U bidiagonal. Obtain expressions for L^{-1} and U^{-1}, and hence deduce that in general A^{-1} will have *no* zero elements.

4.24 If $u(t)$ and $v(t)$ are differentiable functions of t and $y(t) = u(t)v(t)$, use the product rule for differentiation to show that

$$[u, \dot{u}, \ddot{u}]A = [y, \dot{y}, \ddot{y}],$$

where A is an upper triangular matrix. Similarly, if $z = u/v$ use the quotient rule for differentiation to show that

$$[u, \dot{u}, \ddot{u}]B = [z, \dot{z}, \ddot{z}],$$

where B is also an upper triangular matrix. Verify that $B = A^{-1}$.

4.25 The expression for a 3×3 determinant in (4.9) can be obtained as follows:

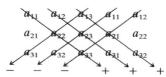

The first two columns of A are repeated on the right of the original array, and products formed along the arrows, with signs attached as shown. Use this method to calculate $\det A$ for A in (4.19).

4.26 The Fibonacci numbers $\{x_k\} = \{1, 1, 2, 3, 5, \ldots\}$ were defined in Problem 2.11. Prove by induction that the associated 2×2 matrix A defined in (2.75) satisfies

$$A^k = \begin{bmatrix} x_{k-1} & x_k \\ x_k & x_{k+1} \end{bmatrix}, \qquad k = 2, 3, 4, \ldots.$$

(a) Use the result in Exercise 4.11(b) to prove that

$$x_{k+1}x_{k-1} - x_k^2 = (-1)^k.$$

(b) Deduce from the identity $A^m A^{k-1} \equiv A^{m+k-1}$ that

$$x_{m+k} = x_m x_{k-1} + x_{m+1} x_k.$$

(c) Set $m = pk$ in the preceding result, and hence prove by induction that x_k is a factor of x_{pk} for all positive integers p.

References

Golub and Van Loan (1989), Mirsky (1963).

5 Rank, non-unique solution of equations, and applications

Apart from an informal discussion in Section 3.1 for the case of two equations and unknowns, we have until now assumed that in the equations

$$Ax = b \qquad (5.1)$$

A is square and the solution is unique. However, in many practical applications A may be rectangular—for example, in linear programming problems (Chapter 1) there are usually fewer equations than unknowns. In this chapter we develop the theory necessary to deal with the situation where the solution of (5.1) is not unique, including the case when A is rectangular. There are two problems—first, to determine whether the equations do indeed have a solution (i.e. whether they are *consistent*); and then, if so, to calculate this solution in its most general form. These problems will be solved completely in Section 5.4.

We begin by restating a result from Chapter 4:

5.1 Unique solution

Equations (5.1) have the unique solution $x = A^{-1}b$ if and only if A is a square non-singular matrix. If $b \equiv 0$ the equations (5.1) are called *homogeneous,* and this unique solution is then $x = A^{-1}0 = 0$, which is called the *trivial* solution. Thus when A is square and $b \equiv 0$, (5.1) can have a non-trivial solution $x \neq 0$ only if A is singular (i.e. $\det A = 0$), but this solution will not be unique.

Example 5.1 The pair of equations

$$x_1 + x_2 = 0, \qquad x_1 + 2x_2 = 0$$

clearly has only the trivial solution $x_1 = x_2 = 0$. However, if the second equation is replaced by $2x_1 + 2x_2 = 0$ then

$$A = \begin{bmatrix} 1 & 1 \\ 2 & 2 \end{bmatrix} \tag{5.2}$$

is singular, and the equations have the solution $x_1 = t$, $x_2 = -t$, where t is an arbitrary parameter.

Exercise 5.1 If x and y are two solutions of (5.1) with A non-singular, prove directly that $x \equiv y$.

5.2 Definition of rank

A new idea we need is that of the *rank* of an $m \times n$ matrix A. This is denoted by $R(A)$, and is defined to be the order of the largest non-singular square submatrix which can be formed by selecting rows and columns of A (as we noted in Section 2.4, a submatrix can include A itself).

Example 5.2

(a) The rank of A in (5.2) is equal to 1, since $\det A = 0$, but $|a_{11}| = 1 \neq 0$.

(b) The rank of

$$A = \begin{bmatrix} 2 & 4 & 8 \\ 1 & 2 & 1 \end{bmatrix} \tag{5.3}$$

is 2, since the submatrix formed by rows 1, 2 and columns 1, 3 is non-singular, i.e.

$$\det \begin{bmatrix} 2 & 8 \\ 1 & 1 \end{bmatrix} = 2 - 8 \neq 0.$$

Notice, however, that the submatrix formed by rows 1, 2 and columns 1, 2 is singular.

(c) Consider

$$A = \begin{bmatrix} 1 & 2 & 4 & 1 \\ 2 & 4 & 8 & 2 \\ 3 & 6 & 2 & 0 \end{bmatrix}. \tag{5.4}$$

Each of the four 3×3 submatrices obtained by dropping one of the columns of A is singular, since the second row is twice the first row (see Exercise 4.8). However, the submatrix formed by rows 1, 3 and columns 1, 3 is non-singular, i.e.

$$\det\begin{bmatrix} 1 & 4 \\ 3 & 2 \end{bmatrix} = 2 - 12 \neq 0$$

so $R(A) = 2$.

We note the following important facts about rank:

(a) The only matrices which have zero rank are those having *all* their elements zero.

(b) When forming submatrices the rows (and columns) need not be adjacent.

(c) If there is at *least one* $r \times r$ non-singular submatrix, but all possible larger-order square submatrices are singular, then $R(A)$ is equal to r—it doesn't matter how many such $r \times r$ non-singular submatrices there are.

(d) If A is $n \times n$ then $R(A) = n$ if and only if A is non-singular.

(e) If A is $m \times n$ then $R(A)$ cannot exceed the smaller of m and n, i.e. $R(A) \leqslant \min(m, n)$.

Exercise 5.2 For what values of k will the matrix

$$A = \begin{bmatrix} 1 & 2 & 3 & 2 \\ 3 & 6 & 9 & 6 \\ 4 & 8 & 12 & k \end{bmatrix}$$

have (a) $R(A) = 1$, (b) $R(A) = 2$, (c) $R(A) = 3$?

Exercise 5.3 Explain why $R(A^\mathsf{T}) = R(A)$ for any matrix A.

5.3 Equivalent matrices

The definition of rank may have seemed to the reader somewhat irrelevant to our proclaimed objective of dealing with linear equations. However, we now link the idea of rank with the fundamental tool of gaussian elimination, which we developed in Chapters 3 and 4.

5.3.1 *Elementary operations*

Let us return to the operations on determinants defined in property PD6, Section 4.1.2:

(a) interchange any two lines of a matrix $[(Li) \leftrightarrow (Lj)]$:
(b) multiply any line by a non-zero scalar k $[(Li) \times k]$;
(c) add an arbitrary multiple of any line to any other line $[(Li) + p(Lj)]$.

These operations applied to rows (or columns) of any matrix A are called *elementary row* (or *column*) *operations*. It is useful to notice how to reverse the effect of an elementary operation: for (a) the two lines are again interchanged; for (b), (Li) is multiplied by $(1/k)$; and the reverse of (c) is $(Li) - p(Lj)$. Thus the inverse of any elementary operation is also an elementary operation of the same type.

The importance of elementary operations for our purposes is the following fact:

Elementary operations do not alter the rank of a matrix.

The justification for this is now given, but the reader who finds the argument difficult to follow can pass on directly to Example 5.3 without loss of continuity. First, recall from Section 4.1.2 that when A is square, property PD6 states that operation (a) alters $\det A$ by a factor -1 and operation (b) alters $\det A$ by a factor k. When A is rectangular it follows that these operations have precisely the same effect on the determinants of *any* square submatrix of A, so singular submatrices remain singular and non-singular submatrices remain non-singular (i.e. (a) and (b) do not alter the rank of A). Next, consider elementary operations of type (c) applied to a rectangular matrix A having rank r. By definition all $(r + 1) \times (r + 1)$ submatrices S_{r+1} of A are singular. After the application of (c), only submatrices containing line i are altered. If S_{r+1} contains both line i *and* line j, then by PD6 its determinant is unaltered, i.e. S_{r+1} remains singular. If S_{r+1} contains only line i, it also remains singular, since an expansion of $\det S_{r+1}$ by the new ith line this determinant is seen to be a sum of minors of order $(r + 1)$ of the original matrix; e.g. if $r = 2$,

$i = 2, j = 4$, then

$$\begin{vmatrix} a_{11} & a_{12} & a_{13} \\ (a_{21} + pa_{41}) & (a_{22} + pa_{42}) & (a_{23} + pa_{43}) \\ a_{31} & a_{32} & a_{33} \end{vmatrix}$$

$$= \begin{vmatrix} a_{11} & a_{12} & a_{13} \\ a_{21} & a_{22} & a_{23} \\ a_{31} & a_{32} & a_{33} \end{vmatrix} + \begin{vmatrix} a_{11} & a_{12} & a_{13} \\ pa_{41} & pa_{42} & pa_{43} \\ a_{31} & a_{32} & a_{33} \end{vmatrix}$$

$$= 0 + p \cdot 0$$

since the two determinants on the right-hand side are 3×3 minors of A. Therefore, in all cases S_{r+1} remains singular after application of operation (c), and the same holds for higher-order square submatrices. Thus any operation of type (c) does not increase the rank of A. Furthermore, an operation of type (c) cannot reduce the rank of A; for if it could, then applying the inverse transformation (which is also an operation of type (c)) would bring back the original matrix, and hence increase the rank back to its original value. However, we have just seen that such an increase is impossible. Therefore, operations of type (c) do not change the rank.

Example 5.3　The matrix A in (5.4) has rank 2. So, therefore, does the matrix obtained from A by applying, for example,

$$(R2) - 2(R1), (R3) - 3(R1), (R3) \times (1 - 1/10) \tag{5.5}$$

namely,

$$\begin{bmatrix} 1 & 2 & 4 & 1 \\ 0 & 0 & 0 & 0 \\ 0 & 0 & 1 & \dfrac{3}{10} \end{bmatrix}. \tag{5.6}$$

To regain (5.4) from (5.6), apply to (5.6):

$$(R3) \times (-10), (R3) + 3(R1), (R2) + 2(R1). \tag{5.7}$$

The operations (5.7) are the inverses of those in (5.5), but notice that the order in which they are applied is reversed. This has a natural correspondence in real life: for example, to get dressed one puts on a shirt and fastens its buttons; to get undressed the operations are inverted and reversed in order, so first undo the buttons and then take off the shirt! This idea can be used to explain why for matrices the reversal of order occurs in $(AB)^{-1} = B^{-1}A^{-1}$.

Any matrix B obtained by applying elementary transformations to A is said to be *equivalent* to A, since B has the same rank as A; A and B are called *equivalent matrices*. Of course, gaussian elimination involves the application of elementary operations to rows, and therefore does not alter the rank of a matrix. It is thus not surprising that we can now show how to calculate rank using the gaussian procedure, suitably modified.

5.3.2 Calculation of rank

It is clearly not feasible to apply the definition as a practical way of calculating rank except for very simple cases, because a large number of determinants have to be evaluated. Instead, we obtain a simple procedure using elementary operations. The basic idea is to reduce any given matrix to an equivalent matrix whose rank can be determined by inspection.

Consider first the case when A is square. We simply carry out the triangularization using gaussian elimination, described in Section 4.2 for evaluating $\det A$. If all the pivots are non-zero, then $\det A \neq 0$, so $R(A) = n$. If at some stage it is impossible to find a non-zero pivot in a column, then, as we remarked in Section 3.2, the direct gaussian elimination procedure cannot be continued. This means that $\det A = 0$, so $R(A) < n$. The modification required is to interchange the offending *column* with a suitable column to the right. That is, using the notation of (3.17), if at the jth step it is not possible to choose a pivot in column j, i.e. $b_{ij} = 0$, for $i = j, j + 1, \ldots, n$, then interchange column j with some column k in which there *is* at least one non-zero element to use as a pivot, i.e. $b_{ik} \neq 0$ for some i such that $j \leq i \leq n$ (see Table 5.1).

A pivot is then selected from this new jth column and the procedure is continued in the usual way. When no further non-zero pivots can be found, $R(A)$ is equal to the total number r of non-zero pivots. This is because the largest non-singular submatrix of the final reduced triangular form is obtained by taking the first r rows and columns. An example should make this clear.

Table 5.1   Interchange of columns j and k

b_{11}	$\cdots$		b_{1j}	$\cdot$	$\cdot$	b_{1k}		$\cdot$	$\cdot$
	b_{22}		b_{2j}		$\cdot$	$\cdot$ b_{2k}		$\cdot$	$\cdot$
		$\cdot$	$\cdot$		$\cdot$ $\cdot$	$\cdot$		$\cdot$ $\cdot$	

$$
\begin{array}{c}
b_{11} \quad \cdots \qquad\qquad\; \overleftrightarrow{b_{1j} \quad\;\; \cdot\;\cdot\;\; b_{1k}} \qquad \cdot\;\cdot \\
\qquad b_{22} \qquad\qquad\quad b_{2j} \qquad \cdot\;\cdot\;\; b_{2k} \qquad\quad \cdot\;\cdot \\
\qquad\quad \cdot \qquad\qquad\qquad \cdot \qquad\;\; \cdot\;\cdot\;\; \cdot \qquad\quad\;\; \cdot\;\cdot \\
\qquad\qquad \cdot \qquad\qquad\quad \cdot \qquad\;\; \cdot\;\cdot\;\; \cdot \qquad\quad \cdot\;\cdot \\
\qquad\qquad b_{j-1,j-1} \quad b_{j-1,j} \quad \cdot\;\cdot\;\; b_{j-1,k} \quad \cdot\;\cdot \\
\qquad\qquad\quad 0 \qquad\qquad 0 \qquad\;\; \cdot\;\cdot\;\; b_{jk} \\
0 \qquad\qquad 0 \qquad\qquad 0 \qquad\;\; \cdot\;\cdot\;\; b_{j+1,k} \\
\qquad\qquad\quad \cdot \qquad\qquad \cdot \qquad\quad \cdot\;\cdot\;\; \cdot \\
\qquad\qquad\quad \cdot \qquad\qquad \cdot \qquad\quad \cdot\;\cdot\;\; \cdot \\
\qquad\qquad\quad 0 \qquad\qquad 0 \qquad\;\; \cdot\;\cdot\;\; b_{nk} \\
\qquad\qquad \text{column } j \qquad\quad \text{column } k
\end{array}
$$

at least one element is non-zero

Example 5.4 Suppose that after reducing the first two columns of a 4×4 matrix in the usual way we obtain an equivalent matrix B, which we then transform as follows:

$$
B = \begin{bmatrix} 1 & 2 & 3 & -1 \\ 0 & 2 & 1 & 4 \\ 0 & 0 & 0 & 6 \\ 0 & 0 & 0 & 3 \end{bmatrix} \rightarrow \begin{bmatrix} 1 & 2 & -1 & 3 \\ 0 & 2 & 4 & 1 \\ 0 & 0 & 6 & 0 \\ 0 & 0 & 3 & 0 \end{bmatrix} \quad (C3) \leftrightarrow (C4)
$$

$$
\rightarrow \begin{bmatrix} 1 & 2 & -1 & 3 \\ 0 & 2 & 4 & 1 \\ 0 & 0 & 6 & 0 \\ 0 & 0 & 0 & 0 \end{bmatrix} \quad (R4) - \frac{1}{2}(R3). \qquad (5.8)
$$

No further non-zero pivots can be found in (5.8). Because of the triangular form we see immediately that the largest non-singular submatrix in (5.8) is formed by the first three rows and columns. Hence $R(A) = 3$, the number of non-zero pivots.

Exercise 5.4 Using the triangularization procedure, calculate the rank of

$$
\begin{bmatrix} 1 & 3 & 4 & 6 \\ 2 & 4 & -1 & 2 \\ 3 & 7 & 3 & 8 \\ 5 & 11 & 2 & 10 \end{bmatrix}.
$$

The above modified gaussian elimination procedure incorporating column interchanges can be applied in exactly the same way if A is $m \times n$, as is now shown in some further examples.

Example 5.5   Using the 3×4 matrix A in (5.4) we obtain

$$A \rightarrow \begin{bmatrix} 1 & 2 & 4 & 1 \\ 0 & 0 & 0 & 0 \\ 0 & 0 & -10 & -3 \end{bmatrix} \quad \begin{array}{l} (R2) - 2(R1) \\ (R3) - 3(R1) \end{array}$$

$$\rightarrow \begin{bmatrix} 1 & 4 & 2 & 1 \\ 0 & 0 & 0 & 0 \\ 0 & -10 & 0 & -3 \end{bmatrix} \quad (C2) \leftrightarrow (C3)$$

$$\rightarrow \begin{bmatrix} 1 & 4 & 2 & 1 \\ 0 & -10 & 0 & -3 \\ 0 & 0 & 0 & 0 \end{bmatrix} \quad (R2) \leftrightarrow (R3). \qquad (5.9)$$

No more non-zero pivots can be found, showing $R(A) = 2$ as before.

Example 5.6   This illustrates the case $m > n$.

$$A = \begin{bmatrix} 1 & 2 & 3 \\ 3 & 6 & 10 \\ 2 & 5 & 7 \\ 1 & 2 & 4 \end{bmatrix} \rightarrow \begin{bmatrix} 1 & 2 & 3 \\ 0 & 0 & 1 \\ 0 & 1 & 1 \\ 0 & 0 & 1 \end{bmatrix} \quad \begin{array}{l} (R2) - 3(R1) \\ (R3) - 2(R1) \\ (R4) - (R1) \end{array}$$

$$\rightarrow \begin{bmatrix} 1 & 2 & 3 \\ 0 & 1 & 1 \\ 0 & 0 & 1 \\ 0 & 0 & 1 \end{bmatrix} \quad (R2) \leftrightarrow (R3)$$

$$\rightarrow \begin{bmatrix} 1 & 2 & 3 \\ 0 & 1 & 1 \\ 0 & 0 & 1 \\ 0 & 0 & 0 \end{bmatrix} \quad (R4) - (R3). \qquad (5.10)$$

There are three non-zero pivots, so $R(A) = 3$.

Thus, irrespective of whether A is square or rectangular, the non-zero pivots can be put into the $(1, 1)$, $(2, 2), \ldots, (r, r)$ positions, showing that $R(A) = r$.

Example 5.7

$$A = \begin{bmatrix} 0 & 1 & 3 & 0 & 0 & 6 \\ 0 & 0 & 0 & 0 & 2 & -3 \\ 0 & 0 & 0 & 5 & 0 & 2 \\ 0 & 0 & 0 & 0 & 0 & 1 \end{bmatrix} \rightarrow \begin{bmatrix} 1 & 0 & 0 & 6 & 3 & 0 \\ 0 & 2 & 0 & -3 & 0 & 0 \\ 0 & 0 & 5 & 2 & 0 & 0 \\ 0 & 0 & 0 & 1 & 0 & 0 \end{bmatrix}$$

by suitable column interchanges, showing $R(A) = 4$.

Exercise 5.5   Determine the rank of the following matrices:

$$\text{(a)} \begin{bmatrix} 0 & 1 & 3 & -2 \\ 2 & 2 & -1 & 1 \\ 4 & 5 & 1 & 0 \end{bmatrix}; \quad \text{(b)} \begin{bmatrix} 3 & 2 & 2 \\ 6 & 4 & 4 \\ -5 & -4 & -6 \\ 10 & 7 & 8 \end{bmatrix},$$

$$\text{(c)} \begin{bmatrix} 1 & 0 & -1 & 1 & 0 & 1 \\ 1 & 1 & 3 & 2 & 4 & 3 \\ 2 & 1 & 2 & 3 & 4 & 4 \\ 1 & -2 & -9 & 1 & -8 & -3 \\ 5 & 4 & 11 & 9 & 16 & 13 \end{bmatrix}.$$

A useful result which can be conveniently stated here without proof is that if A is $m \times n$ and B is $p \times m$ then

$$R(A) + R(B) - m \leqslant R(BA) \leqslant \min[R(A), R(B)]. \quad (5.11)$$

Exercise 5.6   Use (5.11) to prove that if $R(A) = m$ then $R(A^{\mathsf{T}}A) = R(A)$ (in fact it can be shown that this result holds whatever the value of $R(A)$).

5.3.3   Normal form

So far as determination of rank is concerned, we have seen that we need to apply only elementary row transformations and column interchanges. It is both interesting and useful to see what further reduction can be obtained by applying additional elementary operations of types (b) and (c) in Section 5.3.1 to the *columns* of the reduced triangular form.

Example 5.8

(a)   Consider the matrix in (5.8). Reduce the rows so as to have zeros to the *right* of the pivots, by applying the following elementary column operations:

$$(C2) - 2(C1), (C3) + (C1), (C4) - 3(C1) \quad \text{(first row)}$$
$$(C3) - 2(C2), (C4) - \tfrac{1}{2}(C2) \quad \text{(second row)}.$$

This gives the first matrix below, which can be further reduced as shown:

$$\begin{bmatrix} 1 & 0 & 0 & 0 \\ 0 & 2 & 0 & 0 \\ 0 & 0 & 6 & 0 \\ 0 & 0 & 0 & 0 \end{bmatrix} \rightarrow \begin{bmatrix} 1 & 0 & 0 & 0 \\ 0 & 1 & 0 & 0 \\ 0 & 0 & 1 & 0 \\ 0 & 0 & 0 & 0 \end{bmatrix} \quad \begin{array}{l} \tfrac{1}{2}(R2) \\ \tfrac{1}{6}(R3) \end{array}.$$

(b) Similarly, the matrices in (5.9) and (5.10) can be reduced respectively to

$$\begin{bmatrix} 1 & 0 & 0 & 0 \\ 0 & 1 & 0 & 0 \\ 0 & 0 & 0 & 0 \end{bmatrix}, \quad \begin{bmatrix} 1 & 0 & 0 \\ 0 & 1 & 0 \\ 0 & 0 & 1 \\ 0 & 0 & 0 \end{bmatrix}.$$

Generally, any $m \times n$ matrix A can be reduced in this fashion by elementary row *and* column operations to an $m \times n$ matrix N having 1's in the $(1, 1)$, $(2, 2)$, ..., (r, r) positions and zeros everywhere else, where $r = R(A)$. This special matrix N is called the *normal form* of A, and clearly $R(N) = r = R(A)$.

Exercise 5.7 Reduce the three matrices in Exercise 5.5 to normal form.

Exercise 5.8 Write down the normal forms for matrices having rank 2 and orders 2×2, 2×3, 3×2, 4×3 respectively.

It can be shown that the relationship between A and its normal form can be written

$$A = PNQ \tag{5.12}$$

where P $(m \times m)$ and Q $(n \times n)$ are non-singular but not unique. We stated earlier that two matrices A and B are equivalent if they have the same dimensions and rank. In this case it is obvious that A and B have the same normal form, so we can write

$$B = P_1 N Q_1 \tag{5.13}$$

with P_1 and Q_1 non-singular. From (5.13) $N = P_1^{-1} B Q_1^{-1}$ and substituting this into (5.12) gives

$$A = PP_1^{-1} B Q_1^{-1} Q$$
$$= P_2 B Q_2, \tag{5.14}$$

which is an alternative characterization of equivalence of A and B. Notice that P_2 and Q_2 are also non-singular since each is a product of non-singular matrices (see Exercise 4.18).

The converse results also holds, namely, that if A and B satisfy (5.14) for any non-singular P_2 and Q_2 then A and B are equivalent. In particular, if T is an arbitrary $m \times m$ non-singular matrix it follows (by taking $P_2 = T$, $Q_2 = I$) that TA is equivalent to A, i.e. $R(TA) = R(A)$. Similarly, postmultiplication of A by an arbitrary $n \times n$ non-singular matrix does not alter its rank.

It can also be shown that A and B are equivalent if and only if it is possible to pass from one to the other by a sequence of elementary operations. The proof of this result (and some others in this section, e.g. (5.12)) relies on the introduction of 'elementary matrices' (see Problems 5.7 and 5.8), but details lie outside the scope of this book. It is interesting to note, however, that the transformation matrices in (5.12) can be easily obtained from the reduction procedure (see Problem 5.16).

Exercise 5.9   For what matrices A are A and A^T equivalent?

Exercise 5.10   If A and B are equivalent $n \times n$ matrices, determine whether the following pairs are equivalent: (a) A^T and B^T, (b) A^2 and B^2, (c) AB and BA.

Consider the cases when both A and B are non-singular, or both are singular.

5.4   Non-unique solution of equations

We have now developed enough theory to tackle the problems stated at the beginning of this chapter.

5.4.1   Homogeneous equations

As pointed out in Section 5.1, the m homogeneous equations

$$Ax = 0 \tag{5.15}$$

in n unknowns $x_1, \ldots, x_n$ have only the trivial solution $x = 0$ if A is square (i.e. $m = n$) and non-singular. If A is singular or rectangular, express it in normal form as in (5.12). Substituting (5.12) into (5.15) gives $PNQx = 0$, or simply

$$NQx = 0 \tag{5.16}$$

after premultiplying by P^{-1}. Write (5.16) in the form

$$Ny = 0, \tag{5.17}$$

where

$$y = Qx, \qquad x = Q^{-1}y. \tag{5.18}$$

Recalling that N has r 1's in the $(1, 1), \ldots, (r, r)$ positions and zeros elsewhere, where $r = R(A)$, eqns (5.17) when written out are simply

$$y_1 = 0, \qquad y_2 = 0, \ldots, y_r = 0.$$

Since the remaining variables $y_{r+1}, \ldots, y_n$ do not appear in the expansion of (5.17) it follows that they can take arbitrary values. Thus the solution x given by (5.18) contains $n - r$ arbitrary parameters.

We have therefore shown that eqns (5.15) in n unknowns always have a non-trivial solution provided $r = R(A) < n$, and then the most general form of the solution contains $n - r$ arbitrary parameters. If $r = n$, then the only solution is the trivial one, $x = 0$. Notice that if there are fewer equations than unknowns (i.e. $m < n$) then $R(A) \leqslant m < n$, so a non-trivial solution will *always* exist.

The method of determining the solution of (5.15) is to apply the modified gaussian elimination procedure developed in Section 5.3.2 for calculating $R(A)$.

Example 5.9 We find the general solution of three sets of homogeneous equations

(a)
$$x_1 + 2x_2 + 4x_3 + x_4 = 0$$
$$2x_1 + 4x_2 + 8x_3 + 2x_4 = 0 \qquad (5.19)$$
$$3x_1 + 6x_2 + 2x_3 \qquad = 0.$$

The matrix A is that given in (5.4), and the reduction was carried out in Example 5.5, producing the final matrix (5.9). It is important to notice that since columns 2 and 3 were interchanged, (5.9) corresponds to the final equations

$$x_1 + 4x_3 + 2x_2 + x_4 = 0$$
$$- 10x_3 \qquad - 3x_4 = 0. \qquad (5.20)$$

From (5.20) $x_3 = -(3/10)x_4$ and by back substitution

$$x_1 = -2x_2 - 4(-3/10)x_4 - x_4$$
$$= -2x_2 + (1/5)x_4.$$

The general solution of eqns (5.19) is therefore

$$x_1 = -2t_1 + (1/5)t_2, \qquad x_2 = t_1, \qquad x_3 = -(3/10)t_2, \qquad x_4 = t_2,$$
$$(5.21)$$

where t_1 and t_2 are arbitrary parameters. As expected, since we found $R(A) = 2$ in Examples 5.5, there are $4 - 2 = 2$ such parameters.

(b)
$$x_1 + 2x_2 + 3x_3 = 0$$
$$3x_1 + 6x_2 + 10x_3 = 0$$
$$2x_1 + 5x_2 + 7x_3 = 0$$
$$x_1 + 2x_2 + 4x_3 = 0. \tag{5.22}$$

The matrix A was transformed in Example 5.6 into that of (5.10), showing that $R(A) = 3$. Since $r = n$, eqns (5.22) have only the trivial solution $x_1 = x_2 = x_3 = 0$.

(c)
$$x_1 + 2x_2 + 3x_3 - x_4 = 0$$
$$x_1 + 4x_2 + 4x_3 + 3x_4 = 0$$
$$2x_1 + 4x_2 + 6x_3 + 4x_4 = 0$$
$$-x_1 - 2x_2 - 3x_3 + 4x_4 = 0. \tag{5.23}$$

$$A = \begin{bmatrix} 1 & 2 & 3 & -1 \\ 1 & 4 & 4 & 3 \\ 2 & 4 & 6 & 4 \\ -1 & -2 & -3 & 4 \end{bmatrix} \rightarrow \begin{bmatrix} 1 & 2 & 3 & -1 \\ 0 & 2 & 1 & 4 \\ 0 & 0 & 0 & 6 \\ 0 & 0 & 0 & 3 \end{bmatrix} \quad \begin{matrix} (R2) - (R1) \\ (R3) - 2(R1) \\ (R4) + (R1) \end{matrix}$$

$$\begin{matrix} x_1 & x_2 & x_4 & x_3 \end{matrix}$$
$$\rightarrow \begin{bmatrix} 1 & 2 & -1 & 3 \\ 0 & 2 & 4 & 1 \\ 0 & 0 & 6 & 0 \\ 0 & 0 & 3 & 0 \end{bmatrix} \quad (C3) \leftrightarrow (C4). \tag{5.24}$$

It is convenient to record column interchanges by labelling columns with the corresponding variables, as shown in (5.24). The operation $(R4) - \frac{1}{2}(R3)$ on (5.24) then reduces it to the matrix (5.8) in Example 5.4, where it was deduced that $R(A) = 3$. The final equations are

$$x_1 + 2x_2 - x_4 + 3x_3 = 0$$
$$2x_2 + 4x_4 + x_3 = 0$$
$$6x_4 = 0.$$

Hence

$$x_4 = 0, \qquad x_2 = -\frac{1}{2}x_3, \qquad x_1 = -2\left(-\frac{1}{2}x_3\right) - 3x_3 = -2x_3.$$

The general solution of (5.23) contains $4 - 3 = 1$ arbitrary parameter, and can be written

$$x_1 = -2t_1, \qquad x_2 = -\frac{1}{2}t_1, \qquad x_3 = t_1, \qquad x_4 = 0. \qquad (5.25)$$

To summarize, apply gaussian elimination with column interchanges to A until the maximum number $r = R(A)$ of non-zero pivots is obtained in the $(1, 1), \ldots, (r, r)$ positions. If r is equal to n, the number of unknowns, then the only solution is $x = 0$. If $r < n$, the solution is obtained by back substitution using the final reduced form of A, and contains $n - r$ arbitrary parameters. One final point: the actual way in which the parameters are introduced is not unique. For example, in (5.20) we could take $x_4 = -(10/3)x_3$ leading to $x_1 = -2t_1 - (2/3)t_3$, $x_2 = t_1$, $x_3 = t_3$, $x_4 = -(10/3)t_3$ as an alternative *form* of the solution of (5.19); no new values of the x's are obtained, all that has been done is that t_2 in (5.21) has been replaced by $-(10/3)t_3$.

Exercise 5.11 Find the general solution of the equations $Ax = 0$ for each of the three matrices A in Exercise 5.5.

Exercise 5.12 Investigate whether the following equations have a non-trivial solution, and if so determine the solution

(a) $x_1 + 2x_2 - x_3 = 0$ (b) $x_1 + 2x_2 - 3x_3 = 0$

    $2x_1 + 4x_2 + 7x_3 = 0;$ $2x_1 + 5x_2 + 2x_3 = 0$

                              $3x_1 - x_2 - 4x_3 = 0$

                              $7x_1 + 8x_2 - 8x_3 = 0.$

5.4.2 Inhomogenenous equations

We now return to the m equations $Ax = b$ in n unknowns, with $b \neq 0$. We shall see that the solution procedure is very similar to the homogeneous case, but first a little care is needed over the question of consistency. To tackle this we again use the normal form (5.12), which when substituted into (5.1) gives $PNQx = b$. Since we defined $y = Qx$ in (5.18), we can write

$$Ny = P^{-1}b. \qquad (5.26)$$

Recalling that N has r 1's on the principal diagonal, the set of eqns (5.26) when written out is

$$y_1 = \beta_1, \qquad y_2 = \beta_2, \ldots, y_r = \beta_r, \qquad 0 = \beta_{r+1}, \ldots, 0 = \beta_m,$$
$$(5.27)$$

where the β's are the components of the right-hand-side vector $P^{-1}b$. Thus for consistency of (5.27), and hence of (5.1), the last $m - r$ components of $P^{-1}b$ must be zero. In practice this means that for consistency, in the final reduced form of the augmented $m \times (n + 1)$ matrix $B = [A, b]$, any row in which the first n elements are all zero must also have the last column element equal to zero. The elimination process is applied exactly as for the homogeneous case, and again since $y_{r+1}, \ldots, y_n$ do not appear in (5.27), we deduce that the general solution contains $n - r$ arbitrary parameters.

Example 5.10　　We determine consistency of two sets of equations

(a)
$$
\begin{aligned}
x_1 - 2x_2 + x_3 &= -5 \\
x_1 + 5x_2 - 7x_3 &= 2 \\
3x_1 + x_2 - 5x_3 &= 1.
\end{aligned}
\qquad (5.28)
$$

$$
B = \begin{bmatrix} 1 & -2 & 1 & \vdots & -5 \\ 1 & 5 & -7 & \vdots & 2 \\ 3 & 1 & -5 & \vdots & 1 \end{bmatrix} \rightarrow \begin{bmatrix} 1 & -2 & 1 & \vdots & -5 \\ 0 & 7 & -8 & \vdots & 7 \\ 0 & 0 & 0 & \vdots & 9 \end{bmatrix} \qquad (5.29)
$$

after the usual row operations. The last row of B corresponds to the equation $0x_1 + 0x_2 + 0x_3 = 9$, which cannot be satisfied for any values of the x's, so eqns (5.28) are inconsistent.

(b)
$$
\begin{aligned}
2x_1 - 3x_2 + 6x_3 - 5x_4 &= 3 \\
x_2 - 4x_3 + x_4 &= 1 \\
4x_1 - 5x_2 + 8x_3 - 9x_4 &= k.
\end{aligned}
\qquad (5.30)
$$

After the operations $(R3) - 2(R1)$, $(R3) - (R2)$, the augmented matrix becomes

$$
\begin{bmatrix} 2 & -3 & 6 & -5 & \vdots & 3 \\ 0 & 1 & -4 & 1 & \vdots & 1 \\ 0 & 0 & 0 & 0 & \vdots & k-7 \end{bmatrix}. \qquad (5.31)
$$

The last row of (5.31) corresponds to the equation

$$0x_1 + 0x_2 + 0x_3 + 0x_4 = k - 7$$

so eqns (5.30) are consistent if $k = 7$ and inconsistent if $k \neq 7$.

Exercise 5.13 Test for consistency the equations

$$3x_1 - 7x_2 + 14x_3 - 8x_4 = 24$$
$$x_1 - 4x_2 + 3x_3 - x_4 = -2$$
$$x_1 - 3x_2 + 4x_3 - 2x_4 = 4$$
$$2x_1 - 15x_2 - x_3 + 5x_4 = -46.$$

Exercise 5.14 Find the value of k for which the equations

$$3x_1 + 2x_2 + 4x_3 = 3$$
$$x_1 + x_2 + x_3 = k$$
$$5x_1 + 4x_2 + 6x_3 = 15$$

are consistent.

If the equations are found to be consistent, then the general solution is obtained from the final reduced equations by back substitution, as usual, and contains $n - R(A)$ arbitrary parameters.

Example 5.10 (continued) Consider eqns (5.30). These were found to be consistent if $k = 7$. In this case the equations corresponding to (5.31) are

$$2x_1 - 3x_2 + 6x_3 - 5x_4 = 3$$
$$x_2 - 4x_3 + x_4 = 1.$$

Hence

$$x_2 = 1 + 4x_3 - x_4$$

$$x_1 = \frac{3}{2} + \frac{3}{2}(1 + 4x_3 - x_4) - 3x_3 + \frac{5}{2}x_4$$

$$= 3 + 3x_3 + x_4$$

and the general solution can be written

$$x_1 = 3 + 3t_1 + t_2, \qquad x_2 = 1 + 4t_1 - t_2, \qquad x_3 = t_1, \qquad x_4 = t_2, \quad (5.32)$$

where t_1 and t_2 are arbitrary parameters. On inspecting the first four columns of (5.31) we see there are two non-zero pivots, so $R(A) = 2$ and this confirms that the general solution contains $4 - 2 = 2$ arbitrary parameters.

Notice that (5.32) can be written

$$x = [3t_1 + t_2, 4t_1 - t_2, t_1, t_2]^T + [3, 1, 0, 0]^T, \quad (5.33)$$

the sum of a constant vector and one which depends entirely on the parameters. This is true in general: if eqns (5.1) are consistent then the general solution is

$$x = x_h + x_p,$$

where x_h is the general solution of the homogeneous equations in the usual form (5.15) and thus contains no constant terms, and x_p is *any* particular solution of (5.1) and thus is independent of parameters. (The reader who has studied linear differential equations will recognize a parallel with complementary function and particular integral.) For example, in (5.33)

$$x_h = [3t_1 + t_2, \, 4t_1 - t_2, \, t_1, \, t_2]^\mathsf{T}$$

is the general solution of the homogeneous equations obtained by replacing the right-hand sides of (5.30) by zeros.

Example 5.11 We find the general solution of two sets of in-homogeneous equations.

(a)
$$6x_1 + 8x_2 = -1$$
$$3x_1 - x_2 = -3$$
$$3x_1 + 7x_2 = 1.$$

After row operations,

$$B = \begin{bmatrix} 6 & 8 & \vdots & -1 \\ 0 & -5 & \vdots & -\dfrac{5}{2} \\ 0 & 0 & \vdots & 0 \end{bmatrix},$$

showing that the equations are consistent, and $R(A) = 2 = n$, so the solution is unique. The corresponding equations give $-5x_2 = -5/2$, so $x_2 = \frac{1}{2}$, and $x_1 = (-1 - 8 \cdot \frac{1}{2})/6 = -5/6$. Notice that geometrically, this means that the three straight lines corresponding to the equations intersect at a common point.

(b)
$$x_1 + 2x_2 + 3x_3 - x_4 = -1$$
$$x_1 + 4x_2 + 4x_3 + 3x_4 = 5$$
$$2x_1 + 4x_2 + 6x_3 + 4x_4 = 2$$
$$-x_1 - 2x_2 - 3x_4 + 4x_4 = 3.$$

These equations have the same left-hand sides as in (5.23). Applying the same elementary operations as were used in Example

5.9(c), but this time on the augmented matrix B, gives (see (5.24))

$$\begin{array}{cccc} x_1 & x_2 & x_4 & x_3 \end{array}$$
$$\begin{bmatrix} 1 & 2 & -1 & 3 & \vdots & -1 \\ 0 & 2 & 4 & 1 & \vdots & 6 \\ 0 & 0 & 6 & 0 & \vdots & 4 \\ 0 & 0 & 3 & 0 & \vdots & 2 \end{bmatrix} \rightarrow \begin{bmatrix} 1 & 2 & -1 & 3 & \vdots & -1 \\ 0 & 2 & 4 & 1 & \vdots & 6 \\ 0 & 0 & 6 & 0 & \vdots & 4 \\ 0 & 0 & 0 & 0 & \vdots & 0 \end{bmatrix} \quad (R4) - \frac{1}{2}(R3).$$

The equations are consistent, and $r = 3$, $n = 4$. By back substitution

$$x_4 = \frac{2}{3}, \qquad x_2 = \frac{1}{2}(6 - 4x_4 - x_3) = \frac{5}{3} - \frac{1}{2}x_3,$$

$$x_1 = -1 - 2x_2 + x_4 - 3x_3 = -\frac{11}{3} - 2x_3,$$

and the general solution is

$$x_1 = -\frac{11}{3} - 2t_1, \qquad x_2 = \frac{5}{3} - \frac{1}{2}t_1, \qquad x_3 = t_1, \qquad x_4 = \frac{2}{3}.$$

Notice that $x_h = [-2t, -\frac{1}{2}t_1, t_1, 0]^{\mathsf{T}}$, which agrees with (5.25), and $x_p = [-\frac{11}{3}, \frac{5}{3}, 0, \frac{2}{3}]^{\mathsf{T}}$.

Exercise 5.15   Find the general solution of the equations in Exercise 5.13.

Exercise 5.16   Find the general solution of the equations in Exercise 5.14 when they are consistent.

5.4.3   Consistency theorem

We have seen that for consistency of (5.1), if any row of the reduced B matrix has all its first n elements equal to zero then the last element in this row must also be zero. This implies that the number of non-zero pivots is the same for B as for A, so $R(B) = R(A)$.

Example 5.12   Consider again the inconsistent equations (5.28). From the reduced form of A in (5.29) (the first three columns) it is clear that $R(A) = 2$, but for B the operation (C3) $\leftrightarrow$ (C4) produces an extra non-zero pivot, showing $R(B) = 3 \neq R(A)$.

The following general theorem (Mirsky 1963) can be proved:

For the m equations in n unknowns, $Ax = b$ with augmented matrix $B = [A, b]$, then $R(B) \geqslant R(A)$ and the equations possess:

(a) a unique solution if and only if $R(A) = R(B) = n$;
(b) an infinite number of solutions if and only if
     $R(A) = R(B) < n$;
(c) no solution if and only if $R(A) < R(B)$.

This formal result is useful for theoretical purposes, but actual determination of consistency and solutions is carried out as described in Sections 5.4.1 and 5.4.2, using elementary operations.

Exercise 5.17   For the equations

$$kx_1 + x_2 + x_3 = 5$$
$$3x_1 + 2x_2 + kx_3 = 18 - 5k$$
$$x_2 + 2x_3 = 2$$

find the value of k: (a) for which there is no solution; (b) for which there are an infinite number of solutions.

Some readers may find a geometrical interpretation of the theorem helpful. The case $n = 2$ was discussed in Examples 3.1 and 3.2 in Section 3.1. When $n = 3$, an equation $a_{11}x_1 + a_{12}x_2 + a_{13}x_3 = b_1$ describes a plane in three dimensions. Thus for three equations, representing three planes, case (a) of the theorem corresponds to the planes intersecting in a single point; case (b) when $R(A) = 2$ corresponds to the planes intersecting in a straight line (like the pages of an opened book); and case (c) corresponds to the planes having no point in common. An illustration of this last case is provided by the three planes in (5.28).

5.5   Method of least squares

It can happen in a number of situations that an inconsistent system of eqns (5.1) is obtained, but that nevertheless a 'solution' is required which is the 'best possible' in some sense. An important such case is now described.

A common procedure in scientific or other experiments is to measure pairs of values of variables which are associated in some way, and then try to fit some curve to these n points having cartesian coordinates (α_i, β_i), $i = 1, 2, \ldots, n$. A convenient curve is the polynomial expression

$$y = a_0 + a_1 x + a_2 x^2 + \cdots + a_{m-1} x^{m-1}. \tag{5.34}$$

If $m = n$ and the α's are all different, then there is a *unique* curve (5.34) which passes through all the points simultaneously (see Problem 4.17). Otherwise, if $m < n$ then in general not all the given points can lie on the curve.

The simplest, but none the less very important case is to take $m = 2$, so the curve is simply the straight line

$$y = a_0 + a_1 x \tag{5.35}$$

and it is required to fit a 'best' line to a set of points as indicated in Fig. 5.1. If it is assumed that the α_i are all different, and are known exactly, then the error e_i at each point is the difference between the value of y given by the line and the measured value β_i, i.e.

$$e_i = a_0 + a_1 \alpha_i - \beta_i, \qquad i = 1, 2, \ldots, n. \tag{5.36}$$

We wish to choose a_0 and a_1 in (5.35) so that the total of these errors is as small as possible. Since the errors e_i may be positive or negative and we are interested in the *net* total, we must consider $|e_i|$, or more conveniently, e_i^2. The method of least squares is to choose a_0 and a_1 so that the sum of the squares of the errors

$$S = \sum_{i=1}^{n} (a_0 + a_1 \alpha_i - \beta_i)^2 \tag{5.37}$$

is minimized.

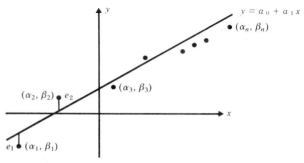

Fig. 5.1   Best straight line through a set of points.

Notice that if all the points in Fig. 5.1 were to lie on the line we should require

$$a_0 + a_1\alpha_i = \beta_i, \qquad i = 1, 2, \ldots, n, \qquad (5.38)$$

which represents n equations in the two unknowns a_0 and a_1. Hence if $n > 2$, unless by some freak all the experimentally determined points do indeed lie on the line, this means that eqns (5.38) are inconsistent.

For simplicity, take $n = 3$ so that (5.37) gives

$$S = (a_0 + a_1\alpha_1 - \beta_1)^2 + (a_0 + a_1\alpha_2 - \beta_2)^2 + (a_0 + a_1\alpha_3 - \beta_3)^2.$$

For S to be a minimum each of the partial derivatives $\partial S / \partial a_0$ and $\partial S / \partial a_1$ must be zero. Hence

$$\frac{\partial S}{\partial a_0} = 0 = 2(a_0 + a_1\alpha_1 - \beta_1) + 2(a_0 + a_1\alpha_2 - \beta_2)$$
$$+ 2(a_0 + a_1\alpha_3 - \beta_3),$$

$$\frac{\partial S}{\partial a_1} = 0 = 2\alpha_1(a_0 + a_1\alpha_1 - \beta_1) + 2\alpha_2(a_0 + a_1\alpha_2 - \beta_2)$$
$$+ 2\alpha_3(a_0 + a_1\alpha_3 - \beta_3),$$

and simplifying these two equations gives respectively

$$\left.\begin{array}{r} 3\alpha_0 + (\alpha_1 + \alpha_2 + \alpha_3)a_1 = \beta_1 + \beta_2 + \beta_3 \\ (\alpha_1 + \alpha_2 + \alpha_3)a_0 + (\alpha_1^2 + \alpha_2^2 + \alpha_3^2)a_1 = \alpha_1\beta_1 + \alpha_2\beta_2 + \alpha_3\beta_3 \end{array}\right\}.$$
$$(5.39)$$

If eqns (5.38) are written in the form

$$Aa = \beta \qquad (5.40)$$

where, when $n = 3$,

$$A = \begin{bmatrix} 1 & \alpha_1 \\ 1 & \alpha_2 \\ 1 & \alpha_3 \end{bmatrix}, \qquad a = \begin{bmatrix} a_0 \\ a_1 \end{bmatrix}, \qquad \beta = \begin{bmatrix} \beta_1 \\ \beta_2 \\ \beta_3 \end{bmatrix}, \qquad (5.41)$$

then it is easy to verify that the pair (5.39) is

$$A^\mathsf{T}Aa = A^\mathsf{T}\beta. \qquad (5.42)$$

Since $R(A) = 2$ (because $\alpha_1 \neq \alpha_2 \neq \alpha_3$), then by Exercise 5.6 it follows that $R(A^\mathsf{T}A) = 2$ also, so the 2×2 matrix $A^\mathsf{T}A$ is non-singular. Hence (5.42) has the unique solution

$$a = (A^\mathsf{T}A)^{-1}A^\mathsf{T}\beta, \tag{5.43}$$

which is the desired least-squares solution of (5.40). It is straightforward to show that the result still applies for $n > 3$.

When the polynomial (5.34) is used with $m > 2$ and $n > m$, then A in (5.40) is $n \times m$ and the preceding argument is easily generalized to show that (5.43) still holds (see Section 10.1.1).

Example 5.13   We determine the least-squares quadratic curve for the points having cartesian coordinates $(-1, 3)$, $(0, 0)$, $(1, 2)$, $(2, 5)$.

Here $m = 3$, $n = 4$ and from (5.34) the curve is

$$y = a_0 + a_1 x + a_2 x^2.$$

Equations (5.40) are

$$a_0 + a_1 \alpha_i + a_2 \alpha_i^2 = \beta_i, \qquad i = 1, 2, 3, 4,$$

which gives

$$A = \begin{bmatrix} 1 & -1 & 1 \\ 1 & 0 & 0 \\ 1 & 1 & 1 \\ 1 & 2 & 4 \end{bmatrix}, \qquad a = \begin{bmatrix} a_0 \\ a_1 \\ a_2 \end{bmatrix}, \qquad \beta = \begin{bmatrix} 3 \\ 0 \\ 2 \\ 5 \end{bmatrix}.$$

A simple calculation gives $A^\mathsf{T}A$ and $A^\mathsf{T}\beta$, and eqns (5.42) can then be solved by gaussian elimination. The solution is $a = \frac{1}{10}[6, -7, 15]^\mathsf{T}$, so the required quadratic is $y = (6 - 7x + 15x^2)/10$. This is shown in Fig. 5.2.

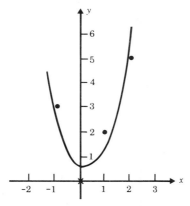

Fig. 5.2   Best quadratic curve for Example 5.13.

Exercise 5.18   Use (5.42) to obtain the best straight line for the points $(1, 2)$, $(2, 4)$, $(3, 5)$, $(4, 7)$. Draw a graph of your result.

Exercise 5.19   Verify that when A in (5.40) is square and non-singular, then (5.43) reduces to the unique solution of the equations (5.40).

5.6   Use of Krokecker product

We now consider a linear equation involving matrices in the form

$$AX = C, \tag{5.44}$$

where A is $n \times n$, C is $n \times m$, and the unknown X to be determined is $n \times m$. Of course, if A is non-singular, we can write the unique solution of (5.44) as $X = A^{-1}C$. However, it is useful to convert (5.44) into the usual matrix-vector form (5.1).

For example, let $n = m = 2$ so (5.44) becomes

$$\begin{bmatrix} a_1 & a_2 \\ a_3 & a_4 \end{bmatrix} \begin{bmatrix} x_1 & x_2 \\ x_3 & x_4 \end{bmatrix} = \begin{bmatrix} c_1 & c_2 \\ c_3 & c_4 \end{bmatrix}. \tag{5.45}$$

Expansion and rearrangement of (5.45) easily show that it is equivalent to

$$\begin{bmatrix} a_1 & 0 & a_2 & 0 \\ 0 & a_1 & 0 & a_2 \\ a_3 & 0 & a_4 & 0 \\ 0 & a_3 & 0 & a_4 \end{bmatrix} \begin{bmatrix} x_1 \\ x_2 \\ x_3 \\ x_4 \end{bmatrix} = \begin{bmatrix} c_1 \\ c_2 \\ c_3 \\ c_4 \end{bmatrix} \tag{5.46}$$

and we recognize the 4×4 matrix in (5.46) as $A \otimes I_2$, so (5.46) can be written

$$(A \otimes I_2)x = c,$$

where x is the column vector in (5.46) obtained by stacking the elements of the rows of X taken in sequence, and c is formed in the same way from C. It is easy to verify that in general (5.44) can be written

$$(A \otimes I_m)x = c, \tag{5.47}$$

where if $X = [x_{ij}]$ then x is the column mn-vector

$$x = [x_{11}, x_{12}, \ldots, x_{1m}, x_{21}, \ldots, x_{2m}, \ldots, x_{nm}]^{\mathsf{T}} \tag{5.48}$$

formed from the rows of X, and similarly for c. The notation

$v(X)$ or vec(X) is often used for x, and some authors apply it to $v(X^T)$, giving the column vector obtained from the columns of X. We can think of $v(\cdot)$ as an operator which transforms a matrix into a column vector. The left-hand side of (5.47) is in fact a special case of the general property (see Problem 5.11)

$$v(AXB^T) = (A \otimes B)v(X), \qquad (5.49)$$

where B is $m \times m$. In particular, this implies that

$$XB = C$$

can be written as

$$(I_n \otimes B^T)x = c, \qquad (5.50)$$

where x and c are as in (5.47). What is of most interest is to combine the two expressions (5.47) and (5.50). On doing this it follows that the matrix equation

$$AX + XB = C \qquad (5.51)$$

(to be solved for X) is equivalent to

$$(A \otimes I_m + I_n \otimes B^T)x = c. \qquad (5.52)$$

Thus (5.51) has been transformed into (5.52), which represents mn equations in the usual form (5.1). For example, it now follows by the consistency theorem of Section 5.4.3 that the solution X of (5.51) is unique if and only if the $mn \times mn$ matrix

$$D = A \otimes I_m + I_n \otimes B^T \qquad (5.53)$$

is non-singular, in which case the elements of X are given by $x = D^{-1}c$.

A simple criterion for determining non-singularity of D will be developed in Section 6.3.5. Equation (5.51) arises in a number of applications, and the Kronecker product has again demonstrated its usefulness by converting the problem into a standard set of linear equations.

Exercise 5.20   Use (5.52) to solve (5.51) when

$$A = \begin{bmatrix} 1 & 2 \\ 0 & 3 \end{bmatrix}, \qquad B = \begin{bmatrix} 2 & 1 \\ 1 & 4 \end{bmatrix}, \qquad C = I_2.$$

Exercise 5.21   Use the result in (2.76), Problem 2.12, to prove that for D in (5.53)

$$D^2 = A^2 \otimes I_m + 2A \otimes B^T + I_n \otimes (B^T)^2. \qquad (5.54)$$

5.7 Linear dependence of vectors

Example 5.14 Consider the following three row vectors

$$u = [1, 4], \qquad v = [2, -1], \qquad w = [4, 7]. \tag{5.55}$$

It is obvious that $2u + v - w = 0$, and the vectors u, v, w are called
linearly dependent. Alternatively,

$$w = 2u + v \tag{5.56}$$

and in (5.56) we have written w as a *linear combination* of u and v, i.e.

$$w = x_1 u + x_2 v \tag{5.57}$$

where x_1 and x_2 are scalars. Writing out (5.57) in full:

$$[4, 3] = [x_1, 4x_1] + [2x_2, -x_2]$$
$$= [x_1 + 2x_2, 4x_1 - x_2].$$

So, on equating components

$$x_1 + 2x_2 = 4, \qquad 4x_1 - x_2 = 3$$

or

$$\begin{bmatrix} 1 & 2 \\ 4 & -1 \end{bmatrix} \begin{bmatrix} x_1 \\ x_2 \end{bmatrix} = \begin{bmatrix} 4 \\ 3 \end{bmatrix}. \tag{5.58}$$

Thus expressing w as in (5.57) is equivalent to solving a pair of linear
equations in the usual form. Notice that the columns of the matrix in
(5.58) are u^T and v^T.

These ideas can be generalized: if we have n row m-vectors

$$v_i = [a_{1i}, a_{2i}, \ldots, a_{mi}], \qquad i = 1, 2, \ldots, n \tag{5.59}$$

then expressing $b = [b_1, b_2, \ldots, b_m]$ as a linear combination of
the v's, i.e.

$$x_1 v_1 + x_2 v_2 + \cdots + x_n v_n = b \tag{5.60}$$

is equivalent to the m linear equations in n unknowns

$$Ax = b^T, \tag{5.61}$$

where $x = [x_1, \ldots, x_n]^T$ and the columns of the $m \times n$ matrix A
are $v_1^T, v_2^T, \ldots, v_n^T$.

The vectors $v_1, v_2, \ldots, v_n$ are *linearly dependent* if it is
possible to find a linear combination of them which gives the zero
vector, i.e. $b \equiv 0$ in (5.60), so

$$x_1 v_1 + x_2 v_2 + \cdots + x_n v_n = 0 \tag{5.62}$$

with not all the x's zero. If the only solution to (5.62) is the trivial one $x_1 = x_2 = \cdots = x_n = 0$, then the vectors are *linearly independent*. Equations (5.62) are equivalent to the set (5.61) with $b \equiv 0$, which is the set of homogeneous equations studied in Section 5.4.1. Thus the problem of determining linear dependence or independence is equivalent to determining whether or not the homogeneous equations $Ax = 0$ have a non-trivial solution, for which the condition is $R(A) < n$. In particular, if $m < n$ then $R(A) \leqslant m < n$, so the vectors are *always* linearly dependent.

Example 5.15

(a) $v_1 = [1, 2, 3]$, $v_2 = [2, 4, 6]$, $v_3 = [4, 8, 2]$, $v_4 = [1, 2, 0]$.

$$(5.63)$$

Here $m = 3$, $n = 4$ and it is easily seen that (5.62) produces the set of eqns (5.19), which in matrix form are

$$\begin{bmatrix} 1 & 2 & 4 & 1 \\ 2 & 4 & 8 & 2 \\ 3 & 6 & 2 & 0 \\ v_1^T & v_2^T & v_3^T & v_4^T \end{bmatrix} \begin{bmatrix} x_1 \\ x_2 \\ x_3 \\ x_4 \end{bmatrix} = 0. \qquad (5.64)$$

In Example 5.9(a) we found that these equations do have a non-trivial solution, so the vectors in (5.63) are linearly dependent. For example, setting $t_1 = 1$, $t_2 = 0$ in (5.21) gives $x_1 = -2$, $x_2 = 1$, $x_3 = 0$, $x_4 = 0$, so

$$-2v_1 + 1v_2 + 0v_3 + 0v_4 = 0.$$

Note that some of the x's in (5.62) can be zero, provided at least one x_i is non–zero.

(b) If we consider only v_1, v_2, and v_3 in (5.63) then (5.62) gives

$$\begin{bmatrix} 1 & 2 & 4 \\ 2 & 4 & 8 \\ 3 & 6 & 2 \end{bmatrix} \begin{bmatrix} x_1 \\ x_2 \\ x_3 \end{bmatrix} = 0. \qquad (5.65)$$

It is easily seen that the matrix in (5.65) has rank 2, so a non-trivial solution exists, implying that v_1, v_2 and v_3 are linearly dependent.

(c) Finally, consider only v_2 and v_3 in (5.63). Equations (5.62) are now

$$\begin{bmatrix} 2 & 4 \\ 4 & 8 \\ 6 & 2 \end{bmatrix} \begin{bmatrix} x_1 \\ x_2 \end{bmatrix} = 0 \qquad (5.66)$$

and since the matrix in (5.66) has rank 2 the only solution of these is $x_1 = x_2 = 0$, showing v_2 and v_3 are linearly independent.

As the preceding example suggests, there is a direct relationship between linear dependence and rank. It can be shown that the rank of *any* matrix A is equal to the maximum number of its columns (regarded as vectors) which are linearly independent. For example, the matrix A in (5.64) has rank 2 (see Example 5.2(c)) and we have seen in Example 5.15 that at most two of its columns are linearly independent. Since $R(A) = R(A^T)$, and the columns of A^T are the rows of A, it follows that $R(A)$ is also equal to the maximum number of its rows which are linearly independent.

Exercise 5.22 For each of the first two matrices in Exercise 5.5 find linear combinations of the columns which gives zero (use Exercise 5.11). Do the same for the rows.

The vectors $v_1, v_2, \ldots, v_n$ in (5.59) are called a *basis* for the set of n-dimensional row vectors if $m = n$ and if they are linearly independent. In this case A is square and has rank n, i.e. A is non-singular, so the solution of (5.61) is unique. That is, *any* row n-vector b can be expressed as a unique linear combination of the basis vectors. Again, identical remarks apply to a set of n column n-vectors.

Example 5.16 The simplest basis consists of the rows (or columns) of I_n. For example, using again the notation introduced in Exercise 2.19, if $n = 3$ then

$$e_1 = [1, 0, 0], \qquad e_2 = [0, 1, 0], \qquad e_3 = [0, 0, 1]$$

form a basis, and if $b = [b_1, b_2, b_3]$ then obviously

$$b = b_1 e_1 + b_2 e_2 + b_e e_3.$$

However, *any* linearly independent set of three vectors will do, e.g.

$$f_1 = [1, 1, 0], \qquad f_2 = [0, 1, 1], \qquad f_3 = [1, 1, 1]$$

(simply verify that the 3×3 matrix having f_1^T, f_2^T, f_3^T as its columns is non-singular) and then

$$b = (b_2 - b_3)f_1 + (b_2 - b_1)f_2 + (b_1 - b_2 + b_3)f_3.$$

In general, the coefficients in the linear combination of the basis vectors are found by solving (5.61) in the usual way. This idea of interpreting the solution of a set of linear equations in terms of linear relationships between row or column vectors is a very important one for the development of the theory of 'vector spaces', but this lies outside the scope of this book.

Exercise 5.23 If $n = 3$ verify that $g_1 = e_1 + e_2 + e_3$, $g_2 = e_1 + 2e_3$, $g_3 = 2e_2 + e_3$ form a basis, and express $b = [b_1, b_2, b_3]$ in terms of this basis.

5.8 Error-correcting codes

When a message is transmitted electronically interference often occurs, so that the signal received may be incorrect. The aim of coding theory is to detect, and if possible, correct errors which occur during transmission. A familiar example of an error-detecting code is the International Standard Book Number (ISBN), where a very simple arithmetical check can be applied to determine whether a given ten-digit number is indeed an ISBN. We shall only consider, however, *binary codes* which consist of a set of *codewords*, each word being a string of n 0's and 1's (each symbol is called a *bit*—short for binary digit).

Example 5.17 Suppose we wish to send the messages *north, south, east, west,* which are coded as follows:

	north	south	east	west
code C_1	00	01	10	11
code C_2	000	101	011	110.

The code C_1 requires transmission of only two bits, but this code cannot detect any errors, since if errors occur in either or both bits, then an incorrect message is received. For example, if 01 (*south*) is sent and there is a transmission error in the first bit so that 11 is received, then this is interpreted as *west*.

The second code C_2 can detect any *single* error. For example if 101 is sent and an error occurs in a single bit, then either 001, 111 or 100 is received. None of these is a codeword, so the receiver is aware of a transmission error. Although this code cannot correct errors, it reveals how redundancy can be added to the original message, in the form of extra bits, in such a way that transmission errors can be detected. In general, the basic aim of coding theory is to append additional bits to the original message so that it can be correctly recovered even if it has been garbled during transmission. Such codes are now widely used in communications and related fields.

We shall assume in what follows that the probability of an error in any single bit is very small, so that if an incorrect message is received then it is much more likely to contain one error rather than several. In practice a code which detects, but does not correct, errors is only useful if the receiver can obtain a repetition of the message in which an error has been detected.

Example 5.18 A simple but useful error-detecting code is the *parity check,* which consists of appending a single bit so that the overall codeword has even *parity,* i.e. an even number of 1's. If the received word has even parity then the most likely situation is that there are no errors. If it has odd parity, then we deduce it contains an error. To be specific, if a message to be sent is 101011 then the additional or *check* bit will be 0, giving the codeword 1010110. If, say, 1011110 is received then since it contains an odd number of 1's we infer that (at least) one error has occurred in transmission. However, there is no way in which we can determine the correct message using this code.

Exercise 5.24 Suppose that the codeword to be transmitted is $x_1x_2\cdots x_{n-1}x_n$, where each symbol is 0 or 1, and $x_n = 1$ if $s = x_1 + \cdots + x_{n-1}$ is odd and $x_n = 0$ if s is even. If the received word is $X_1X_2\cdots X_n$ show that there must be an even or odd number of errors according to whether parity(X_n) is either equal or not equal to parity $(X_1X_2\cdots X_{n-1})$.

On the assumption that the probability of more than one error is extremely small, the most likely circumstances in this example are therefore either no error or one error, respectively.

Exercise 5.25 A message transmitted consists of a single bit, 0 or 1 ('Yes' or 'No'). The *repetition code* of length 3 simply consists of transmitting the message three times:

$$\begin{array}{lcc} \text{Message:} & 0 & 1 \\ \text{Codeword:} & 000 & 111 \end{array}.$$

For example, if 100 is received, then again assuming that at most one error has occurred in transmission, we deduce that 000 was sent. By considering all the other seven cases of possible received messages, show that this code corrects all single errors.

We now introduce arithmetic modulo 2, that is

$$\begin{array}{lll} 0+0=0, & 0+1=1+0=1, & 1+1=0 \\ 0\times 0=0, & 0\times 1=1\times 0=0, & 1\times 1=0. \end{array} \tag{5.67}$$

To make sense of these rules, think of '1' as representing an odd number and '0' an even number, so that for example $1 + 1 = 0$ means that the sum of two odd numbers is even. Subtraction can also be defined: there is no difficulty with $0 - 0 = 0$, $1 - 1 = 0$ and

$1 - 0 = 1$, but we also have

$$0 - 1 = (1 + 1) - 1 = 1.$$

The *sum* of two codewords $a = a_1 a_2 \cdots a_n$, $b = b_1 b_2 \cdots b_n$ is the codeword having ith bit $a_i + b_i$, $i = 1, \ldots, n$, where addition follows the rules in (5.67). Then a *linear code C* is defined by the condition that for any two codewords a, b in C their sum $a + b$ also belongs to C.

Example 5.19 The code $C = \{00, 01, 10, 11\}$ in Example 5.17 is a linear code, since

$$01 + 10 = 11, \qquad 01 + 11 = 10, \qquad 10 + 11 = 01, \qquad 00 + 01 = 01, \text{ etc.,}$$

so the sum of all possible pairs of codewords remains in C_1.

Exercise 5.26 If a is any codeword in a linear code, prove that $a + a = 0$ (the codeword consisting of n zeros). Hence deduce that 0 is always a codeword for any linear code.

Exercise 5.27 Determine whether each of the following sets of codewords forms a linear code.

(a) 000, 110, 100, 010;

(b) 000, 100, 011, 101, 110;

(c) 00000, 01110, 10111, 11001.

If $x = x_1 x_2 \cdots x_n$ is a codeword of *length n*, write x' to denote the column vector $[x_1, x_2, \ldots, x_n]^T$. Let H be an $m \times n$ *binary* matrix with $n > m$, all of whose elements are 0 or 1. Then a linear code C is defined as the set of codewords for which $Hx' = 0$, where the matrix–vector multiplication is carried out using the rules (5.67): this follows because if $Ha' = 0$ and $Hb' = 0$ then obviously $H(a' + b') = 0$. The matrix H is known as the *parity check* (or simply the *check*) *matrix*.

Example 5.20 Consider the check matrix

$$H = \begin{bmatrix} 1 & 0 & 0 & 1 & 1 \\ 0 & 1 & 0 & 0 & 1 \\ 0 & 0 & 1 & 1 & 0 \end{bmatrix} \tag{5.68}$$

which, since it has five columns, defines a linear code of length 5. If $x_1 x_2 x_3 x_4 x_5$ is a codeword then $Hx' = 0$ gives

$$
\begin{aligned}
x_1 \qquad\quad + x_4 + x_5 &= 0 \\
x_2 \qquad\qquad + x_5 &= 0 \\
x_3 + x_4 \qquad\quad &= 0.
\end{aligned}
\tag{5.69}
$$

Because the x_i are binary digits subject to (5.67), equations (5.69) give

$$x_1 = -x_4 - x_5 = x_4 + x_5$$

$$x_2 = -x_5 = x_5, \qquad x_3 = -x_4 = x_4.$$

The bits x_4, x_5 are called the *information bits* because they can be selected independently, and x_1, x_2, x_3 are the *check bits*, which are uniquely determined by the information bits. Thus when $x_4 = 1$, $x_5 = 0$ we obtain $x_1 = 1$, $x_2 = 0$, $x_3 = 1$ giving the codeword 10110. Similarly, the other three codewords are 00000 ($x_4 = x_5 = 0$), 11001 ($x_4 = 0$, $x_5 = 1$) and 01111 ($x_4 = 1$, $x_5 = 1$).

Exercise 5.28 What is the check matrix for the code in Example 5.18?

Exercise 5.29 Determine all the codewords for linear codes having check matrices

(a) $\begin{bmatrix} 1 & 0 & 0 & 0 & 1 & 1 \\ 0 & 1 & 0 & 1 & 0 & 1 \\ 0 & 0 & 1 & 1 & 1 & 1 \end{bmatrix}$;   (b) $\begin{bmatrix} 1 & 0 & 0 & 0 & 1 & 1 & 1 \\ 0 & 1 & 0 & 0 & 0 & 1 & 1 \\ 0 & 0 & 1 & 0 & 1 & 1 & 1 \\ 0 & 0 & 0 & 1 & 0 & 0 & 1 \end{bmatrix}.$

The check matrices in Example 5.20 and Exercise 5.29 are examples of the general form

$$H = [I_m, A], \tag{5.70}$$

where $A = [a_{ij}]$ is an arbitrary $m \times (n - m)$ binary matrix. Generalizing what we did for (5.69), writing out $Hx' = 0$ gives

$$x_1 = a_{11}x_{m+1} + \cdots + a_{1,n-m}x_n$$

$$\vdots \tag{5.71}$$

$$x_m = a_{m1}x_{m+1} + \cdots + a_{m,n-m}x_n,$$

where $x_1, \ldots, x_m$ are the check bits and $x_{m+1}, \ldots, x_n$ are the (independent) information bits. The equations (5.71) can be written as

$$\begin{bmatrix} x_1 \\ \vdots \\ x_m \end{bmatrix} = A \begin{bmatrix} x_{m+1} \\ \vdots \\ x_n \end{bmatrix} \tag{5.72}$$

and (5.72) gives a convenient way of determining $x_1, \ldots, x_m$ for any given values of the information bits $x_{m+1}, \ldots, x_n$. Since there are 2^{n-m} ways of choosing the latter (each bit can be 0 or 1) there are a total of 2^{n-m} codewords. The number $n - m$ of information bits is often called the *dimension* of the code. In general for an arbitrary H the dimension is equal to $n - R(H)$

(see Problem 5.19). Notice that the order of the columns in H is unimportant—rearranging the columns of H simply leads to a reordering of the bits in each codeword.

Example 5.20 *(continued)* For the check matrix in (5.68) we have

$$A = \begin{bmatrix} 1 & 1 \\ 0 & 1 \\ 1 & 0 \end{bmatrix}. \tag{5.73}$$

When $x_4 = 1$, $x_5 = 0$, premultiplying $[1, 0]^\mathsf{T}$ by A produces $[1, 0, 1]^\mathsf{T}$ as the vector of check bits, giving the codeword is 10110 as before.

Exercise 5.30 If, say, columns 2 and 4 in (5.68) are interchanged, this is equivalent to interchanging x_2 and x_4 in the codewords. Write down the check matrix for Example 5.20 modified in this way and determine the codewords.

It can be shown that provided H does not have a zero column, nor two identical columns, then it will be the check matrix for a code which corrects *all* single errors. If a codeword c is transmitted, and a single error occurs in transmission, then the received word will have the form $r = c + e_i$, where e_i consists of zeros except for a single 1 in the ith position. Then

$$Hr' = Hc' + He_i'$$
$$= He_i', \quad \text{since } Hc' = 0,$$

and from the way e_i' is defined, He_i' is just the ith column of H. We have therefore obtained the following scheme for *decoding* the received message r to give the correct transmitted codeword c. Form the so-called *syndrome* Hr': if it is zero, then r is a codeword. Otherwise, compare it with the columns of H. If the syndrome is equal to the ith column of H, then a single transmission error has occurred and the ith bit is incorrect; if the syndrome is not equal to a column of H, then we have detected that more than one error has occurred in transmission.

Example 5.21 Consider the linear code whose check matrix H is given in (5.68), and suppose that a received word is $r = 01101$. The syndrome is

$$Hr' = H\begin{bmatrix} 0 \\ 1 \\ 1 \\ 0 \\ 1 \end{bmatrix} = \begin{bmatrix} 1 \\ 0 \\ 1 \end{bmatrix},$$

which by comparison with (5.68) is seen to be the fourth column of H. Hence we deduce that there is a single error which occurs in the fourth bit, so the correctly decoded message is 01111. Alternatively, if $r = 00011$ then

$$Hr' = \begin{bmatrix} 0 \\ 1 \\ 1 \end{bmatrix},$$

which is not equal to a column of H, so more than one error must have occurred in transmission.

Exercise 5.31   For a linear code with check matrix

$$H = \begin{bmatrix} 1 & 0 & 1 & 0 & 0 \\ 1 & 1 & 0 & 1 & 0 \\ 0 & 1 & 0 & 0 & 1 \end{bmatrix}$$

decode each of the following received messages (a) 11110; (b) 11101; (c) 10001.

An ingenious way of constructing a check matrix for a single-error-correcting code, named after *Hamming,* is to select as the columns of H the binary number representations of the integers 1 to n.

Example 5.22   When $n = 6$ the binary representations are

$$\begin{array}{cccccc} 1 & 2 & 3 & 4 & 5 & 6 \\ 001 & 010 & 011 & 100 & 101 & 110 \end{array}$$

and the corresponding check matrix is obtained by writing these binary numbers as columns, to give

$$H = \begin{bmatrix} 0 & 0 & 0 & 1 & 1 & 1 \\ 0 & 1 & 1 & 0 & 0 & 1 \\ 1 & 0 & 1 & 0 & 1 & 0 \end{bmatrix}. \tag{5.74}$$

The crucial property of a Hamming code is that if there is a single error in the ith position, then the syndrome Hr' is equal to the ith column of H, which is precisely the number i in binary form, so there is no need to compare the syndrome with the columns of H.

Example 5.22 (continued)   If a received message is $r = 101110$ then using H in (5.74) gives

$$Hr' = \begin{bmatrix} 0 \\ 1 \\ 1 \end{bmatrix}$$

and 011 is the binary representation of 3, showing that the error is in the third bit. The transmitted codeword is therefore 100110.

It is convenient to use those columns of the Hamming code check matrix containing a single 1 as the check bits (as was done for the non-Hamming code in Example 5.20) i.e. x_1, x_2 and x_4 in this example. Using H in (5.74), $Hx' = 0$ gives

$$x_1 = x_3 + x_5, \qquad x_2 = x_3 + x_6, \qquad x_4 = x_5 + x_6.$$

Exercise 5.32 Determine the eight codewords for the linear code with check matrix (5.74).

When $n = 2^m - 1$, where m (>1) is the number of rows of H, the Hamming code is called *perfect* and every syndrome (except 0) represents a single correctable error. When $n < 2^m - 1$, some syndromes represent multiple errors. For example, the code defined by (5.74) is not perfect ($m = 3$, $n = 6$). If a received word is $r = 000111$, then the syndrome is

$$Hr' = \begin{bmatrix} 1 \\ 1 \\ 1 \end{bmatrix}$$

and 111 is the binary representation of 7. Since the wordlength is 6, this means that more than one transmission error has occurred.

Exercise 5.33 Write down the check matrix for the perfect Hamming code with $m = 3$. Decode the received word 1101011.

It is clear that for the Hamming check matrix in (5.74) the columns can be interchanged so as to put H into the standard form (5.70), and this is true for a Hamming matrix of any dimensions. When H has this form we have seen that $Hx' = 0$ is equivalent to (5.72), which can be extended to read

$$\begin{bmatrix} x_1 \\ \vdots \\ x_n \end{bmatrix} = \begin{bmatrix} A \\ I_{n-m} \end{bmatrix} \begin{bmatrix} x_{m+1} \\ \vdots \\ x_n \end{bmatrix}.$$

Transposing this equation gives (see Section 2.4)

$$[x_1, \ldots, x_n] = [x_{m+1}, \ldots, x_n][A^\mathsf{T}, I_{n-m}]. \tag{5.75}$$

Equation (5.75) expresses any codeword in terms of the information bits. Since these are independent, it follows that the codewords are given by all possible binary combinations of the rows of the matrix $G = [A^T, I_{n-m}]$, which is called the *generator matrix* for the code. In other words, the rows of G form a basis for the set of codewords.

Example 5.23   For the code in Example 5.20, with check matrix (5.68) and A given in (5.73), the generator matrix is

$$G = [A^T, \quad I_2]$$

$$= \begin{bmatrix} 1 & 0 & 1 & 1 & 0 \\ 1 & 1 & 0 & 0 & 1 \end{bmatrix},$$

so the codewords have the form

$$\alpha_1(10110) + \alpha_2(11001)$$

for all possible $\alpha_i = 0, 1$. The four codewords are therefore 00000, 10110 ($\alpha_1 = 1$, $\alpha_2 = 0$), 11001 ($\alpha_1 = 0$, $\alpha_2 = 1$) and 01111 ($\alpha_1 = 1$, $\alpha_2 = 1$), as before.

Exercise 5.34   For each case in Exercise 5.29 determine the generator matrix, and use this to find the codewords.

We have only been able to give a very brief introduction to coding theory, and books such as Hill (1986) or MacWilliams and Sloane (1977) should be consulted for further development of the subject.

Problems

5.1   The idea of controllability, introduced in Example 4.10, can be extended to the case where there are m control variables, so that the set of differential equations (4.36) is replaced by

$$\frac{dx}{dt} = Ax + Bu,$$

where B is an $n \times m$ matrix and u is an m-vector. It can be shown that the controllability matrix (4.37) is replaced by the $n \times nm$ matrix

$$\mathscr{C} = [B, AB, A^2B, \ldots, A^{n-1}B] \qquad (5.76)$$

and in this case the condition for controllability is that $R(\mathscr{C}) = n$.

If

$$A = \begin{bmatrix} 1 & 1 & 0 \\ 0 & 1 & 0 \\ 0 & 0 & 1 \end{bmatrix}, \qquad B = \begin{bmatrix} 0 & 1 \\ 1 & 0 \\ 1 & 0 \end{bmatrix}$$

determine whether in this case the system is controllable.

5.2 Suppose that a certain product is stored in two warehouses, in which there are 17 and 11 units of the product available respectively. It is required to deliver the product to three supermarkets, where the requirements are 9, 12, and 7 units respectively. Let x_{ij} be the number of units transported from warehouse number i to supermarket number j ($i = 1, 2$; $j = 1, 2, 3$). Thus, for example, considering the first supermarket, we must have $x_{11} + x_{21} = 9$, and so on. Write down the remaining four equations which must be satisfied. Show that the total set of equations is consistent and find the general solution.

This is a very simple example of a *transportation problem*, an important special type of LP problem. In practice, if c_{ij} is the unit transport cost from warehouse i to supermarket j, the aim is to find the transport schedule which minimizes the total cost $\sum_i \sum_j c_{ij} x_{ij}$.

5.3 Prove that for the equations

$$x_1 + 2x_2 + 3x_3 - 3x_4 = k_1$$
$$2x_1 - 5x_2 - 3x_3 + 12x_4 = k_2$$
$$7x_1 + x_2 + 8x_3 + 5x_4 = k_3$$

to be consistent then $37k_1 + 13k_2 - 9k_3 = 0$. Find the general solution when $k_1 = 2$, $k_2 = 4$.

5.4 Determine the values of k such that the equations

$$x_1 + 2x_2 = x_3$$
$$2x_1 + (3 + k)x_2 = 3x_3$$
$$(k - 1)x_1 + 4x_2 = 3x_3$$

have a non-trivial solution. Find the general solution in these cases.

5.5 Find the values of λ for which the equations

$$2x_1 + (4 - \lambda)x_2 + 7 = 0$$
$$(2 - \lambda)x_1 + 2x_2 + 3 = 0$$
$$2x_1 + 5x_2 + 6 - \lambda = 0$$

are consistent. Find the general solution in these cases.

5.6   Find the rank of the matrix

$$A = \begin{bmatrix} 0 & -2 & 4 & 3 \\ 1 & 1 & 1 & 1 \\ 3 & 5 & -1 & a \\ 2 & 1 & 4 & b \end{bmatrix}$$

for all possible values of the parameters a and b. Determine also in each case the general solution of the corresponding equations $Ax = 0$.

5.7   An *elementary matrix* is a matrix obtained from I_n by applying a single elementary operation to it. For example, $(R1) - 2(R2)$ applied to I_3 gives the elementary matrix

$$E = \begin{bmatrix} 1 & -2 & 0 \\ 0 & 1 & 0 \\ 0 & 0 & 1 \end{bmatrix}. \qquad (5.77)$$

For simplicity consider only elementary row operations. Show that if E is obtained from I_n by any elementary row operation, then forming the product EA produces the matrix A with the same operation applied to its rows (for example, with E in (5.77) then EA is equal to the matrix obtained by applying $(R1) - 2(R2)$ to A).

5.8   The matrix E in (5.77) corresponds to $(R1) - 2(R2)$ applied to I_3; the inverse of this operation is $(R1) + 2(R2)$, which corresponds to the elementary matrix

$$F = \begin{bmatrix} 1 & 2 & 0 \\ 0 & 1 & 0 \\ 0 & 0 & 1 \end{bmatrix}.$$

Notice that $F = E^{-1}$; why?

By considering the inverse of each elementary operation (listed in Section 5.3.1), show that in general the inverse of any elementary matrix is itself an elementary matrix.

5.9   If

$$A = \begin{bmatrix} \overset{m}{A_1} & \overset{n}{0} \\ 0 & A_2 \end{bmatrix} \begin{matrix} p \\ q \end{matrix}$$

and $R(A_1) = r_1$, $R(A_2) = r_2$, prove that $R(A) = r_1 + r_2$.

5.10 If A is $m \times n$ and B is $n \times m$ with $m < n$, use (5.11) to prove that BA is singular.

5.11 By using the method which was applied to (5.44), verify (5.49). Hence deduce that the equation

$$AXB^\mathsf{T} = C, \tag{5.78}$$

where A is $n \times n$, B is $m \times m$ and X and C are $n \times m$, can be written in the form

$$(A \otimes B)v(X) = v(C), \tag{5.79}$$

where $v(X)$ denotes the column vector x in (5.48).

5.12 Using the notation of the preceding problem, reduce that

$$v(C^\mathsf{T}) = Pv(C), \qquad v(X) = Qv(X^\mathsf{T}),$$

where P, Q are constant $mn \times mn$ matrices whose rows are a permutation of those of I_{mn} (such matrices are called *permutation matrices*—see Section 13.4.5). In other words the elements of $v(C^\mathsf{T})$ are simply a reordering of those of $v(C)$, and similarly for $v(X)$ and $v(X^\mathsf{T})$.

By considering (5.78) and its transpose, show that

$$B \otimes A = P(A \otimes B)Q.$$

5.13 Consider the equation

$$A^2 X + 2AXB + XB^2 = C, \tag{5.80}$$

where the matrices have the same dimensions as in (5.51). Using (5.47), (5.50), (5.54), and (5.78), show that (5.80) can be written in the form $D^2 x = c$, where D is defined in (5.53) and x, c are as in (5.47).

This shows that the solution of (5.80) can be obtained from that of (5.51).

5.14 It can be shown that if A and B are two matrices having the same dimensions then

$$R(A + B) \leqslant R(A) + R(B).$$

Use this result, together with (5.11), to prove that if X is an idempotent matrix (defined in Problem 4.14) then

$$R(X) + R(I_n - X) = n.$$

5.15 If X is an idempotent matrix as defined in Problem 4.14, use (5.12) and the result (b) of Problem 2.7 to prove that $R(X) = \mathrm{tr}(X)$.

5.16 Write the relationship (5.12) between an $m \times n$ matrix A and its normal form N as $RAT = N$, where $R = P^{-1}$, $T = Q^{-1}$. Then R and

T can be obtained by recording the elementary transformations used to transform A into N via the array (not a matrix!):

$$\frac{I_n}{A \mid I_m} \longrightarrow \frac{T}{N \mid R}.$$

For example, if

$$A = \begin{bmatrix} 1 & 2 & 3 \\ 2 & 4 & 6 \end{bmatrix},$$

then reduction of A proceeds as follows:

$$
\begin{array}{ccc|cc}
1 & 0 & 0 \\
0 & 1 & 0 \\
0 & 0 & 1 \\
\hline
1 & 2 & 3 & 1 & 0 \\
2 & 4 & 6 & 0 & 1
\end{array}
\xrightarrow{(R2)-2(R1)}
\begin{array}{ccc|cc}
1 & 0 & 0 \\
0 & 1 & 0 \\
0 & 0 & 1 \\
\hline
1 & 2 & 3 & 1 & 0 \\
1 & 0 & 0 & -2 & 1
\end{array}
$$

$$
\xrightarrow[(C3)-3(C1)]{(C2)-2(C1)}
\begin{array}{ccc|cc}
1 & -2 & -3 \\
0 & 1 & 0 \\
0 & 0 & 1 \\
\hline
1 & 0 & 0 & 1 & 0 \\
0 & 0 & 0 & -2 & 1.
\end{array}
$$

The row operations on A are also applied to the rows in the lower right block, and the column operations on A are also applied to the upper left block. In this example

$$R = \begin{bmatrix} 1 & 0 \\ -2 & 1 \end{bmatrix}, \quad T = \begin{bmatrix} 1 & -2 & -3 \\ 0 & 1 & 0 \\ 0 & 0 & 1 \end{bmatrix}, \quad N = \begin{bmatrix} 1 & 0 & 0 \\ 0 & 0 & 0 \end{bmatrix}$$

and it is easy to check that $RAT = N$.

Carry out this procedure for the matrix A in (5.4), and check that your answer is correct.

5.17 For a set of n real row m-vectors $v_1, \ldots, v_n$ it can be shown that they are linearly independent if and only if the $n \times n$ real symmetric *Gram* matrix $G = [g_{ij}]$, with $g_{ij} = v_i^T v_j$, is non-singular. Use this result to test the vectors in (5.55).

5.18 By considering the equations $Hx' = 0$, prove that $111000\ldots0$ is always a codeword for any Hamming code with $n \geqslant 4$.

5.19 By considering the equations $Hx' = 0$, prove that if H is the check matrix for a linear code of length n, then the dimension of the code is $n - R(H)$.

5.20 Prove that if a linear code has check matrix H and generator matrix G, then $GH^{\mathsf{T}} = 0$.

5.21 The Faculty of Difficult Studies contains 1920 students, and each is to be given a unique identity number in the form of a binary codeword.

(a) Show that there must be at least eleven information bits.

(b) If the code is to be capable of correcting any single error, show that there must be at least four check bits.

5.22 Let u, v be arbitrary column n-vectors, and set $B = uv^{\mathsf{T}}$ (assumed $\neq 0$). Deduce from (5.11) that $R(B) = 1$, and show that

$$B^2 = [\mathrm{tr}(B)]B$$

(use property (b) in Problem 2.7).

5.23 If A is a non-singular $n \times n$ matrix, and B is the matrix defined in the previous problem, show that $BA^{-1}BA^{-1} = BA^{-1}\mathrm{tr}(BA^{-1})$. Hence verify that

$$(A + B)^{-1} = A^{-1} - \frac{A^{-1}BA^{-1}}{1 + \mathrm{tr}(BA^{-1})}, \tag{5.81}$$

assuming $A + B$ is non-singular.

This show how the inverse can be easily modified when A is perturbed by a rank-one addition.

5.24 Verify that provided the appropriate matrices are non-singular

$$(A - UBV)^{-1} = A^{-1} + A^{-1}U(B^{-1} - VA^{-1}U)^{-1}VA^{-1}. \tag{5.82}$$

This result is a generalization of (5.81), and equivalent expressions are

$$(A - UBV)^{-1} = A^{-1} + A^{-1}UB(I - VA^{-1}UB)^{-1}VA^{-1}$$
$$= A^{-1} + A^{-1}U(I - BVA^{-1}U)^{-1}BVA^{-1}.$$

5.25 The expressions in the preceding problem are useful in linear systems theory. For example, use (5.82) to prove

$$[I + G(s)]^{-1} = I - C(sI - A + BC)^{-1}B,$$

where $G(s) = C(sI - A)^{-1}B$ is the *transfer function matrix* associated with the control system described in Problem 5.1, with a vector of measurable output variables $y = Cx$.

5.26 Let A_1, A_2 be $m \times m$ matrices, B_1, B_2 be $n \times n$ matrices and define

$$C = A_1 \otimes B_1 + A_2 \otimes B_2.$$

Assume A_1, B_1 and C are non-singular and $R(B_2) = 1$. Use the first result in Problem 5.23 to verify that

$$C^{-1} = A_1^{-1} \otimes B_1^{-1} - D \otimes B_1^{-1}B_2B_1^{-1}$$

where $D = [A_1 + A_2 \operatorname{tr}(B_2B_1^{-1})]^{-1}A_2A_1^{-1}$.

5.27 Let U be the matrix in Problem 4.7, where W is non-singular. Verify that

$$U = \begin{bmatrix} I_n & 0 \\ YW^{-1} & I_m \end{bmatrix} \begin{bmatrix} W & 0 \\ 0 & (U/W) \end{bmatrix} \begin{bmatrix} I_n & W^{-1}X \\ 0 & I_m \end{bmatrix}$$

and hence deduce using Problem 5.9 that

$$R(U) = R(W) + R(U/W),$$

where $(U/W) = Z - YW^{-1}X$ is the Schur complement of W in U.

5.28 By appropriate choices for A, U, B, V in (5.82), show that the Schur complement (U/W) has inverse

$$(Z - YW^{-1}X)^{-1} = Z^{-1} + Z^{-1}Y(W - XZ^{-1}Y)^{-1}XZ^{-1}.$$

References

Hill (1986), MacWilliams and Sloane (1977), Mirsky (1963).

6 Eigenvalues and eigenvectors

The underlying problem in much of the theory discussed so far in this book is that of solving sets of linear algebraic equations. To be able to deal with other applications, such as solution of sets of linear differential equations, we need the ideas of this chapter. Indeed, these are essential for much further development of matrices.

6.1 Definitions

The basic problem is to find values of the scalar λ and corresponding column n-vectors x ($\neq 0$) which satisfy the equation

$$Ax = \lambda x, \tag{6.1}$$

where A is a given $n \times n$ matrix. Rewriting (6.1) as

$$(\lambda I_n - A)x = 0, \tag{6.2}$$

we see that (6.2) represents a system of homogeneous equations, so, as stated in Section 5.1, for a non-trivial solution x to exist we must have

$$\det(\lambda I_n - A) = 0. \tag{6.3}$$

When the $n \times n$ determinant in (6.3) is expanded, it produces an nth-degree polynomial in λ, called the *characteristic polynomial* $k(\lambda)$ of A, and (6.3) is called the *characteristic equation* of A, i.e.

$$\begin{aligned} k(\lambda) &\equiv \det(\lambda I_n - A) \\ &\equiv \lambda^n + k_1\lambda^{n-1} + k_2\lambda^{n-2} + \cdots + k_{n-1}\lambda + k_n = 0. \end{aligned} \tag{6.4}$$

The n roots $\lambda_1, \lambda_2, \ldots, \lambda_n$ of this equation are called the *eigenvalues* of A (the term *characteristic roots* is also used). When λ in (6.2) is equal to one of the λ_i, then a solution $x \neq 0$ of (6.2) will exist, but as we have seen in Section 5.4.1 this solution will not be unique. Any such vector x satisfying (6.2) when $\lambda = \lambda_i$ is called an *eigenvector* (or *characteristic vector*) corresponding to λ_i. The set $\lambda_1, \lambda_2, \ldots, \lambda_n$ is called the *spectrum* of A. Although the Anglo-German hybrid terms involving 'eigen-' are rather ugly, they are in widespread use and we shall therefore adopt these from now on.

Example 6.1 If

$$A = \begin{bmatrix} 1 & 3 \\ 2 & 2 \end{bmatrix}, \tag{6.5}$$

then

$$\det(\lambda I_2 - A) = \begin{vmatrix} \lambda - 1 & -3 \\ -2 & \lambda - 2 \end{vmatrix}$$
$$= (\lambda - 1)(\lambda - 2) - (-3)(-2)$$
$$= \lambda^2 - 3\lambda - 4$$
$$= (\lambda + 1)(\lambda - 4),$$

so the eigenvalues of A are $\lambda_1 = -1$, $\lambda_2 = 4$. When $\lambda = -1$ the equations (6.2) are

$$\begin{bmatrix} -2 & -3 \\ -2 & -3 \end{bmatrix} \begin{bmatrix} x_1 \\ x_2 \end{bmatrix} = 0 \tag{6.6}$$

and the solution of (6.6) is easily found to be $x_1 = t_1$, $x_2 = -\frac{2}{3}t_1$, with t_1 an arbitrary parameter. Similarly, when $\lambda = 4$ the solution of (6.2) is $x_1 = t_2$, $x_2 = t_2$. Thus eigenvectors corresponding to the eigenvalues $-1, 4$ respectively have the form

$$u = t_1 \begin{bmatrix} 1 \\ -\dfrac{2}{3} \end{bmatrix}, \qquad v = t_2 \begin{bmatrix} 1 \\ 1 \end{bmatrix} \tag{6.7}$$

for arbitrary non-zero t_1 and t_2.

As expected (from the theory of homogeneous equations in Chapter 5) the eigenvectors are not unique. However, it is common practice to choose the parameters so that the eigenvectors are *unit vectors*, i.e. they have unit length, where the *length* (or *euclidean norm*) of an n-vector x with real or complex components is defined to be the real non-negative number

$$\|x\| = (|x_1|^2 + |x_2|^2 + \cdots + |x_n|^2)^{1/2} = (x^*x)^{1/2}. \tag{6.8}$$

This definition agrees with the usual idea of length of real vectors in two or three dimensions (see (4.35)). When a vector has been converted to a unit vector it is said to be *normalized*. To carry out this normalization, simply divide x by its length $l = \|x\|$, for by the definition (6.8) the length of x/l is

$$\|x/l\| = (|x_1|^2/l^2 + |x_2|^2/l^2 + \cdots + |x_n|^2/l^2)^{1/2}$$
$$= (|x_1|^2 + |x_2|^2 + \cdots + |x_n|^2)^{1/2}/l$$
$$= l/l = 1,$$

so x/l is a unit vector.

For example, the lengths of the vectors in (6.7) are respectively

$$l_1 = \left(t_1^2 + \frac{4}{9}t_1^2\right)^{1/2} = \frac{\sqrt{13}}{3}t_1, \qquad l_2 = (t_2^2 + t_2^2)^{1/2} = (\sqrt{2})t_2,$$

so normalized eigenvectors are u/l_1, v/l_2, i.e.

$$\frac{3}{\sqrt{13}}\begin{bmatrix} 1 \\ -\dfrac{2}{3} \end{bmatrix}, \qquad \frac{1}{\sqrt{2}}\begin{bmatrix} 1 \\ 1 \end{bmatrix},$$

corresponding to eigenvalues -1 and 4 respectively. Notice that there is an ambiguity over the sign, since $-u/l_1$, $-v/l_2$ are also normalized eigenvectors. In practice, however, this causes no difficulty.

Some further terminology: the vectors x in (6.1) are often called *right* eigenvectors, to distinguish them from *left* eigenvectors y, which are *row* n-vectors satisfying

$$yA = \lambda y. \qquad (6.9)$$

Notice that since (6.9) gives $y(\lambda I_n - A) = 0$, the condition for a non-trivial y to exist is still (6.3), so the values of λ which satisfy (6.9) are exactly the same as those satisfying (6.1). In this book, if no qualification is made then the reader can assume that right eigenvectors are being used. Before studying properties of eigenvalues and eigenvectors, we devote a section to describing a few out of the multitude of applications in which they arise. This will serve to illustrate that (6.1) is definitely not a merely abstract definition.

Exercise 6.1 Determine the eigenvalues and associated normalized eigenvectors for the matrix

$$A = \begin{bmatrix} 1 & 3 \\ 3 & 1 \end{bmatrix}.$$

Exercise 6.2 If $A = \mathrm{diag}[a_{11}, a_{22}, \ldots, a_{nn}]$ show using property PD4 of Section 4.1.2 that $\lambda_i = a_{ii}$ for $i = 1, 2, \ldots, n$. Show also that this result still holds if A is a triangular matrix. Hence give an example of a 3×3 matrix which has all its eigenvalues equal to zero but is not itself a zero matrix.

Exercise 6.3 Consider again the 2×2 matrix $A^{(1)}$ in (2.23) associated with the complex number $z_1 = a_1 + ib_1$. Show that the eigenvalues of $A^{(1)}$ are z_1 and $\bar{z}_1$.

Exercise 6.4 Prove that, for any non-zero scalar c,

$$\|cx\| = |c|\,\|x\|\,.$$

6.2 Some applications

Example 6.2 Consider a simplified version of the mass–spring system shown in Fig. 1.4, in which it is assumed that the damping forces are negligible, i.e. $d_1 = d_2 = 0$, and that $m_1 = m_2 = 1$. Equations (1.8) and (1.9) can then be written as

$$\begin{bmatrix} \ddot{x}_1 \\ \ddot{x}_2 \end{bmatrix} = -\begin{bmatrix} k_1 & -k_1 \\ -k_1 & (k_1 + k_2) \end{bmatrix}\begin{bmatrix} x_1 \\ x_2 \end{bmatrix}$$

or in matrix form

$$\ddot{x} = -Ax, \tag{6.10}$$

where in (6.10), $x = [x_1, x_2]^{\mathrm{T}}$. Intuitively we feel that the motion has an oscillatory nature, which suggests we try

$$x = y \sin \omega t \tag{6.11}$$

as a solution of (6.10), where y is a constant vector. Then differentiating (6.11) gives $\ddot{x} = -\omega^2 y \sin \omega t$, which on substitution into (6.10) produces

$$-\omega^2 y \sin \omega t = -Ay \sin \omega t. \tag{6.12}$$

The parameter ω represents the frequency of oscillations, so we reject $\omega = 0$ in (6.12), which then reduces to

$$(\omega^2 I - A)y = 0,$$

which has precisely the form (6.2) with $\lambda = \omega^2$. Thus the solution of the set of second-order linear differential equations (6.10) depends on the eigenvalues and eigenvectors of A, and the eigenvalues determine the frequencies of the oscillations.

Example 6.3 As was done in (1.10), we can convert the second-order equations (6.10) into a first-order set by defining new variables $x_3 = \dot{x}_1$, $x_4 = \dot{x}_2$, to obtain

$$\dot{x} = Bx, \tag{6.13}$$

where B is the 4×4 matrix obtained from (1.10) by setting $d_1 = d_2 = 0$, $m_1 = m_2 = 1$, and now $x = [x_1, x_2, x_3, x_4]^\mathsf{T}$. An approach to solving (6.13) is suggested by the *scalar* differential equation

$$\dot{z} = bz. \tag{6.14}$$

On writing dz/dt for $\dot{z}$ and separating the variables, the reader should confirm that the solution of (6.14) is

$$z(t) = e^{bt}\alpha, \tag{6.15}$$

where α is a constant. It therefore seems worthwhile to use the vector

$$x(t) = e^{\lambda t}v \tag{6.16}$$

as a trial solution of (6.14), where v is a constant vector and λ a parameter to be determined. Differentiation of (6.16) gives $\dot{x} = \lambda e^{\lambda t}v$, and substituting into (6.13):

$$\lambda e^{\lambda t}v = Be^{\lambda t}v$$

whence, since $e^{\lambda t} \neq 0$,

$$(\lambda I - B)v = 0. \tag{6.17}$$

Again we have obtained an equation in the form (6.2), this time showing that the solution of the set of first-order linear differential equations (6.13) depends on the eigenvalues and eigenvectors of B.

Example 6.4 A stretched string has its ends A and B fixed, and four equal particles m are attached at equal distances d apart, the whole resting on a smooth horizontal table, as shown in Fig. 6.1. The tension T in the string is assumed large so that the portions of the string can be assumed straight, and the angles made with AB assumed small. If the

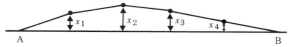

Fig. 6.1 Stretched string for Example 6.4.

displacements of the particles are x_1, x_2, x_3, x_4, it can be shown that the equations of motion are

$$\begin{bmatrix} \ddot{x}_1 \\ \ddot{x}_2 \\ \ddot{x}_3 \\ \ddot{x}_4 \end{bmatrix} = -\frac{T}{md} \begin{bmatrix} 2 & -1 & 0 & 0 \\ -1 & 2 & -1 & 0 \\ 0 & -1 & 2 & -1 \\ 0 & 0 & -1 & 2 \end{bmatrix} \begin{bmatrix} x_1 \\ x_2 \\ x_3 \\ x_4 \end{bmatrix}, \qquad (6.18)$$

which can be written in the matrix form (6.10), so the same conclusion about dependence of the solution on eigenvalues and eigenvectors can be drawn. (In passing, notice that the matrix in (6.18) is tridiagonal, as defined in Problem 3.2.)

The preceding examples illustrate the general fact that the solution of a set of linear differential equations can be expressed in terms of eigenvalues and eigenvectors. Such equations arise in many areas, including mechanical vibrations, electrical networks, and control systems, and will be studied further in Section 6.5 together with linear difference equations, which exhibit similar properties.

Example 6.5 A geometrical interpretation in two and three dimensions can be given. Let P be an arbitrary point in two dimensions having cartesian coordinates x_1 and x_2. If we regard a given 2×2 matrix A as the matrix of a transformation of coordinates (see Example 1.6), then after the transformation has been applied P moves to some point P' having coordinates

$$x' = \begin{bmatrix} x_1' \\ x_2' \end{bmatrix} = A \begin{bmatrix} x_1 \\ x_2 \end{bmatrix} = Ax,$$

Thus the condition (6.1) shows that $x' = Ax = \lambda x$. This implies that x' is a scalar multiple of x, which is equivalent to requiring that P' lies on the straight line OP, and its distance from the origin is $OP' = \|x'\| = \|\lambda x\| = |\lambda| \|x\| = |\lambda|$ (OP) (see Exercise 6.4).

The eigenvectors of A can therefore be thought of as those vectors which are left unaltered in direction after the transformation has been applied, and the eigenvalues measure the ratio of the lengths of these vectors before and after the transformation. For example, the eigenvectors u and v in (6.7) with $t_1 = t_2 = 1$ for the matrix A in (6.5) are shown before and after transformation in Fig. 6.2(a) and Fig. 6.2(b) respectively.

Exercise 6.5 Deduce that for the transformation in Problem 1.5(a), normalized eigenvectors are unit vectors along the two coordinate axes.

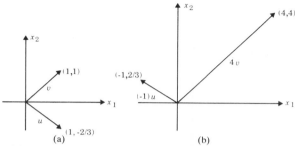

Fig. 6.2 (a) Eigenvectors of A in equation (6.5). (b) Eigenvectors after transformation.

6.3 Properties

6.3.1 The characteristic equation

If A is a real matrix then obviously all the coefficients k_i in the characteristic equation (6.4) will be real, but some of the roots λ_i may be complex. From a well-known result in elementary algebra, any complex roots of (6.4) will occur in conjugate pairs. Eigenvectors corresponding to real eigenvalues will have real elements because the equations (6.2) will then have all coefficients real; however, eigenvectors associated with complex eigenvalues may also have complex elements. If A is a complex matrix then complex eigenvalues will not in general form conjugate pairs.

Exercise 6.6 Calculate the eigenvalues of:

(a) $\begin{bmatrix} 1 & -10 \\ 3 & -5 \end{bmatrix}$; (b) $\begin{bmatrix} 0 & -1 & 1 \\ 2 & 3 & 3 \\ -2 & 1 & 1 \end{bmatrix}$.

Since the roots of (6.4) are $\lambda_1, \ldots, \lambda_n$, we have

$$\det(\lambda I_n - A) \equiv \lambda^n + k_1\lambda^{n-1} + \cdots + k_n$$
$$\equiv (\lambda - \lambda_1)(\lambda - \lambda_2)\cdots(\lambda - \lambda_n). \quad (6.19)$$

Setting $\lambda = 0$ in the identity (6.19) gives

$$\det(-A) = k_n = (-\lambda_1)(-\lambda_2)\cdots(-\lambda_n). \quad (6.20)$$

It follows from property PD2 of Section 4.1.2 that $\det(-A) = (-1)^n \det A$, so the expressions in (6.20) give

$$\det A = \lambda_1 \lambda_2 \cdots \lambda_n = (-1)^n k_n. \qquad (6.21)$$

The result expressed by (6.21) is an interesting one: it shows that the determinant of A is equal to the product of the eigenvalues of A; thus in particular A is singular if and only if it has at least one zero eigenvalue. Next consider the coefficients of λ^{n-1} in (6.19). For simplicity take the case $n = 3$. The right-hand side of (6.19) is then

$$(\lambda - \lambda_1)(\lambda - \lambda_2)(\lambda - \lambda_3) = \lambda^3 - \lambda^2(\lambda_1 + \lambda_2 + \lambda_3) + \cdots$$

and the left-hand side of (6.19) is

$$\det(\lambda I_3 - A) = \begin{vmatrix} \lambda - a_{11} & -a_{12} & -a_{13} \\ -a_{21} & \lambda - a_{22} & -a_{23} \\ -a_{31} & -a_{32} & \lambda - a_{33} \end{vmatrix} \qquad (6.22)$$

$$= (\lambda - a_{11})A_{11} + (-a_{12})A_{12} + (-a_{13})A_{13}$$

on expanding by the first row using (4.23) (the A_{1i} are here the cofactors of the first row of $\lambda I - A$). The only term involving λ^2 in (6.22) is contained in

$$(\lambda - a_{11})A_{11} = (\lambda - a_{11}) \begin{vmatrix} \lambda - a_{22} & -a_{23} \\ -a_{32} & \lambda - a_{33} \end{vmatrix}$$

$$= (\lambda - a_{11})(\lambda^2 - a_{22}\lambda - a_{33}\lambda + \cdots)$$

$$= \lambda^3 - (a_{11} + a_{22} + a_{33})\lambda^2 + \cdots.$$

Hence when $n = 3$, comparing the coefficients of λ^2 in (6.19) gives

$$-(a_{11} + a_{22} + a_{33}) = k_1 = -(\lambda_1 + \lambda_2 + \lambda_3),$$

showing that the sum of the eigenvalues of A is equal to the sum of the elements on the principal diagonal, i.e. the *trace* of A. A similar argument applies for the general case, giving

$$\lambda_1 + \lambda_2 + \cdots + \lambda_n = \text{tr}(A) = -k_1. \qquad (6.23)$$

For example, for the matrix A in (6.5), $\text{tr}(A) = 1 + 2 = 3$ and $\lambda_1 + \lambda_2 = -1 + 4 = 3$.

Expressions involving minors of A can be derived for the other coefficients $k_2, \ldots, k_{n-1}$ in the characteristic equation, but these are more complicated. In fact when n is large it is difficult to calculate the characteristic equation and practical procedures for determining eigenvalues avoid this (see Section 6.6). For 2×2 and 3×3 matrices we can find $\det(\lambda I - A)$ using the cofactor formulae (4.23) or (4.24).

Example 6.6 The characteristic polynomial of the matrix

$$A = \begin{bmatrix} -1 & 0 & 2 \\ 0 & 1 & 2 \\ 2 & 2 & 0 \end{bmatrix} \tag{6.24}$$

is

$$\det(\lambda I_3 - A) = \begin{vmatrix} \lambda + 1 & 0 & -2 \\ 0 & \lambda - 1 & -2 \\ -2 & -2 & \lambda \end{vmatrix} \tag{6.25}$$

$$= (\lambda + 1)[(\lambda - 1)\lambda - 4] - 2 \cdot 2(\lambda - 1)$$
$$= \lambda^3 - 9\lambda = \lambda(\lambda - 3)(\lambda + 3)$$

(expanding the determinant by the first row). Thus the eigenvalues of A are $\lambda_1 = 0$, $\lambda_2 = 3$, $\lambda_3 = -3$. To obtain an eigenvector corresponding to λ_1, set $\lambda = 0$ in the right-hand side of (6.25) so as to obtain the coefficients in the equations (6.2), which are therefore

$$\begin{aligned} x_1 \qquad\quad - 2x_3 &= 0 \\ - \; x_2 - 2x_3 &= 0 \\ - 2x_1 - 2x_2 \qquad\quad &= 0. \end{aligned}$$

It is easily confirmed that the general solution of these equations is $x_1 = 2t_1$, $x_2 = -2t_1$, $x_3 = t_1$ for arbitrary non-zero t_1. Thus an eigenvector of A corresponding to λ_1 is $x = t[2, -2, 1]^{\mathsf{T}}$. Since by (6.8), $\|x\| = (4t_1^2 + 4t_1^2 + t_1^2)^{1/2} = 3t_1$, a normalized eigenvector corresponding to λ_1 is $x/3t_1 = \frac{1}{3}[2, -2, 1]^{\mathsf{T}}$.

In general, the set of homogeneous equations (6.2) in the n unknowns $x_1, \ldots, x_n$ is solved for each value λ_i of λ by the method of Section 5.4.1. Since by construction $\lambda_i I_n - A$ is singular, these equations will always have a non-trivial (and non-unique) solution. If all the λ_i are all different from each other, then $R(\lambda_i I_n - A) = n - 1$, and each eigenvector will contain just one arbitrary parameter.

Exercise 6.7 Determine normalized eigenvectors: (a) for the eigenvalues λ_2 and λ_3 of A in (6.24); (b) for the matrix in Exercise 6.6(b).

Exercise 6.8 For each of the following matrices A expand $\det(\lambda I_3 - A)$ using cofactors, and hence calculate the characteristic equation and eigenvalues:

(a) $\begin{bmatrix} -2 & -1 & 0 \\ 1 & 2 & 3 \\ 4 & 5 & 6 \end{bmatrix}$; (b) $\begin{bmatrix} 1 & 0 & -1 \\ 1 & 2 & 1 \\ 2 & 2 & 3 \end{bmatrix}$.

6.3.2 Hermitian and symmetric matrices

It was remarked in Section 6.3.1 that even real matrices may in general possess some complex eigenvalues (and eigenvectors). However, an important exception to this is provided by hermitian and real symmetric matrices (defined in Section 2.3.2) which always have all their eigenvalues real.

To prove this, let A be an $n \times n$ hermitian matrix, so from the definition (2.56)

$$A^* = A \tag{6.26}$$

and let λ and u be associated eigenvalue and eigenvector of A, so that

$$Au = \lambda u. \tag{6.27}$$

At this stage we cannot assume that either λ or u is real, so if we take the complex conjugate of both sides of (6.27) and then transpose, we obtain

$$u^*A^* = \bar{\lambda}u^*. \tag{6.28}$$

Postmultiplying (6.28) by u and applying (6.26) results in

$$u^*Au = \bar{\lambda}u^*u. \tag{6.29}$$

However, premultiplying (6.27) by u^* gives

$$u^*Au = \lambda u^*u$$

and subtracting (6.29) from this last equation produces

$$(\lambda - \bar{\lambda})u^*u = 0. \tag{6.30}$$

Now u^*u is the scalar product of $u^* = [\bar{u}_1, \ldots, \bar{u}_n]$ and u, which by (2.44) is

$$u^*u = \bar{u}_1 u_1 + \bar{u}_2 u_2 + \cdots + \bar{u}_n u_n$$
$$= |u_1|^2 + |u_2|^2 + \cdots + |u_n|^2 = \|u\|^2, \tag{6.31}$$

and the expression in (6.31) is non-zero since $u \neq 0$. Hence we can conclude from (6.30) that $\bar{\lambda} = \lambda$, showing that λ is real, as required. Since a symmetric matrix is the special case of a hermitian matrix when all the elements are real, the result also applies in this case. In fact we can also see that when A in (6.27) is real and symmetric, then since all the numbers occurring in (6.27) will be real the eigenvectors will also be real.

Exercise 6.9 Calculate the eigenvalues and corresponding normalized eigenvectors for the hermitian matrix

$$A = \begin{bmatrix} 3 & 1+i \\ 1-i & 2 \end{bmatrix}.$$

Exercise 6.10 Prove that the determinant of any hermitian matrix is a real number.

Exercise 6.11 If A is a skew hermitian matrix (defined in (2.57)) show that all its eigenvalues have zero real part.

6.3.3 *Matrix polynomials and the Cayley–Hamilton theorem*

We now show how to obtain eigenvalues and eigenvectors of matrices derived from A. Let λ_i and u_i be associated eigenvalues and right eigenvectors of an arbitrary $n \times n$ matrix A, so that

$$Au_i = \lambda_i u_i, \qquad i = 1, 2, \ldots, n. \tag{6.32}$$

Notice that we are here using u_i to stand for a column n-vector rather than an ith component—it will always be clear from the context when this is meant. Premultiply both sides of (6.32) by A to give

$$A^2 u_i = \lambda_i A u_i = \lambda_i (\lambda_i u_i) = \lambda_i^2 u_i,$$

which shows that A^2 has eigenvalues λ_i^2 and eigenvectors u_i, for $i = 1, 2, \ldots, n$. Repeated premultiplication by A similarly shows that A^m has eigenvalues λ_i^m and eigenvectors u_i for any positive integer m, i.e.

$$A^m u_i = \lambda_i^m u_i. \tag{6.33}$$

A slightly different argument shows that A^T has the same eigenvalues as those of A. For we have

$$\det(\lambda I - A^T) = \det(\lambda I - A^T)^T, \quad \text{by property PD1, Section 4.1.2,}$$
$$= \det(\lambda I - A), \qquad \text{by (2.38),}$$

showing that the characteristic equation of A^T is identical to that of A.

Exercise 6.12 Show that for u_i defined in (6.32), u_i^T is a left eigenvector of A^T.

Exercise 6.13 For A in (6.32), prove: (a) the eigenvectors of A^* are $\bar{\lambda}_i$; (b) if A is non-singular, the eigenvalues and eigenvectors of A^{-1} are $1/\lambda_i$ and u_i for $i = 1, 2, \ldots, n$. What are the eigenvalues of adj A?

Exercise 6.14 Let A and B be two $n \times n$ matrices with B non-singular, and let the eigenvalues and eigenvectors of AB be μ_i and v_i, $i = 1, 2, \ldots, n$. Prove that the eigenvalues and eigenvectors of BA are μ_i and Bv_i.

We now generalize the result in (6.33). Let

$$p(\lambda) = \lambda^r + p_1\lambda^{r-1} + \cdots + p_{r-1}\lambda + p_r \qquad (6.34)$$

be an arbitrary polynomial of degree r. Then we define the matrix *polynomial* $p(A)$ in the matrix A to be the $n \times n$ matrix obtained by replacing λ by A, i.e.

$$p(A) = A^r + p_1A^{r-1} + \cdots + p_{r-1}A + p_rI_n. \qquad (6.35)$$

If u_i is defined by (6.32) then

$$\begin{aligned}
p(A)u_i &= A^r u_i + p_1 A^{r-1}u_i + \cdots + p_{r-1}Au_i + p_r u_i \\
&= \lambda_i^r u_i + p_1\lambda_i^{r-1}u_i + \cdots + p_{r-1}\lambda_i u_i + p_r u_i, \quad \text{by (6.33)} \\
&= p(\lambda_i)u,
\end{aligned}$$

showing that the eigenvalues and eigenvectors of $p(A)$ are $p(\lambda_i)$ and u_i, for $i = 1, 2, \ldots, n$.

Exercise 6.15 For A in (6.32), if c is a scalar prove that the eigenvalues of $A + cI_n$ are $\lambda_i + c$, $i = 1, 2, \ldots, n$: (a) by using the preceding result; (b) by using the definition (6.3). Prove also that if $\lambda_i + c \neq 0$ for all i, then the eigenvalues and eigenvectors of $(A + cI)^{-1}$ are $1/(\lambda_i + c)$ and u_i.

Exercise 6.16 Prove that $p(A)$ commutes with A for any polynomial $p(\lambda)$ and any square matrix A.

Notice that if λ_i is a root of $p(\lambda)$ then the preceding argument shows that $p(A)u_i \equiv 0$. In particular, a most important result arises when we take $p(\lambda)$ to be the characteristic polynomial of A defined in (6.4), i.e.

$$k(\lambda) = \det(\lambda I_n - A) = \lambda^n + k_1\lambda^{n-1} + \cdots + k_{n-1}\lambda + k_n. \quad (6.36)$$

We shall prove:

Cayley–Hamilton theorem Every matrix satisfies its own characteristic equation, i.e.

$$k(A) \equiv A^n + k_1 A^{n-1} + \cdots + k^{n-1}A + k_n I_n \equiv 0. \quad (6.37)$$

Example 6.7 The characteristic polynomial of A in (6.5) was found to be $\lambda^2 - 3\lambda - 4$, and

$$A^2 - 3A - 4I_2 = \begin{bmatrix} 7 & 9 \\ 6 & 10 \end{bmatrix} - 3 \begin{bmatrix} 1 & 3 \\ 2 & 2 \end{bmatrix} - 4 \begin{bmatrix} 1 & 0 \\ 0 & 1 \end{bmatrix}$$

$$= \begin{bmatrix} 0 & 0 \\ 0 & 0 \end{bmatrix},$$

thereby verifying (6.37) in this case.

It must be stressed that (6.37) is *not* an obvious result: the equation $k(\lambda) = 0$ is satisfied only for the n values $\lambda = \lambda_1, \lambda_2, \ldots, \lambda_n$, whereas (6.37) states that the matrix polynomial $k(A)$ is identically equal to the $n \times n$ zero matrix.

For ease of understanding we give a proof of the Cayley–Hamilton theorem for $n = 3$, but the method generalizes directly for any value of n. First, from (4.46) we have

$$(\lambda I_3 - A)^{-1} = \text{adj}(\lambda I_3 - A)/\det(\lambda I_3 - A)$$
$$= \text{adj}(\lambda I - A)/k(\lambda), \quad (6.38)$$

where for convenience the suffix on I has been dropped. Recall next that $\text{adj}(\lambda I - A)$ is the transposed matrix of cofactors of $\lambda I - A$, so each of its elements is a 2×2 determinant which will be a polynomial in λ of degree at most 2. For example, from the array in (6.22) we see that the 2, 2 element of $\text{adj}(\lambda I - A)$ is

$$\begin{vmatrix} \lambda - a_{11} & -a_{13} \\ -a_{31} & \lambda - a_{33} \end{vmatrix} = (\lambda - a_{11})(\lambda - a_{33}) - a_{13}a_{31}.$$

Collecting together powers of λ, we can therefore write

$$\text{adj}(\lambda I - A) = \lambda^2 B_1 + \lambda B_2 + B_3, \quad (6.39)$$

where B_1, B_2, and B_3 are constant 3×3 matrices. Combining together (6.38) and (6.39) gives

$$I = (\lambda I - A)(\lambda I - A)^{-1}$$
$$= (\lambda I - A)(\lambda^2 B_1 + \lambda B_2 + B_3)/k(\lambda) \quad (6.40)$$

and multiplying both sides of (6.40) by $k(\lambda)$ produces

$$(\lambda^3 + k_1\lambda^2 + k_2\lambda + k_3)I = (\lambda I - A)(\lambda^2 B_1 + \lambda B_2 + B_3)$$
$$= \lambda^3 B_1 + \lambda^2(B_2 - AB_1) + \lambda(B_3 - AB_2) - AB_3. \quad (6.41)$$

Equating coefficients of powers of λ in (6.41):

$$\begin{aligned}
\lambda^3: & \quad I \equiv B_1 \\
\lambda^2: & \quad k_1 I \equiv B_2 - AB_1 \\
\lambda: & \quad k_2 I \equiv B_3 - AB_2 \\
\lambda^0: & \quad k_3 I \equiv -AB_3.
\end{aligned} \qquad (6.42)$$

Finally, premultiply the identities in (6.42) by A^3, A^2, A, and I respectively, and add the resulting expressions to obtain

$$\begin{aligned}
A^3 + k_1 A^2 + k_2 A + k_3 I = A^3 B_1 &+ (A^2 B_2 - A^3 B_1) \\
&+ (AB_3 - A^2 B_2) - AB_3 \\
&\equiv 0,
\end{aligned}$$

which is the desired result. The only differences for $n > 3$ are that (6.39) is replaced by $\lambda^{n-1} B_1 + \cdots + B_n$, and there are $n + 1$ identities like (6.42), but otherwise the proof proceeds in the same way.

Exercise 6.17 Verify the Cayley–Hamilton theorem for the two matrices in Exercise 6.8 (i.e. evaluate A^3 and A^2 and substitute into the characteristic polynomial).

Exercise 6.18 Obtain the characteristic equation of the 3×3 matrix A in (2.25), and hence write down (6.37) in this case. Notice that it was verified in Exercise 2.9 that for this matrix A we have $A^2 - 3A + 2I_3 \equiv 0$.

Exercise 6.18 illustrates that it is possible for some $n \times n$ matrices to have $q(A) \equiv 0$, where $q(\lambda)$ is a polynomial having degree *less* than n. Such a matrix is called *derogatory*. The unique monic polynomial $m(\lambda)$ of lowest degree such that $m(A) \equiv 0$ is called the *minimum polynomial* of A. When $m(\lambda)$ is the same as the characteristic polynomial $k(\lambda)$, then A is called *non-derogatory*. An important condition which ensures that A is non-derogatory is that all its eigenvalues are *distinct* (i.e. all different from each other). However, not all matrices possessing repeated eigenvalues are derogatory.

Equation (6.37) can be rearranged to give

$$A^n = -k_1 A^{n-1} - k_2 A^{n-2} - \cdots - k_{n-1} A - k_n I, \qquad (6.43)$$

which expresses A^n as a linear combination of A^{n-1}, $A^{n-2}, \ldots, A, I$. If (6.43) is premultiplied by A, we get A^{n+1} expressed as a linear combination of A^n, $A^{n-1}, \ldots, A, I$, and hence it follows that A^{n+1} can be expressed as a linear combination of $A^{n-1}, A^{n-2}, \ldots, I$. Similarly, by repeated premultiplication of (6.43) by A it follows that *any* power A^{n+t}, $t = 0, 1, 2, \ldots$, can be expressed as a linear combination of powers of A up to A^{n-1}.

Example 6.8 Consider again the 2×2 matrix A in (6.5). In Example 6.7 we found that

$$A^2 = 3A + 4I. \qquad (6.44)$$

Hence

$$A^3 = 3A^2 + 4A = 3(3A + 4I) + 4A = 13A + 12I$$

$$= \begin{bmatrix} 25 & 39 \\ 26 & 38 \end{bmatrix}$$

and similarly

$$A^4 = 13A^2 + 12A = 13(3A + 4I) + 12A = 51A + 52I$$

and so on.

On the other hand, if A is non-singular, we can premultiply (6.37) by A^{-1} and rearrange to get

$$k_n A^{-1} = -(A^{n-1} + k_1 A^{n-2} + \cdots + k_{n-1} I). \qquad (6.45)$$

Since A is non-singular, (6.21) shows that $k_n \neq 0$, so we can divide in (6.45) to give

$$A^{-1} = -(A^{n-1} + k_1 A^{n-2} + \cdots + k_{n-1} I)/k_n. \qquad (6.46)$$

Example 6.8 (continued) Premultiplying (6.44) by A^{-1} gives

$$A = 3I + 4A^{-1},$$

so that

$$A^{-1} = \frac{1}{4}(A - 3I) = \frac{1}{4} \begin{bmatrix} -2 & 3 \\ 2 & -1 \end{bmatrix}.$$

In general, calculation of A^{-1} via (6.46) is not recommended because of the previously mentioned difficulty of calculating the k's.

Exercise 6.19 Determine A^4 and A^{-1} for the matrix in Exercise 6.8(b) using the Cayley–Hamilton theorem.

The use of the Cayley–Hamilton theorem to evaluate powers of A can be extended to evaluation of $p(A)$, where $p(\lambda)$ is an arbitrary polynomial (6.34) having degree *greater* than n. To see how this is done, first recall the fact from elementary algebra that if $p(\lambda)$ is divided by $k(\lambda)$ to give some quotient $q(\lambda)$, say, then the remainder $r(\lambda)$ has degree *less* than n, i.e.

$$p(\lambda) \equiv q(\lambda)k(\lambda) + r(\lambda). \tag{6.47}$$

Now (6.47) is an identity, not an equation, so it still holds if we replace λ by A, which gives

$$p(A) \equiv q(A)k(A) + r(A). \tag{6.48}$$

However, by the Cayley–Hamilton theorem $k(A) \equiv 0$, so (6.48) reduces to

$$p(A) \equiv r(A). \tag{6.49}$$

Thus to evaluate $p(A)$, simply divide $p(\lambda)$ by the characteristic polynomial, and the desired result is obtained by substituting A into the remainder polynomial.

Example 6.9 Continuing with the 2×2 matrix A in (6.5) having $k(\lambda) = \lambda^2 - 3\lambda - 4$, if

$$p(\lambda) = \lambda^5 + 2\lambda^4 + \lambda^3 - 3\lambda^2 + \lambda - 1,$$

then division of $p(\lambda)$ by $k(\lambda)$ produces the remainder $r(\lambda) = 312\lambda + 307$. Hence by (6.49)

$$\begin{aligned} p(A) &= A^5 + 2A^4 + A^3 - 3A^2 + A - I \\ &= 312A + 307I \\ &= \begin{bmatrix} 619 & 936 \\ 624 & 931 \end{bmatrix}. \end{aligned}$$

Exercise 6.20 Determine $A^5 - 7A^4 - 18A^3 + 54A^2 + 116A + I_3$ for the matrix in Exercise 6.8(a).

6.3.4 Companion matrix

We now introduce a matrix with a very simple form which has the useful property that its characteristic polynomial is the same as that of A, so both matrices have the same eigenvalues. For example, when $n = 2$ define

$$C = \begin{bmatrix} 0 & 1 \\ -k_2 & -k_1 \end{bmatrix}, \tag{6.50}$$

where the k's are the coefficients in (6.4). It is very easy to see that

$$\det(\lambda I_2 - C_2) = \begin{vmatrix} \lambda & -1 \\ k_2 & \lambda + k_1 \end{vmatrix}$$
$$= \lambda^2 + k_1\lambda + k_2,$$

which is the same as $k(\lambda)$ in (6.4) with $n = 2$. Similarly, if

$$C_3 = \begin{bmatrix} 0 & 1 & 0 \\ 0 & 0 & 1 \\ -k_3 & -k_2 & -k_1 \end{bmatrix}, \tag{6.51}$$

then

$$\det(\lambda I_3 - C_3) = \lambda^3 + k_1\lambda^2 + k_2\lambda + k_3 \tag{6.52}$$

and the pattern is now emerging. We define C_n to be the $n \times n$ matrix having 1's along the diagonal immediately above the principal diagonal, having last row $[-k_n, -k_{n-1}, \ldots, -k_2, -k_1]$, and zeros everywhere else. Then

$$\det(\lambda I_n - C_n) = \lambda^n + k_1\lambda^{n-1} + \cdots + k_n \equiv k(\lambda) \equiv \det(\lambda I_n - A),$$
$$\tag{6.53}$$

and for this reason C_n is called the *companion matrix* associated with the polynomial $k(\lambda)$. Because of its simple form C_n is useful in various applications (alternative forms are sometimes used— see Problem 6.12).

Exercise 6.21 Verify that if λ_i is an eigenvalue of C_3 then an associated eigenvector is $[1, \lambda_i, \lambda_i^2]^\mathrm{T}$.

Similarly, verify that $m_i = [1, \lambda_i, \lambda_i^2, \ldots, \lambda_i^{n-1}]^\mathrm{T}$ is an eigenvector of C_n (notice that m_i is the ith column of the Vandermonde matrix defined in Problem 4.16).

As an illustration of one of the uses of the companion matrix, we now show that determination of polynomials in C_n is especially easy. For simplicity we consider C_3 in (6.51), but the argument is easily generalized for any value of n. We wish to compute

$$p(C_3) = C_3^2 + p_1 C_3 + p_2 I_3. \tag{6.54}$$

First, if e_i denotes the ith row of I_3 then it is obvious from (6.51) that the first and second rows of C_3 are e_2 and e_3 respectively.

Next, if X is *any* 3×3 matrix, then the product $e_i X$ simply gives the ith row of X (see Exercise 2.19), so in particular we can write (dropping for convenience the suffix on C):

$$e_1 C = e_2, \qquad e_2 C = e_3. \tag{6.55}$$

Hence the first row of C^2 is given by

$$e_1 C^2 = (e_1 C) C = e_2 C = e_3.$$

We can now determine the rows ρ_1, ρ_2, ρ_3 of $p(C)$ in (6.54). Since the first rows of C^2, C, and I are e_3, e_2, and e_1 respectively, then clearly

$$\begin{aligned}
\rho_1 &= e_3 + p_1 e_2 + p_2 e_1 \\
&= [0, 0, 1] + p_1 [0, 1, 0] + p_2 [1, 0, 0] \\
&= [p_2, p_1, 1].
\end{aligned} \tag{6.56}$$

Also,

$$\begin{aligned}
\rho_2 &= e_2 p(C) \\
&= (e_1 C) p(C), \quad \text{by (6.55)}, \\
&= e_1 p(C) C
\end{aligned} \tag{6.57}$$

since $p(C)$ commutes with C (see Exercise 6.16). By definition $e_1 p(C) = \rho_1$, so (6.57) reduces to

$$\rho_2 = \rho_1 C. \tag{6.58}$$

Similarly,

$$\begin{aligned}
\rho_3 &= e_3 p(C) \\
&= e_2 C p(C), \quad \text{by (6.55)}, \\
&= [e_2 p(C)] C \\
&= \rho_2 C \\
&= \rho_1 C^2, \quad \text{by (6.58)},
\end{aligned}$$

so we have shown that

$$p(C_3) = \begin{bmatrix} \rho_1 \\ \rho_1 C \\ \rho_1 C^2 \end{bmatrix}, \tag{6.59}$$

where ρ_1 is given by (6.56). In general, if C is the $n \times n$

companion matrix, and we form the matrix polynomial

$$p(C) = C^r + p_1 C^{r-1} + p_2 C^{r-2} + \cdots + p_{r-1}C + p_r I \quad (6.60)$$

with $r < n$, then an extension of the argument used to obtain (6.59) shows that

$$p(C) = \begin{bmatrix} \rho_1 \\ \rho_1 C \\ \rho_1 C^2 \\ \vdots \\ \rho_1 C^{n-1} \end{bmatrix}, \quad (6.61)$$

where

$$\rho_1 = [p_r, p_{r-1}, \ldots, p_1, 1, 0, \ldots, 0]. \quad (6.62)$$

Thus, for any polynomial $p(\lambda)$ having degree less than n, the rows $\rho_1, \rho_2, \ldots, \rho_n$ of the matrix polynomial $p(C)$ can be obtained from (6.62) and the recurrence formula

$$\rho_{i+1} = \rho_i C, \quad i = 1, 2, \ldots, n-1 \quad (6.63)$$

and this is much easier in general than constructing $p(A)$ for an arbitrary matrix A.

We can note three interesting points about (6.61). First, it follows by inspection that $p(C) \equiv 0$ only if $\rho_1 = 0$. That is, there is no polynomial of degree less than n such that $p(C) \equiv 0$, so according to the definition in Section 6.3.3, C is non-derogatory. Second, we saw in Section 6.3.3 that the eigenvalues of $p(C)$ are $p(\lambda_1), p(\lambda_2), \ldots, p(\lambda_n)$, where the λ's are the eigenvalues of C, i.e. the roots of the polynomial $k(\lambda)$. Also, (6.21) shows that

$$\det p(C) = p(\lambda_1)p(\lambda_2) \cdots p(\lambda_n),$$

so $p(C)$ is non-singular if and only if each $p(\lambda_i) \neq 0$. This latter condition means that none of the roots of $k(\lambda)$ can also be a root of $p(\lambda)$, or in other words $k(\lambda)$ and $p(\lambda)$ must have no common factor. Thus $p(C)$ is non-singular if and only if $k(\lambda)$ and $p(\lambda)$ are relatively prime. In the terminology of Problem 4.1, $\det p(C)$ is a *resultant* for the two polynomials $k(\lambda)$ and $p(\lambda)$.

The third feature of $p(C)$ which is worth noting is that if we transpose (6.61) we get

$$[p(C)]^T = [\rho_1^T, C^T \rho_1^T, (C^T)^2 \rho_1^T, \ldots, (C^T)^{n-1} \rho_1^T],$$

which has precisely the same form as the controllability matrix introduced in (4.37), with b replaced by ρ_1^T and A by C^T. This can be developed to produce some interesting results in control theory.

Exercise 6.22 Let C be the companion matrix for $k(\lambda) = \lambda^3 + 6\lambda^2 + 11\lambda + 6$. Use (6.62) and (6.63) to construct $p(C)$, where $p(\lambda) = \lambda^2 - \lambda - 2$. Hence determine whether $k(\lambda)$ and $p(\lambda)$ have a common factor.

6.3.5 Kronecker product expressions

If A and B are arbitrary $n \times n$ matrices, then in general there is no way of determining the eigenvalues either of the product AB or of the sum $A + B$ in terms of the eigenvalues of A and B. However, as we have seen in other situations, these difficulties can be avoided if we use the Kronecker product.

Let A be $n \times n$ with eigenvalues and eigenvectors λ_i, u_i ($i = 1, 2, \ldots, n$) and let B be $m \times m$ with eigenvalues and eigenvectors μ_j, y_j ($j = 1, 2, \ldots, m$), so by definition $Au_i = \lambda_i u_i$, $By_j = \mu_j y_j$. Hence, using (2.76) we have

$$
\begin{aligned}
(A \otimes B)(u_i \otimes y_j) &= (Au_i) \otimes (By_j) \\
&= (\lambda_i u_i) \otimes (\mu_j y_j) \\
&= (\lambda_i \mu_j)(u_i \otimes y_j), \quad (6.64)
\end{aligned}
$$

showing that the mn eigenvalues and eigenvectors of $A \otimes B$ are simply $\lambda_i \mu_j$ and $u_i \otimes y_j$, for $i = 1, \ldots, n$ and $j = 1, \ldots, m$.

A similar result can be obtained for a Kronecker 'sum' involving A and B. We use the matrix

$$
D = A \otimes I_m + I_n \otimes B^T, \quad (6.65)
$$

which arose in (5.52) in connection with a linear matrix equation. We shall show that the eigenvalues of the $mn \times mn$ matrix D are all possible *sums* $\lambda_i + \mu_j$. (The occurrence of B^T instead of B in (6.65) is of no consequence—we saw at the beginning of Section 6.3.3 that the eigenvalues of B^T are the same as those of B.)

The method of proof is quite ingenious. Let ε be an arbitrary scalar parameter, and consider the product

$$
\begin{aligned}
(I_n + \varepsilon A) \otimes (I_m + \varepsilon B^T) &= I_n \otimes I_m + \varepsilon(A \otimes I_m + I_n \otimes B^T) \\
&\qquad + \varepsilon^2 A \otimes B^T \\
&= I_n \otimes I_m + \varepsilon D + \varepsilon^2 A \otimes B^T. \quad (6.66)
\end{aligned}
$$

By the result of Exercise 6.15 the eigenvalues of $I + \varepsilon A$ and $I + \varepsilon B^{\mathsf{T}}$ are $1 + \varepsilon\lambda_i$ and $1 + \varepsilon\mu_j$ respectively. By the foregoing argument, expressed in (6.64), the eigenvalues of the matrix on the left-hand side of (6.66) are

$$(1 + \varepsilon\lambda_i)(1 + \varepsilon\mu_j) = 1 + \varepsilon(\lambda_i + \mu_j) + \varepsilon^2\lambda_i\mu_j. \tag{6.67}$$

Since ε is *arbitrary*, it follows by comparing terms in ε on the right-hand sides of (6.66) and (6.67) that D has eigenvalues $\lambda_i + \mu_j$.

Since $\det D$ is equal to the product of the eigenvalues of D (see (6.21)) it follows that D is non-singular if and only if $\lambda_i + \mu_j \neq 0$, for all i and j, and this is the condition for the solution X of the matrix equation (5.51) (i.e. $AX + XB = C$) to be unique.

Exercise 6.23 Write down in terms of A and B a matrix whose eigenvalues are the *mn* numbers $\lambda_i^2 + \mu_j^2$, $i = 1, \ldots, n$; $j = 1, \ldots, m$.

Exercise 6.24 If E is a $p \times p$ matrix having eigenvalues and eigenvectors v_k, z_k, what are the eigenvalues and eigenvectors of $A \otimes B \otimes E$?

6.4 Similarity

We shall assume *throughout* this section that all the eigenvalues λ_i of A are distinct (i.e. $\lambda_i \neq \lambda_j$ for all $i \neq j$). This is often the case in practical problems.

6.4.1 Definition

Two $n \times n$ matrices A and B are called *similar* if

$$B = P^{-1}AP, \tag{6.68}$$

where P is an arbitrary non-singular $n \times n$ matrix. The relationship between A and B is called that of *similarity*. The characteristic equation of B is

$$0 = |\lambda I_n - P^{-1}AP| \tag{6.69}$$

and we now do some juggling with (6.69) to prove that the eigenvalues of B are the same as those of A. First, replace I_n in (6.69) by $P^{-1}P$ and factorize inside the determinant sign:

$$0 = |P^{-1}(\lambda I - A)P|. \tag{6.70}$$

Then use the result (4.33) on the determinant of a product, thereby reducing (6.70) to

$$0 = |P^{-1}|\,|\lambda I - A|\,|P|$$
$$= |\lambda I - A|\,,$$

since $|P^{-1}| = 1/|P|$ (see (4.42)). This shows that the characteristic equations of B and A are identical, as required. Furthermore, if u_i are the eigenvectors of A, then

$$B(P^{-1}u_i) = P^{-1}AP(P^{-1}u_i)$$
$$= P^{-1}Au_i$$
$$= P^{-1}\lambda_i u_i = \lambda_i(P^{-1}u_i),$$

showing that the eigenvectors of B are $P^{-1}u_i$, for $i = 1, 2, \ldots, n$.

The converse result also holds, namely, that if B is a matrix having the same eigenvalues as A then (6.68) holds for some P. (It is worth remarking that this does not apply in general if any of the λ's are repeated—for further discussion see Section 8.1.)

Exercise 6.25   Prove that if A and B are similar then (a) $\det A = \det B$; (b) $\mathrm{tr}(A) = \mathrm{tr}(B)$.

Exercise 6.26   Prove that if all the n eigenvalues of A are distinct then there exists an $n \times n$ non-singular matrix X such that $A^{\mathsf{T}}X - XA = 0$.

Exercise 6.27   If A and B are similar, determine which of the following pairs is similar: (a) A^{T} and B^{T}; (b) A^k and B^k (k = positive integer); (c) A^{-1} and B^{-1} (assuming A is non-singular).

6.4.2   Diagonalization

We now show how to choose P in (6.68) so that B is a diagonal matrix. The development depends on the fact (stated here without proof), that provided the λ's are distinct then the eigenvectors $u_1, \ldots, u_n$ of A are linearly independent. From the discussion in Section 5.7 this means that the matrix formed by putting the u's side by side:

$$T = [u_1, u_2, \ldots, u_n] \qquad (6.71)$$

has rank n, i.e. is non-singular. It is T which we use for P in

(6.68), which becomes

$$
\begin{aligned}
T^{-1}AT &= T^{-1}A[u_1, u_2, \ldots, u_n] \\
&= T^{-1}[Au_1, Au_2, \ldots, Au_n] \\
&= T^{-1}[\lambda_1 u_1, \lambda_2 u_2, \ldots, \lambda_n u_n], \quad \text{by (6.32)}, \\
&= [\lambda_1 T^{-1}u_1, \lambda_2 T^{-1}u_2, \ldots, \lambda_n T^{-1}u_n].
\end{aligned} \tag{6.72}
$$

However, we also have

$$
\begin{aligned}
I_n &= T^{-1}T = T^{-1}[u_1, u_2, \ldots, u_n] \\
&= [T^{-1}u_1, T^{-1}u_2, \ldots, T^{-1}u_n]
\end{aligned} \tag{6.73}
$$

and comparison of (6.72) and (6.73) shows that the ith column in (6.72) is simply λ_i times the ith column in (6.73), and this latter is the ith column of I_n. Therefore (6.72) reduces to

$$
T^{-1}AT = \text{diag}[\lambda_1, \lambda_2, \ldots, \lambda_n] \equiv \Lambda, \tag{6.74}
$$

showing that when all the λ's are distinct then A is similar to Λ, the diagonal matrix of its eigenvalues. The similarity transformation (6.74) of A into Λ is called *diagonalization* and has many important applications.

Example 6.10 Return to the matrix A in (6.5) having $\lambda_1 = -1$, $\lambda_2 = 4$, and use the eigenvectors obtained in (6.7):

$$
u_1 = t_1 \begin{bmatrix} 1 \\ -\dfrac{2}{3} \end{bmatrix}, \qquad u_2 = t_2 \begin{bmatrix} 1 \\ 1 \end{bmatrix}.
$$

For convenience set $t_1 = 3$, $t_2 = 1$, so from (6.71)

$$
T = \begin{bmatrix} 3 & 1 \\ -2 & 1 \end{bmatrix}, \qquad T^{-1} = \frac{1}{5}\begin{bmatrix} 1 & -1 \\ 2 & 3 \end{bmatrix}, \tag{6.75}
$$

the inverse being obtained using (4.7). It is easily verified that $T^{-1}AT = \text{diag}[-1, 4]$.

Notice that any suitable vectors u_i in (6.71) will do—there is no need to use normalized eigenvectors.

It should be noted that if some of the λ's are repeated then A is similar to the diagonal matrix Λ only if A possesses n linearly independent eigenvectors (see Problem 6.32).

Exercise 6.28 Determine eigenvectors for the matrix A in Exercise 6.8(b). Calculate T^{-1}, and hence verify (6.74) in this case.

Exercise 6.29 Use (6.74) to show that $A = T\Lambda T^{-1}$, and hence prove that

$$A^k = T\Lambda^k T^{-1} \tag{6.76}$$

for any positive integer k. Use (6.76) to calculate A^3 for the matrix A in Exercise 6.28. Similarly, show that for the matrix polynomial in (6.35), $p(A) = Tp(\Lambda)T^{-1}$.

Exercise 6.30 With the notation of (6.74), if A is non-singular, define

$$X = T\{\text{diag}[\pm\lambda_1^{1/2}, \ \pm\lambda_2^{1/2}, \ldots, \ \pm\lambda_n^{1/2}]\}T^{-1} \tag{6.77}$$

and show that $X^2 = A$. Thus X can be thought of as a 'square root' of A.

It is worth mentioning here that for T in (6.71) (the columns of which are right eigenvectors of A) then the rows $v_1, \ldots, v_n$ of its inverse T^{-1} are *left* eigenvectors of A (defined in (6.9)). To show this, rearrange (6.74) as $T^{-1}A = \Lambda T^{-1}$ and compare the ith rows on each side to obtain $v_i A = \lambda_i v_i$, as required. As an illustration, the rows of T^{-1} in Example 6.10 are

$$v_1 = \frac{1}{5}[1, -1], \qquad v_2 = \frac{1}{5}[2, 3]$$

and using A in (6.5) it is easy to verify that

$$v_1 A = (-1)v_1, \qquad v_2 A = 4v_2,$$

showing that v_1, v_2 are left eigenvectors. For a converse result, see Problem 6.15.

Exercise 6.31 Verify that in Exercise 6.28 the rows of T^{-1} are left eigenvectors of A.

We close this section by pointing out that similarity in (6.68) is a special case of equivalence of matrices, defined in (5.14), since in (6.68) both P^{-1} and P are non-singular. Thus all the properties of equivalent matrices also apply to similar matrices. For example, two similar matrices have the same rank.

6.4.3 Hermitian and symmetric matrices

When A is hermitian (or real symmetric) the diagonalization formula (6.74) can be simplified. To develop this, let

$$Au_i = \lambda_i u_i, \qquad Au_j = \lambda_j u_j, \qquad i \neq j. \tag{6.78}$$

We saw in Section 6.3.2 that all the λ's are real. Premultiply the first equation in (6.78) by u_j^*:

$$u_j^* A u_i = \lambda_i u_j^* u_i. \tag{6.79}$$

Take the conjugate transpose of the second equation in (6.78) and then postmultiply by u_i to get

$$u_j^* A u_i = \lambda_j u_j^* u_i \tag{6.80}$$

(using $A^* = A$, $\bar{\lambda}_j = \lambda_j$). Comparison of (6.79) and (6.80) reveals that

$$\lambda_i u_j^* u_i = \lambda_j u_j^* u_i \tag{6.81}$$

and since $\lambda_i \neq \lambda_j$ (by assumption) we conclude from (6.81) that

$$u_j^* u_i = 0, \qquad \text{for all } i \neq j. \tag{6.82}$$

Any two vectors whose scalar product is zero are called *mutually orthogonal*, by analogy with the geometrical result in two and three dimensions. Thus (6.82) shows that all pairs of eigenvectors of any hermitian matrix (having distinct eigenvalues) are mutually orthogonal. If the u's are normalized, i.e. by (6.8),

$$u_i^* u_i = 1, \qquad i = 1, \ldots, n, \tag{6.83}$$

then for T in (6.71) we have

$$T^* T = \begin{bmatrix} u_1^* \\ \vdots \\ u_n^* \end{bmatrix} [u_1, \ldots, u_n] = I_n$$

using (6.82) and (6.83). This shows that T in this case is a unitary matrix (see Problem 4.11), and $T^{-1} = T^*$. Hence the diagonalization (6.74) becomes

$$T^* A T = \Lambda \tag{6.84}$$

and A is said to be *unitarily* similar to Λ. When A is real symmetric all the u's are real vectors, and T is orthogonal (i.e. $T^{-1} = T^T$) so that (6.84) is replaced by *orthogonal* similarity

$$T^T A T = \Lambda. \tag{6.85}$$

The transformations (6.84) and (6.85) are simpler to apply than the general case (6.74) since there is no need to calculate T^{-1}.

Example 6.11   For the real symmetric matrix

$$A = \begin{bmatrix} 1 & -2 \\ -2 & -2 \end{bmatrix} \qquad (6.86)$$

the usual calculations give $\lambda_1 = 2$, $\lambda_2 = -3$, with corresponding normalized eigenvectors

$$u_1 = \frac{1}{\sqrt{5}} \begin{bmatrix} 2 \\ -1 \end{bmatrix}, \qquad u_2 = \frac{1}{\sqrt{5}} \begin{bmatrix} 1 \\ 2 \end{bmatrix}, \qquad T = [u_1, u_2].$$

Hence in this case (6.85) is

$$\underbrace{\frac{1}{\sqrt{5}} \begin{bmatrix} 2 & -1 \\ 1 & 2 \end{bmatrix}}_{T^\mathsf{T}} \underbrace{\begin{bmatrix} 1 & -2 \\ -2 & -2 \end{bmatrix}}_{A} \underbrace{\frac{1}{\sqrt{5}} \begin{bmatrix} 2 & 1 \\ -1 & 2 \end{bmatrix}}_{T} = \underbrace{\begin{bmatrix} 2 & 0 \\ 0 & -3 \end{bmatrix}}_{\Lambda}.$$

Exercise 6.32   Show that the matrix

$$A = \begin{bmatrix} 11 & 2 & 8 \\ 2 & 2 & -10 \\ 8 & -10 & 5 \end{bmatrix} \qquad (6.87)$$

has one eigenvalue equal to -9, and calculate the other eigenvalues, and normalized eigenvectors. Hence carry out the diagonalization (6.85).

It is important to note that it can be shown that *every* hermitian (or real symmetric) matrix can be reduced by unitary (or orthogonal) similarity to a diagonal matrix, even if some of the λ's occur more than once (in fact, unitary similarity also holds for *normal* matrices, defined in Problem 2.10). As remarked at the end of the preceding section, because A is similar to Λ we have $R(A) = R(\Lambda)$, and since Λ is diagonal its rank is equal to the number of non-zero diagonal elements. Thus, for *any* hermitian (or real symmetric) matrix, its rank is equal to the total number of its non-zero eigenvalues, including repetitions.

Exercise 6.33   Let A be an $n \times n$ matrix having all its elements equal to unity. By considering $R(A)$ deduce that A has only one non-zero eigenvalue. Determine its value by applying (6.23).

6.4.4   Transformation to companion form

If A has characteristic polynomial $k(\lambda)$ in (6.53), then the companion matrix C defined in Section 6.3.4 has the same eigenvalues as A, so if all these values are distinct then A and C

are similar. In fact the eigenvectors of C were found in Exercise 6.21 to be the columns m_i of the Vandermonde matrix

$$V_n = [m_1, m_2, \ldots, m_n], \tag{6.88}$$

where

$$m_i = [1, \lambda_i, \lambda_i^2, \ldots, \lambda_i^{n-1}]^{\mathsf{T}}.$$

From (4.81) V_n is non-singular provided all the λ's are distinct, so in this case the similarity transformation between C and Λ is

$$V_n^{-1} C V_n = \Lambda. \tag{6.89}$$

Combining (6.89) with (6.74) gives

$$\begin{aligned} C = V_n \Lambda V_n^{-1} &= V_n T^{-1} A T V_n^{-1} \\ &= (T V_n^{-1})^{-1} A (T V_n^{-1}). \end{aligned} \tag{6.90}$$

In (6.90) T is the matrix (6.71) of eigenvectors of A, and V_n is defined by (6.88) and (6.89), so the similarity transformation from A to C can be determined.

Example 6.12 For A in (6.5) we found $\lambda_1 = -1$, $\lambda_2 = 4$ so

$$V_2 = \begin{bmatrix} 1 & 1 \\ -1 & 4 \end{bmatrix}, \quad V_2^{-1} = \frac{1}{5} \begin{bmatrix} 4 & -1 \\ 1 & 1 \end{bmatrix},$$

the inverse being obtained from (4.7). The matrix T is given in (6.75) so

$$T V_2^{-1} = \frac{1}{5} \begin{bmatrix} 13 & -2 \\ -7 & 3 \end{bmatrix}$$

and substitution into (6.90) gives

$$(T V_2^{-1})^{-1} A (T V_2^{-1}) = \begin{bmatrix} 0 & 1 \\ 4 & 3 \end{bmatrix},$$

which is the companion matrix associated with the characteristic polynomial $\lambda^2 - 3\lambda - 4$ of A.

Exercise 6.34 Carry out the transformation (6.90) for A in Exercise 6.1.

When some of the λ's are repeated, the above method is no longer valid since V_n is singular. In general, A is similar to the companion matrix C associated with its characteristic polynomial if and only if A is non-derogatory (in Problem 6.21, take B to be this matrix C).

6.5   Solution of linear differential and difference equations

An important use of (6.74) is in solving a set of linear differential equations

$$\dot{x} = Ax, \tag{6.91}$$

where A is a constant $n \times n$ matrix and $x = [x_1(t), \ldots, x_n(t)]^T$. Two examples of such sets of equations were given in Section 6.2. Let

$$y = T^{-1}x, \qquad x = Ty, \tag{6.92}$$

where T is the matrix of eigenvectors in (6.71). Then

$$\dot{y} = T^{-1}\dot{x} = T^{-1}Ax$$
$$= T^{-1}ATy$$
$$= \Lambda y, \tag{6.93}$$

the last step following by (6.74). Since Λ is diagonal the equations represented by (6.93) have the very simple form

$$\dot{y}_i = \lambda_i y_i, \qquad i = 1, 2, \ldots, n. \tag{6.94}$$

As in (6.15) in Section 6.2, the solution of (6.94) is

$$y_i = \alpha_i e^{\lambda_i t}, \qquad i = 1, 2, \ldots, n, \tag{6.95}$$

where the α's are arbitrary constants of integration. Hence by (6.92) the general solution of (6.91) is

$$x(t) = T[\alpha_1 e^{\lambda_1 t}, \ \alpha_2 e^{\lambda_2 t}, \ \ldots, \ \alpha_n e^{\lambda_n t}]^T. \tag{6.96}$$

Example 6.13   If A is the matrix in (6.5), then eqns (6.91) are

$$\frac{dx_1}{dt} = x_1 + 3x_2, \qquad \frac{dx_2}{dt} = 2x_1 + 2x_2. \tag{6.97}$$

In this case $\lambda_1 = -1$, $\lambda_2 = 4$ and T is given in (6.75), so by (6.96) the general solution of (6.97) is

$$\begin{bmatrix} x_1(t) \\ x_2(t) \end{bmatrix} = \begin{bmatrix} 3 & 1 \\ -2 & 1 \end{bmatrix} \begin{bmatrix} \alpha_1 e^{-t} \\ \alpha_2 e^{4t} \end{bmatrix}$$

or

$$x_1 = 3\alpha_1 e^{-t} + \alpha_2 e^{4t}, \qquad x_2 = -2\alpha_1 e^{-t} + \alpha_2 e^{4t}, \tag{6.98}$$

where α_1 and α_2 are arbitrary constants whose values can be determined from given initial conditions $x_1(0)$ and $x_2(0)$.

Exercise 6.35 Verify by differentiation that (6.98) is indeed the solution of (6.97).

Exercise 6.36 Solve (6.91) when A is the matrix used in Exercise 6.8(b) (use the results obtained in Exercise 6.28).

A very similar development holds for linear difference equations of the form

$$X(k + 1) = AX(k), \qquad k = 0, 1, 2, \ldots, \tag{6.99}$$

where $X(k) = [X_1(k), X_2(k), \ldots, X_n(k)]^\mathsf{T}$. For examples of applications of (6.99), see Example 1.8 and Problems 1.6, 2.11 and 6.10. Using the transformation $Y(k) = T^{-1}X(k)$, $X(k) = TY(k)$, leads to

$$Y(k + 1) = \Lambda Y(k)$$

or in component form

$$Y_i(k + 1) = \lambda_i Y_i(k), \qquad i = 1, 2, \ldots, n. \tag{6.100}$$

It is easily verified that eqns (6.100) have the general solution

$$Y_i(k) = \beta_i \lambda_i^k, \qquad i = 1, 2, \ldots, n,$$

where the β's are arbitrary constants determined by the initial conditions; so the general solution of (6.99) is

$$X(k) = T[\beta_1 \lambda_1^k, \beta_2 \lambda_2^k, \ldots, \beta_n \lambda_n^k]^\mathsf{T}. \tag{6.101}$$

Example 6.14 Consider the difference equations

$$X_1(k + 1) = X_1(k) + 3X_2(k), \qquad X_2(k + 1) = 2X_1(k) + 2X_2(k).$$

Once again A is the matrix in (6.5) with $\lambda_1 = -1$, $\lambda_2 = 4$ and T is given by (6.75). Thus from (6.101) the general solution of this pair of difference equations is

$$\begin{bmatrix} X_1(k) \\ X_2(k) \end{bmatrix} = \begin{bmatrix} 3 & 1 \\ -2 & 1 \end{bmatrix} \begin{bmatrix} \beta_1(-1)^k \\ \beta_2(4)^k \end{bmatrix}.$$

The reader can easily check by substitution into the equations that this solution is indeed correct.

Exercise 6.37 Show that if $X(0)$ is an eigenvector of A corresponding to an eigenvalue λ_i, then the solution of (6.99) is $X(k) = \lambda_i^k X(0)$. Apply this result to the last part of the beetle model in Problem 2.15.

Exercise 6.38 Determine the eigenvalues of the matrix A in (2.74) (notice that A is in companion form). Apply (6.101) to obtain the general expression for the solution of (2.74). Using the conditions $X_1(1) = 1$, $X_2(1) = 1$ determine β_1 and β_2, and hence show that the kth Fibonacci number is

$$x_k = X_1(k) = \frac{1}{\sqrt{5}}\left[\left(\frac{1 + \sqrt{5}}{2}\right)^k - \left(\frac{1 - \sqrt{5}}{2}\right)^k\right]$$

$$= \frac{1}{2^{k-1}}\left[k + \binom{k}{3}5 + \binom{k}{5}5^2 + \binom{k}{7}5^3 + \ldots\right]$$

where $\binom{k}{r}$ denotes the binomial coefficient.

6.6   Calculation of eigenvalues and eigenvectors

As remarked previously, it is not recommended that the eigenvalues of a matrix be calculated by solving the characteristic equation. This is because of difficulties both in computing the coefficients in the characteristic polynomial, and in applying root-finding methods (these may also be very sensitive to errors in the coefficients). Instead, some iterative process is applied to the matrix itself. We confine ourselves to describing one simple procedure which can easily be carried out with a pocket calculator. Details of computer methods used in real-life problems can be found in numerical textbooks such as Golub and van Loan (1989).

6.6.1   Power method

Suppose that all the eigenvalues of a real matrix A are distinct, and that the eigenvalues having largest modulus is real. Let this value (termed the *dominant* eigenvalue) be denoted by λ_1, so that the remaining eigenvalues $\lambda_2, \ldots, \lambda_n$ satisfy

$$|\lambda_i| < |\lambda_1|, \qquad i = 2, 3, \ldots, n. \tag{6.102}$$

Since corresponding eigenvectors $u_1, \ldots, u_n$ are linearly independent they form a basis, so any arbitrary column n-vector X_0 can be expressed as a linear combination

$$X_0 = \alpha_1 u_1 + \alpha_2 u_2 + \cdots + \alpha_n u_n. \tag{6.103}$$

If $X_1 = AX_0$, then since $Au_i = \lambda_i u_i$ we have

$$X_1 = \alpha_1 A u_1 + \alpha_2 A u_2 + \cdots + \alpha_n A u_n$$
$$= \alpha_1 \lambda_1 u_1 + \alpha_2 \lambda_2 u_2 + \cdots + \alpha_n \lambda_n u_n.$$

Similarly if $X_2 = AX_1$, $X_3 = AX_2$, and so on, we obtain

$$X_k = \alpha_1 \lambda_1^k u_1 + \alpha_2 \lambda_2^k u_2 + \cdots + \alpha_n \lambda_n^k u_n, \qquad k = 1, 2, 3, \ldots \quad (6.104)$$

and hence

$$\frac{1}{\lambda_1^k} X_k = \alpha_1 u_1 + \alpha_2 \left(\frac{\lambda_2}{\lambda_1}\right)^k u_2 + \cdots + \alpha_n \left(\frac{\lambda_n}{\lambda_1}\right)^k u_n. \quad (6.105)$$

Since from (6.102) $|\lambda_i/\lambda_1| < 1$ for each $i > 1$, it follows that as $k \to \infty$ the terms $(\lambda_i/\lambda_1)^k$ tend to zero, so (6.105) gives

$$\frac{1}{\lambda_1^k} X_k \to \alpha_1 u_1, \quad \text{as } k \to \infty. \quad (6.106)$$

Thus provided $\alpha_1 \neq 0$, (6.106) shows that X_k tends to a multiple of the eigenvector u_1 as $k \to \infty$. However, as k increases the elements of X_k can become large, so to avoid this it is preferable to scale X_k at each step so that its largest element is unity. We therefore define a modified sequence $X_1, X_2, X_3, \ldots$, as follows:

$$\left. \begin{array}{l} Y_{k+1} = AX_k, \qquad k = 0, 1, 2, \ldots \\[4pt] \beta_{k+1} = \text{element of } Y_{k+1} \text{ having largest modulus} \\[4pt] X_{k+1} = \dfrac{1}{\beta_{k+1}} Y_{k+1}. \end{array} \right\} \quad (6.107)$$

As $k \to \infty$ we still have X_k tending to some multiple of u_1, so we can write $X_k \to pu_1$ and then (6.107) implies

$$Y_k \to A(pu_1) = p(Au_1) = p(\lambda_1 u_1) = \lambda_1(pu_1),$$

showing that $Y_k \to \lambda_1 X_k$ as $k \to \infty$. Because the largest element of X_k is unity, if follows that the element of Y_k having largest modulus approaches λ_1, i.e. $\beta_k \to \lambda_1$ as $k \to \infty$. Thus β_k and X_k defined by (6.107) provide successively better approximations to λ_1 and u_1, and the process is terminated when X_{k+1} and X_k are sufficiently close.

Example 6.15 For the matrix A in (6.5) start by choosing an arbitrary vector $X_0 = [1, \frac{1}{2}]^T$. Then (6.107) gives

$$Y_1 = AX_0 = \begin{bmatrix} 2.5 \\ 3 \end{bmatrix}, \qquad \beta_1 = 3, \qquad X_1 = \frac{1}{3} Y_1 = \begin{bmatrix} 0.833 \\ 1 \end{bmatrix},$$

$$Y_2 = AX_1 = \begin{bmatrix} 3.833 \\ 3.667 \end{bmatrix}, \qquad \beta_2 = 3.833, \qquad X_2 = \begin{bmatrix} 1 \\ 0.957 \end{bmatrix},$$

$$Y_3 = AX_2 = \begin{bmatrix} 3.871 \\ 3.914 \end{bmatrix}, \qquad \beta_3 = 3.914, \qquad X_3 = \begin{bmatrix} 0.989 \\ 1 \end{bmatrix},$$

and so on. Working to three decimal places, continuation of this process gives $X_6 = [1, 1]^T$ and $\beta_7 = 4$, which agree with the exact values found in Example 6.1.

Clearly in general the speed of convergence will depend on how quickly the terms in (6.105) tend to zero, that is, upon the magnitudes of the ratios $|\lambda_i/\lambda_1|$ for $i > 1$. The method breaks down if by chance α_1 in (6.103) is zero, but in any case this can be overcome simply by choosing a different initial vector and starting again.

Once λ_1 is known, an $(n-1) \times (n-1)$ matrix A' can be constructed having eigenvalues $\lambda_2, \lambda_3, \ldots, \lambda_n$ (see Problem 6.31). If λ_2 is real and $|\lambda_2| > |\lambda_i|$, for all $i > 2$, then the power method can be used to estimate λ_2 and u_2, and so on.

Exercise 6.39   Use the power method with $X_0 = [1, 1]^T$ to estimate the domainant eigenvalue and corresponding eigenvector for $A = \begin{bmatrix} 4 & 1 \\ 2 & 5 \end{bmatrix}$.

Exercise 6.40   Let the eigenvalue λ_n of A having smallest modulus be real. Prove that the dominant eigenvalue of A^{-1} is $1/\lambda_n$ (hint: see Exercise 6.13).

Hence compute an estimate of the smaller eigenvalue of A in Exercise 6.39 by applying the power method to A^{-1}, with $X_0 = [1, 0]^T$.

6.6.2   Other methods

A great deal of effort has been devoted to devising effective computer algorithms for calculating eigenvalues and eigenvectors. One important class of methods is based on the use of similarity transformations of the form $A_{k+1} = Q_k^T A_k Q_k$, $k = 0, 1, 2, \ldots$, with $A_0 = A$ and Q_k orthogonal. The matrices Q_k are chosen so that A_k approaches a simple form (e.g. diagonal) whose eigenvalues are readily obtainable. For example, in the so-called QR-type methods, which are widely used, a sequence is generated by

$$A_k = Q_k R_k, \qquad R_k Q_k = A_{k+1}, \qquad k = 0, 1, 2, \ldots \quad (6.108)$$

with Q_k orthogonal and R_k upper triangular, and then A_k tends to upper triangular form as $k \to \infty$. These transformation methods are usually convergent, and produce a full set of eigenvalues and eigenvectors.

As in the case of linear equations, any large practical problem will require the use of a computer and a proven library program.

Exercise 6.41 Show that in (6.108) A_{k+1} is similar to A_k.

6.7 Iterative solution of linear equations

We are now able to consider some other methods for solving the n linear equations in n unknowns in our usual form

$$Ax = b. \tag{6.109}$$

The procedures we shall describe are different in nature from those considered in Chapter 3, where the solution was obtained after a *fixed* finite number of operations (such methods are termed 'direct'). Instead, with an iterative approach, an initial guess $x^{(0)}$ for the solution of (6.109) is used to obtain a better approximation $x^{(1)}$, which in turn is used to produce $x^{(2)}$, and so on. The aim is that the sequence of vectors $\{x^{(k)}\}$, $k = 1, 2, 3, \ldots$ converges to the exact solution x of (6.109), by which we mean that each element of $x^{(k)}$ converges to the corresponding element of the exact solution x as $k \to \infty$.

6.7.1 Gauss–Seidel and Jacobi methods

The first step is to split up A into two parts

$$A = B - C, \tag{6.110}$$

where B is chosen to be non-singular and to have an appropriate form which will be described later. Substitution of (6.110) into (6.109) gives

$$Bx = b + Cx \tag{6.111}$$

and the iterative procedure is defined by

$$Bx^{(k+1)} = Cx^{(k)} + b, \qquad k = 0, 1, 2, \ldots. \tag{6.112}$$

It is, of course, necessary to determine conditions under which $x^{(k)}$ in (6.112) tends to x as $k \to \infty$, and this problem provides an interesting application of diagonalization. Substract (6.111) from

(6.112) to obtain

$$B(x^{(k+1)} - x) = C(x^{(k)} - x)$$

or

$$x^{(k+1)} - x = L(x^{(k)} - x), \qquad (6.113)$$

where $L = B^{-1}C$. Substituting $k = 0, 1, 2, \ldots$ in (6.113) gives in turn

$$x^{(1)} - x = L(x^{(0)} - x)$$
$$x^{(2)} - x = L(x^{(1)} - x) = L^2(x^{(0)} - x)$$
$$x^{(3)} - x = L(x^{(2)} - x) = L^3(x^{(0)} - x)$$
$$\vdots$$
$$x^{(k)} - x = L^k(x^{(0)} - x). \qquad (6.114)$$

We require $x^{(k)} - x$ to tend to zero as $k \to \infty$. For arbitrary $x^{(0)}(\neq x)$, (6.114) therefore shows that we must have $L^k \to 0$ as $k \to \infty$ (i.e. every element of L^k must tend to zero as $k \to \infty$). Now suppose the eigenvalues of L are $l_1, l_2, \ldots, l_n$ and are distinct. Then by the diagonalization formula (6.74)

$$L = T[\text{diag}(l_1, l_2, \ldots, l_n)]T^{-1}$$

so as in (6.76)

$$L^k = T[\text{diag}(l_1^k, l_2^k, \ldots, l_n^k)]T^{-1}. \qquad (6.115)$$

Since T in (6.115) is a constant matrix it follows that for $L^k \to 0$ we must have each term $l_i^k \to 0$, and the condition for this is that $|l_i| < 1$ for each i. Thus we have shown that provided all the eigenvalues of L have modulus less than unity, then the process defined in (6.112) converges to the solution of (6.109) (the argument can be extended to cover the case when some l's are repeated—see Exercise 9.8).

Unfortunately this test for convergence is of limited value, since calculation of the l's may involve more effort than solving the original equations (6.109)! A more useful criterion, which, however, only gives a *sufficient* condition for convergence of (6.114), is that

$$\|L\| < 1, \qquad (6.116)$$

where $\|L\|$, the *euclidean norm* of the matrix L, is defined in a similar way to the norm of a vector in (6.8):

$$\|L\| = \left(\sum_{i=1}^{n} \sum_{j=1}^{n} |l_{ij}|^2 \right)^{1/2}. \tag{6.117}$$

Notice (see Problem 2.7(c)) that $\|L\| = [\text{tr}(AA^*)]^{1/2}$. Although (6.116) is too restrictive it has the advantage of being very easy to apply.

We now give two ways of choosing B in (6.110). Let $A = [a_{ij}]$, and assume all $a_{ii} \neq 0$. Define

$$D = \text{diag}[a_{11}, a_{22}, \dots, a_{nn}] \tag{6.118}$$

and a lower triangular matrix $E = [e_{ij}]$ with all $e_{ii} = 0$ and $e_{ij} = -a_{ij}$, $j < i$, and an upper triangular matrix $F = [f_{ij}]$ with all $f_{ii} = 0$ and $f_{ij} = -a_{ij}$, $j > i$. In other words, E is the 'lower part' of $-A$, F is the 'upper part' of $-A$, and D is the 'diagonal part' of A, so that

$$A = D - E - F. \tag{6.119}$$

For example, if

$$A = \begin{bmatrix} 4 & 1 & -2 \\ -1 & 6 & -1 \\ 1 & -3 & 8 \end{bmatrix}, \tag{6.120}$$

then

$$D = \text{diag}[4, 6, 8] \tag{6.121}$$

$$E = \begin{bmatrix} 0 & 0 & 0 \\ 1 & 0 & 0 \\ -1 & 3 & 0 \end{bmatrix}, \qquad F = \begin{bmatrix} 0 & -1 & 2 \\ 0 & 0 & 1 \\ 0 & 0 & 0 \end{bmatrix}. \tag{6.122}$$

If in (6.110) we take

$$B = D - E, \qquad C = F \tag{6.123}$$

the scheme (6.112) is called the *Gauss–Seidel* method. If we take

$$B = D, \qquad C = E + F \tag{6.124}$$

the scheme (6.112) is called *Jacobi's* method. The reason for these choices is that in both cases B has a simple form (respectively triangular and diagonal) so that the equations (6.112) are easily solved at each iteration. In fact some simple explicit formulae can be derived as follows. For Jacobi's method

(6.112) and (6.124) give

$$x^{(k+1)} = B^{-1}(Cx^{(k)} + b)$$
$$= D^{-1}[(E + F)x^{(k)} + b] \tag{6.125}$$

and since D is diagonal we have $D^{-1} = \text{diag}[1/a_{11}, \ldots, 1/a_{nn}]$. Also, $E + F$ is simply the matrix $(-A)$ with all the elements on the principal diagonal replaced by zeros. Thus when (6.125) is written out in component form we obtain

$$x_i^{(k+1)} = \left(-\sum_{\substack{j=1 \\ j \neq i}}^{n} a_{ij}x_j^{(k)} + b_i \right) \Big/ a_{ii}, \quad i = 1, 2, \ldots, n. \tag{6.126}$$

It is illuminating to compare (6.126) with the ith equation in (6.109), which is

$$\sum_{j=1}^{n} a_{ij}x_j = b_i$$

or, on rearranging to express x_i in terms of the other x's (assuming $a_{ii} \neq 0$):

$$x_i = \left(-\sum_{\substack{j=1 \\ j \neq i}}^{n} a_{ij}x_j + b_i \right) \Big/ a_{ii}. \tag{6.127}$$

Clearly (6.126) is the same as (6.127), except for the superscripts. This shows that the Jacobi process computes the value of x_i at the $(k + 1)$th step by simply substituting into the ith equation the values of the x's obtained at the *previous* kth stage.

For the Gauss–Seidel method, (6.112) and (6.123) give

$$(D - E)x^{(k+1)} = Fx^{(k)} + b$$

or

$$x^{(k+1)} = D^{-1}[Ex^{(k+1)} + Fx^{(k)} + b],$$

which in component form becomes

$$x_i^{(k+1)} = \left(-\sum_{j=1}^{i-1} a_{ij}x_j^{(k+1)} - \sum_{j=i+1}^{n} a_{ij}x_j^{(k)} + b_i \right) \Big/ a_{ii}. \tag{6.128}$$

Again, a comparison with (6.127) shows that $x_i^{(k+1)}$ is obtained by substituting previously calculated values into the ith equation, but this time $x_i^{(k+1)}$ is evaluated using the values $x_1^{(k+1)}$, $x_2^{(k+1)}, \ldots, x_{i-1}^{(k+1)}$ as soon as they are obtained. The following example will serve to clarify this. (For both methods it is convenient to start with $x^{(0)} = 0$.)

Example 6.16 Consider the equations

$$4x_1 + x_2 - 2x_3 = 4$$
$$-x_1 + 6x_2 - x_3 = -9.5$$
$$x_1 - 3x_2 + 8x_3 = 17,$$

so that A is the matrix in (6.120). For Jacobi's method, using (6.121) and (6.122)

$$L = D^{-1}(E + F)$$

$$= \begin{bmatrix} 0 & -\dfrac{1}{4} & \dfrac{1}{2} \\ \dfrac{1}{6} & 0 & \dfrac{1}{6} \\ -\dfrac{1}{8} & \dfrac{3}{8} & 0 \end{bmatrix}$$

and from (6.117)

$$\|L\| = \left(\frac{1}{4^2} + \frac{1}{2^2} + \frac{1}{6^2} + \frac{1}{6^2} + \frac{1}{8^2} + \frac{3^2}{8^2}\right)^{1/2}$$

$$= 0.72 < 1,$$

so the process converges. From (6.126) with $k = 0$ and $x^{(0)} = 0$ we have

$$x_1^{(1)} = b_1/a_{11} = 4/4 = 1$$
$$x_2^{(1)} = b_2/a_{22} = -9.5/6 = -1.583$$
$$x_3^{(1)} = b_3/a_{33} = 17/8 = 2.125.$$

Next, set $k = 1$ in (6.126) to give

$$x_1^{(2)} = (-a_{12}x_2^{(1)} - a_{13}x_3^{(1)} + b_1)/a_{11}$$
$$= (-x_2^{(1)} + 2x_3^{(1)} + 4)/4 = 2.458,$$
$$x_2^{(2)} = (-a_{21}x_1^{(1)} - a_{23}x_3^{(1)} + b_2)/a_{22}$$
$$= (x_1^{(1)} + x_3^{(1)} - 9.5)/6 = -1.062 \qquad (6.129)$$
$$x_3^{(2)} = (-x_1^{(1)} + 3x_2^{(1)} + 17)/8 = 1.406$$

and similarly

$$x^{(3)} = [1.969, -0.939, 1.420]^{\mathsf{T}}$$
$$x^{(4)} = [1.945, -1.018, 1.527]^{\mathsf{T}}$$
$$x^{(5)} = [2.018, -1.005, 1.500]^{\mathsf{T}},$$

which is converging to the exact solution $x = [2, -1, \frac{3}{2}]^{\mathsf{T}}$. The expressions for $x_i^{(2)}$ have been written out in full to emphasize the simplicity of the method. There is no need to remember any formulae: compute x_1 from the first given equation, x_2 from the second, x_3 from the third, etc., by using the values of the x's found at the previous stage. The matrix representation is necessary only for investigation of convergence.

With the Gauss–Seidel method, using (6.121) and (6.122) it is easy to compute $L = (D - E)^{-1}F$, and (6.117) gives $\|L\| = 0.61 < 1$, so again the process converges. From (6.128) with $k = 0$ and $x^{(0)} = 0$ we obtain

$$x_1^{(1)} = b_1/a_{11} = 4/4 = 1$$
$$x_2^{(1)} = (-a_{21}x_1^{(1)} + b_2)/a_{22} = (x_1^{(1)} - 9.5)/6 = -1.417$$
$$x_3^{(1)} = (-a_{31}x_1^{(1)} - a_{32}x_2^{(1)} + b_3)/a_{33} = (-x_1^{(1)} + 3x_2^{(1)} + 17)/8 = 1.469.$$

Next, set $k = 1$ in (6.128) to give

$$
\begin{aligned}
x_1^{(2)} &= (-a_{12}x_2^{(1)} - a_{13}x_3^{(1)} + b_1)/a_{11} \\
&= (-x_2^{(1)} + 2x_3^{(1)} + 4)/4 = 2.089 \\
x_2^{(2)} &= (-a_{21}x_1^{(2)} - a_{23}x_3^{(1)} + b_2)/a_{22} \\
&= (x_1^{(2)} + x_3^{(1)} - 9.5)/6 = -0.990 \\
x_3^{(2)} &= (-x_1^{(2)} + 3x_2^{(2)} + 17)/8 = 1.493
\end{aligned}
\tag{6.130}
$$

and similarly

$$x^{(3)} = [1.994, \ -1.002, \ 1.500]^{\mathsf{T}}.$$

We stress again that the only difference between the Jacobi and Gauss–Seidel methods is that in the latter the latest available values of $x_1, x_2, \ldots, x_{i-1}$ are used to compute x_i—for example, compare the two expressions for $x_2^{(2)}$ in (6.129) and (6.130). In the above example, the Gauss–Seidel method converges to the exact solution more rapidly than Jacobi's method, and this is often, but not always, the case.

Exercise 6.42 Find an approximate solution of the equations

$$
\begin{aligned}
6x_1 + \ \ x_2 - \ 3x_3 &= \ \ 17.5 \\
x_1 + 8x_2 + \ 3x_2 &= \ \ 10 \\
-x_1 + 4x_2 + 12x_3 &= -12
\end{aligned}
$$

using the methods of (a) Jacobi, (b) Gauss–Seidel.

Notice that for the matrix A in (6.120) which was used in Example 6.16, in each row the modulus of the diagonal element is greater than the sum of the moduli of the off-diagonal elements, i.e.

$$
\begin{aligned}
|a_{11}| &= 4 > |a_{12}| + |a_{13}| = 1 + 2 \\
|a_{22}| &= 6 > |a_{21}| + |a_{23}| = 1 + 1 \\
|a_{33}| &= 8 > |a_{31}| + |a_{32}| = 1 + 3.
\end{aligned}
$$

In general, any $n \times n$ matrix A having this property,

$$|a_{ii}| > \sum_{\substack{j=1 \\ j \neq i}}^{n} |a_{ij}|, \qquad i = 1, 2, \ldots, n, \qquad (6.131)$$

is called *diagonal dominant*, and it can be shown that both the Jacobi and Gauss–Seidel methods converge when A in (6.109) has this property (although again this provides only a sufficient condition for convergence). Other sufficient conditions for convergence of the Gauss–Seidel method are when A is real, symmetric, and 'positive definite' (see Section 7.4); or when A is an 'M-matrix' (see Section 14.2.2).

Exercise 6.43 If

$$A = \begin{bmatrix} 2 & a & 0 \\ 1 & 3 & b \\ 0 & 1 & 2 \end{bmatrix},$$

where a and b are parameters, compute the matrix L for the Gauss–Seidel procedure and hence prove that the method converges if and only if $|a + b| < 6$. Obtain sufficient conditions for convergence by applying (6.131).

Generally speaking, iterative methods are most useful as compared to direct methods when a very good initial approximation $x^{(0)}$ is known, or when a large proportion of the elements of A is zero (A is then called *sparse*).

6.7.2 Newton–Raphson type method

A somewhat different iterative procedure can be developed for calculating A^{-1} by generalizing the Newton–Raphson formula for finding the root of a scalar equation $f(z) = 0$. If z_0 is a sufficiently close initial approximation then (see Dahlquist and Björck (1974))

$$z_{k+1} = z_k - f(z_k)/f'(z_k), \qquad k = 0, 1, 2, \ldots \qquad (6.132)$$

provides successively better approximations. Let $f(z) = a - z^{-1}$, so the exact root is $z = a^{-1}$. Then $f' = z^{-2}$ and (6.132) becomes

$$\begin{aligned} z_{k+1} &= z_k - (a - z_k^{-1})/z_k^{-2} \\ &= z_k(2 - az_k) \end{aligned} \qquad (6.133)$$

and $z_k \to a^{-1}$ as $k \to \infty$. By analogy with (6.133) we define a sequence of matrices $X_1, X_2, X_3, \ldots$ by

$$X_{k+1} = X_k(2I - AX_k), \qquad k = 0, 1, 2, \ldots \qquad (6.134)$$

and show that $X_k \to A^{-1}$ as $k \to \infty$. To prove this, define an 'error' matrix

$$E_k = I - AX_k; \qquad (6.135)$$

our task is to show that $E_k \to 0$ as $k \to \infty$, for then $AX_k \to I$ as $k \to \infty$, which is the desired result. Clearly (6.135) implies

$$
\begin{aligned}
E_{k+1} &= I - AX_{k+1} \\
&= I - AX_k(2I - AX_k), \qquad \text{by (6.134)}, \\
&= I - (I - E_k)[2I - (I - E_k)], \qquad \text{by (6.135)}, \\
&= I - (I - E_k)(I + E_k) = E_k^2.
\end{aligned}
$$

Hence $E_1 = E_0^2$, $E_2 = E_1^2 = E_0^4$, $E_3 = E_2^2 = E_0^8, \ldots, E_k = E_0^p$, where $p = 2^k$. By the same argument as was applied to L in the preceding section (see (6.115)) it then follows that $E_0^p \to 0$ as $k \to \infty$ provided $E_0 = I - AX_0$ has all its eigenvalues with modulus less than unity. Also, as previously for L in (6.116), a *sufficient* condition for convergence of (6.134) is that $\|E_0\| = \|I - AX_0\| < 1$, and this will be satisfied provided the initial approximation X_0 is sufficiently close to A^{-1}.

Example 6.17 Let

$$A = \begin{bmatrix} 2 & 1 \\ 3 & 4 \end{bmatrix}, \qquad (6.136)$$

which has the exact inverse

$$A^{-1} = \begin{bmatrix} 0.8 & -0.2 \\ -0.6 & 0.4 \end{bmatrix}. \qquad (6.137)$$

Take

$$X_0 = \begin{bmatrix} 0.5 & -0.1 \\ -0.3 & 0.2 \end{bmatrix},$$

so that

$$E_0 = I - AX_0 = \begin{bmatrix} 0.3 & 0 \\ -0.3 & 0.5 \end{bmatrix}$$

and from (6.117), $\|E_0\| = 0.66 < 1$, so convergence is assured.

From (6.134)

$$X_1 = X_0(2I - AX_0) = \begin{bmatrix} 0.68 & -0.15 \\ -0.45 & 0.30 \end{bmatrix}$$

$$X_2 = X_1(2I - AX_1) = \begin{bmatrix} 0.78 & -0.19 \\ -0.56 & 0.38 \end{bmatrix},$$

which is a reasonable approximation to (6.137).

Exercise 6.44 If

$$A = \begin{bmatrix} 5 & 2 \\ 3 & -1 \end{bmatrix}, \qquad X_0 = \begin{bmatrix} 0.1 & 0.2 \\ 0.3 & -0.4 \end{bmatrix},$$

calculate X_1 and X_2 using (6.134), and compare with the exact expression for A^{-1}.

Exercise 6.45 If E_k is the matrix defined in (6.135), prove that provided the convergence condition is satisfied then

$$A^{-1} = X_k(1 + E_k + E_k^2 + E_k^3 + \cdots)$$

(the sum being taken to infinity).

Exercise 6.46 Prove that the sequence defined by

$$Y_{k+1} = (2I - Y_kA)Y_k, \qquad k = 0, 1, 2, \ldots$$

converges to A^{-1} provided all the eigenvalues of $I - Y_0A$ have modulus less than unity.

Problems

This chapter is the one of the most important in the book. A fairly large number of exercises is therefore given below, some of which develop further aspects of the theory.

6.1 Let A be a normal matrix (defined in Problem 2.10) and U be a unitary matrix.

   (a) Show that $U^{-1}AU$ is also normal. (It can be shown that every normal matrix is unitarily similar to a *diagonal* matrix—see Problem 8.5.)

   (b) If λ is any eigenvalue of U prove that $1/\bar{\lambda}$ is also an eigenvalue of U, and that $|\lambda| = 1$.

   (c) Show that $\|Ux\| = \|x\|$ for any vector x.

6.2   If S is an arbitrary $n \times n$ skew hermitian matrix, use the results of
      Exercises 6.11 and 6.15 to show that $I_n + S$ is non-singular. (This
      result was used in Problem 4.11.)

6.3   Use the result of Problem 4.5 to show that the eigenvalues of the
      matrix M in (4.78) are the eigenvalues of A together with those of
      D.

6.4   *Gershgorin's theorem* states that the eigenvalues of an arbitrary
      $n \times n$ real or complex matrix A lie in the region of the complex
      z-plane consisting of the n discs having centres a_{ii}, radii $\rho_i =$
      $\sum_{\substack{j=1 \\ j \neq i}}^{n} |a_{ij}|$, $i = 1, 2, \ldots, n$ (thus ρ_i is the sum of the moduli of the
      off-diagonal elements in the ith row of A). The discs can be written
      $|z - a_{ii}| \leqslant \rho_i$, $i = 1, \ldots, n$.
      For example, application of the theorem to the matrix A in (6.5)
      gives the two discs $|z - 1| \leqslant 3$, $|z - 2| \leqslant 2$. The eigenvalues lie in
      the union of these two discs, which in this example is simply the
      larger disc (the actual values were found in Example 6.1 to be
      $\lambda_1 = -1$, $\lambda_2 = 4$). The theorem is useful because it enables ap-
      proximate bounds for the eigenvalues to be determined easily.
      Apply Gershgorin's theorem to the matrices in Exercise 6.6, and
      compare with the actual values of the λ's.

6.5   The definition of a diagonal dominant matrix was given in (6.131).
      Use Gershgorin's theorem (Problem 6.4) to prove: (a) any
      diagonal dominant matrix is non-singular (use (6.21)); (b) if all the
      diagonal elements of a diagonal dominant matrix are negative, then
      all the eigenvalues have negative real parts.

6.6   Let A_n denote an $n \times n$ *tridiagonal* matrix in the form given in
      Problem 3.2, so that the matrix in (3.42) is A_4. If φ_n is the
      characteristic polynomial of A_n, expand $\det(\lambda I_n - A_n)$ by the last
      row to obtain the recurrence formula

$$\varphi_n = (\lambda - a_n)\varphi_{n-1} - b_{n-1}c_{n-1}\varphi_{n-2}, \qquad n = 2, 3, \ldots \quad (6.138)$$

      with $\varphi_0 = 1$, $\varphi_1 = \lambda - a_1$.
      Hence obtain the characteristic polynomial of

$$\begin{bmatrix} 1 & 2 & 0 & 0 \\ 2 & -1 & 1 & 0 \\ 0 & 3 & -3 & -1 \\ 0 & 0 & 7 & 4 \end{bmatrix}.$$

6.7   Apply (6.138) to obtain the characteristic equation of the 4×4
      matrix in (6.18). By setting $(\lambda - 2)^2 = \mu$, or otherwise, determine
      the four eigenvalues of this matrix.

6.8 Let ψ_n denote the characteristic polynomial of the $n \times n$ tridiagonal matrix

$$
\begin{bmatrix}
0 & 1 & 0 & 0 & & \\
\frac{1}{2} & 0 & \frac{1}{2} & 0 & & \mathbf{0} \\
0 & \frac{1}{2} & 0 & \frac{1}{2} & & \\
& & & \ddots & \ddots & \frac{1}{2} \\
& \mathbf{0} & & & \frac{1}{2} & 0
\end{bmatrix}.
$$

Use (6.138) to show that if $T_0 = 1$, $T_n = 2^{n-1}\psi_n (n \geq 1)$, then

$$T_n = 2\lambda T_{n-1} - T_{n-2}, \qquad n \geq 2.$$

These polynomials $T_0 = 1$, $T_1 = \lambda$, $T_2 = 2\lambda^2 - 1$, $T_3 = 4\lambda^3 - 3\lambda$, ... , are called *Chebyshev* polynomials and are very useful in numerical analysis.

6.9 Using the notation of Problem 6.6, assume that in (3.42) we have $b_i c_i > 0$, $i = 1, 2, 3$. If

$$D = \text{diag}[1, (b_1/c_1)^{1/2}, (b_1 b_2/c_1 c_2)^{1/2}, (b_1 b_2 b_3/c_1 c_2 c_3)^{1/2}]$$

show that $DA_4 D^{-1}$ is a real symmetric matrix.

This result can be extended to any value of n, showing that when $b_i c_i > 0$, for $i = 1, \ldots, n-1$, then all the eigenvalues of A_n are real, since A_n is similar to a real symmetric matrix.

6.10 Consider the infinitely long resistor ladder network shown in Fig. 6.3. It can be shown that the current i_k in the $(k+1)$th loop satisfies the difference equation

$$i_{k+2} - 3i_{k+1} + i_k = 0, \qquad k = 0, 1, 2, \ldots$$

and $i_k \to 0$ as $k \to \infty$. Define $X_1(k) = i_k$, $X_2(k) = i_{k+1}$, and hence

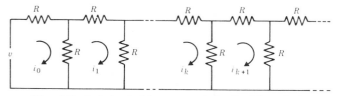

Fig. 6.3 Infinite ladder network for Problem 6.10.

write the equation in the matrix form (6.99). Use (6.101) to show that $i_k = i_0[\frac{1}{2}(3 - \sqrt{5})]^k$.

6.11 If K_n is the matrix defined in Problem 2.4, prove that the eigenvalues of $K_n A K_n$ are the same as those of A. What are its eigenvectors?

6.12 Application of the preceding exercise to the companion matrix C_n shows that the characteristic polynomial of $K_n C_n K_n$ is $k(\lambda)$ in (6.53), and similarly for C_n^T and $K_n C_n^T K_n$. These three matrices are sometimes used as companion forms. Write out each one in full.

6.13 Let the companion matrix C_n defined in Section 6.3.4 be non-singular. Define an $n \times n$ matrix F_n having the first row

$$[-k_{n-1}/k_n, -k_{n-2}/k_n, \ldots, -k_1/k_n, -1/k_n],$$

all the elements on the diagonal immediately below the principal diagonal equal to unity, and all other elements equal to zero. By calculating the product $C_n F_n$ verify that $F_n = C_n^{-1}$.
   Determine the characteristic equation of C_n^{-1} without evaluating $\det(\lambda I - C_n^{-1})$.

6.14 Deduce that the polynomial $k(\lambda)$ in (6.53) has a repeated root if and only if the matrix $D = k'(C_n)$ is singular, where $k'(\lambda) = dk/d\lambda$ ($\det D$ is called the *discriminant* of $k(\lambda)$).

6.15 Prove that if left and right eigenvectors of A are denoted by $v_1, \ldots, v_n$ and $u_1, \ldots, u_n$ respectively, then $v_j u_i = 0$, for $i \neq j$ (assuming as usual that all the eigenvalues of A are distinct). Hence prove that if S is the matrix having rows $v_1, v_2, \ldots, v_n$ and T is the matrix in (6.71), then ST is diagonal.

6.16 Let A be an arbitrary $n \times n$ non-singular matrix and B any other $n \times n$ matrix. By considering $\det(\lambda I_n - AB) = 0$ and using (4.33), prove that AB and BA have the same eigenvalues. Hence, by applying the first result of Exercise 6.15 to the matrix $AB - I_n$, prove that there cannot exist a matrix B such that $AB - BA = I_n$.

6.17 By setting $Z = I_m$ and $W = \lambda I_n$ in Problem 4.7, prove that

$$\det(\lambda I_n - XY) = \lambda^{n-m} \det(\lambda I_m - YX),$$

where X is an arbitrary $n \times m$ matrix and Y an arbitrary $m \times n$ matrix ($n > m$).
   This shows that the eigenvalues of XY are those of YX together with $n - m$ zeros, thus generalizing the first part of the preceding problem.

6.18 Construct a real symmetric 3×3 matrix having eigenvalues $\lambda_1 = -2$, $\lambda_2 = 1$, $\lambda_3 = 3$ and corresponding eigenvectors $[1, 2, 2]^\mathsf{T}$, $[2, -2, 1]^\mathsf{T}$, $[-2, -1, 2]^\mathsf{T}$ (hint: use (6.85)).

6.19 Prove that: (a) any idempotent matrix (defined in Problem 4.14) has its eigenvalues equal to 0 or 1; (b) any nilpotent matrix (defined in Problem 4.15) has all its eigenvalues equal to zero.

6.20 If A is an $n \times n$ matrix having only one non-zero eigenvalue, write down its characteristic equation. Hence show that $\det(I_n + A) = 1 + \mathrm{tr}(A)$. Also, deduce from Problem 5.23 that if $I + A$ is non-singular then $(I + A)^{-1} = I - A/[1 + \mathrm{tr}(A)]$.

6.21 If A and B satisfy (6.68) and $p(\lambda)$ is an arbitrary polynomial, prove that $p(B) = P^{-1}p(A)P$. Hence deduce that the minimum polynomials of A and B are identical.

6.22 Let M and S be the symmetric and skew-symmetric parts of a real $n \times n$ matrix A, and let the eigenvalues of AA^T, M and S be a_i, m_i and s_i respectively ($i = 1, 2, \ldots, n$). Use the results in Problem 2.7 to prove that

$$\sum_{i=1}^{n} a_i = \sum_{i=1}^{n} (m_i^2 - s_i^2).$$

6.23 From (6.96) it is clear that the solution $x(t)$ of the differential equations (6.91) tends to zero as $t \to \infty$ provided each λ_i has negative real part, for then $\exp(\lambda_i t) \to 0$. The system (6.91) is then called *asymptotically stable* (the result still holds even if some λ's are repeated). If

$$A = \begin{bmatrix} 2 & -1 \\ c & -3 \end{bmatrix},$$

calculate the eigenvalues of A and hence show that in this case the condition for asymptotic stability of (6.91) is $c > 6$.

An alternative method, for application to the characteristic polynomial of A, was given in Problem 4.2.

6.24 Consider an $n \times n$ matrix having the form

$$A = \begin{bmatrix} a_1 & a_2 & a_3 & \cdots & a_n \\ a_n & a_1 & a_2 & \cdots & a_{n-1} \\ a_{n-1} & a_n & a_1 & \cdots & a_{n-2} \\ \vdots & \vdots & \vdots & & \vdots \\ a_2 & a_3 & a_4 & \cdots & a_1 \end{bmatrix}.$$

Such matrices are called *circulants,* and arise in a number of applications. By considering the product Au, where $u = [1, \theta, \theta^2, \ldots, \theta^{n-1}]^{\mathsf{T}}$ and θ is an nth root of unity (i.e. $\theta^n = 1$), show that u is an eigenvector of A corresponding to an eigenvalue $a_1 + a_2\theta + a_3\theta^2 + \cdots + a_n\theta^{n-1}$. The n eigenvalues and eigenvectors of A are obtained by taking the n values of θ. Circulants will be studied further in Section 13.2.

6.25 Consider two linear control systems in the form (4.36), i.e.

$$\dot{x}_1 = A_1 x_1 + b_1 u, \qquad \dot{x}_2 = A_2 x_2 + b_2 u,$$

where x_1, x_2, b_1, b_2 are column n-vectors. Assume that the corresponding controllability matrices $\mathscr{C}_1$ and $\mathscr{C}_2$ defined in (4.37) are both non-singular. If the similarity transformation $x_1 = Tx_2$ transforms the first system into the second, prove that

$$Tb_2 = b_1, \qquad TA_2 b_2 = A_1 b_1, \qquad TA_2^2 b_2 = A_1^2 b_1, \ldots$$

and hence show that $T = \mathscr{C}_1 \mathscr{C}_2^{-1}$.

6.26 Let T be an $n \times n$ matrix having rows $t, tA, tA^2, \ldots, tA^{n-1}$, where t is some row n-vector such that T is non-singular. By comparing elements in the products TT^{-1} and TAT^{-1}, show that TAT^{-1} has the companion form defined in Section 6.3.4.

This gives a way of transforming a non-derogatory matrix A to companion form C without using the eigenvalues of A: select b such that $\mathscr{C}$ defined in (4.37) is non-singular, and solve $t\mathscr{C} = d$, where $d^{\mathsf{T}} = [0, 0, \ldots, 0, 1]$. Show that in this case $w = Tx$ transforms the differential equations (4.36) into

$$\frac{dw}{dt} = Cw + du.$$

This transformation is used in control theory.

6.27 Using (5.79), apply the method of Section 5.6 to show that the matrix equation

$$AXB - X = C$$

has a unique solution X if and only if $\lambda_i \mu_j \neq 1$, for all i and j (A, B have dimensions $n \times n$ and $m \times m$ and eigenvalues λ_i, μ_j respectively).

6.28 If A is $n \times n$ and B is $m \times m$, use (6.64) to prove

(a) $\operatorname{tr}(A \otimes B) = \operatorname{tr}(A) \operatorname{tr}(B)$;

(b) $\det(A \otimes B) = (\det A)^m (\det B)^n$.

6.29 By considering the identity

$$(\lambda I - A)\, \mathrm{adj}(\lambda I - A) = k(\lambda)I,$$

where $k(\lambda)$ is defined in (6.36), generalize the formulae (6.42) to obtain

$$\mathrm{adj}(\lambda I - A) = \lambda^{n-1}I + \lambda^{n-2}B_1 + \cdots + B_{n-1},$$

$$B_1 = A + k_1 I, \qquad B_i = AB_{i-1} + k_i I, \qquad i = 2, 3, \ldots, n-1.$$

It can also be shown that

$$k_1 = -\mathrm{tr}(A), \qquad k_i = -\frac{1}{i}\mathrm{tr}(AB_{i-1}), \qquad i = 2, 3, \ldots, n.$$

This scheme is often called *Leverrier's algorithm*. Use it to determine the characteristic polynomial and $\mathrm{adj}(\lambda I - A)$ for the matrix in Exercise 6.6(b).

6.30 Use the power method with $X_0 = [1, 1, 1]^T$ to estimate the dominant eigenvalue and corresponding eigenvector of

$$A = \begin{bmatrix} 0 & 5 & -6 \\ -4 & 12 & -12 \\ -2 & -2 & 10 \end{bmatrix}.$$

6.31 Let P be any non-singular $n \times n$ matrix such that $Pu_1 = f$, where u_1 is an eigenvector associated with the eigenvalues λ_1 of an $n \times n$ matrix A, and f is the first column of I_n. By considering $PAP^{-1}f$ ($\equiv$ the first column of PAP^{-1}) show that

$$PAP^{-1} = \begin{bmatrix} \lambda_1 & b \\ 0 & A' \end{bmatrix} \begin{matrix} 1 \\ n-1 \end{matrix}$$

(where the elements of b are unimportant). Hence deduce that the eigenvalues of the $(n-1) \times (n-1)$ matrix A' are the remaining $n-1$ eigenvalues of A.

   Determine such a P for the matrix A in Problem 6.30, and by obtaining A' find the other eigenvalues of A.

6.32 If $A = \begin{bmatrix} 1 & 2 \\ 0 & 1 \end{bmatrix}$, show by considering the equation $AP = PD$ with D diagonal, that it is impossible to find a non-singular P such that A is similar to a diagonal matrix.

   Notice that A has eigenvalues 1, 1 but only one independent eigenvector. Further study of such problems involves introduction of the *Jordan form* (see Section 8.1).

6.33 Let U be the matrix in Problem 4.7. Suppose that there exist an $n \times n$ matrix B and an $(n + m) \times n$ matrix

$$C = \begin{bmatrix} C_1 \\ C_2 \end{bmatrix} \begin{matrix} n \\ m \end{matrix},$$

such that $UC = CB$ and C_1 is non-singular. By considering $P^{-1}UP$, where

$$P = \begin{bmatrix} C_1 & 0 \\ C_2 & I_m \end{bmatrix},$$

show that the eigenvalues of U consist of the eigenvalues of B together with those of the Schur complement (D/C_1) (defined in Problem 4.7), where

$$D = \begin{bmatrix} C_1 & X \\ C_2 & Z \end{bmatrix}.$$

6.34 Let $x(t)$ be the solution of the differential equations (6.91). Show that

$$\frac{\mathrm{d}}{\mathrm{d}t}[x(t) \otimes x(t)] = Bx(t) \otimes x(t),$$

where $B = A \otimes I + I \otimes A$ (use (2.71) and (2.76)). Hence deduce that

$$\int_\alpha^\beta x(t) \otimes x(t) \, \mathrm{d}t = B^{-1}[x(\beta) \otimes x(\beta) - x(\alpha) \otimes x(\alpha)].$$

References

Dahlquist and Björck (1974), Gantmacher (1959, Vol. 1), Golub and Van Loan (1989), Mirsky (1963), Wilkinson (1965).

7 Quadratic and hermitian forms

The expression

$$q = ax_1^2 + 2bx_1x_2 + cx_2^2, \qquad (7.1)$$

where a, b, c are real constants, is called a *quadratic form* in the two variables x_1 and x_2, since each term has total degree *two* in the x's. We can write (7.1) as

$$q = [x_1, x_2]\begin{bmatrix} ax_1 + bx_2 \\ bx_1 + cx_2 \end{bmatrix}$$

$$= [x_1, x_2]\begin{bmatrix} a & b \\ b & c \end{bmatrix}\begin{bmatrix} x_1 \\ x_2 \end{bmatrix}$$

$$= x^{\mathsf{T}}Ax. \qquad (7.2)$$

The 2×2 matrix A in (7.2) is called the *matrix of the form q*; notice that it is symmetric. Suppose we change coordinates in (7.2) according to $x = Py$. Then $q = (Py)^{\mathsf{T}}APy = y^{\mathsf{T}}P^{\mathsf{T}}APy$, and we know from Section 6.4.3 that it is always possible to choose an orthogonal matrix P such that $P^{\mathsf{T}}AP = B$ is *diagonal*. In this case $q = y^{\mathsf{T}}By$ reduces simply to $b_{11}y_1^2 + b_{22}y_2^2$, a sum of squares. It is interesting to give a simple geometrical interpretation of this result. The ellipse shown in Fig. 7.1(a) has equation $ax_1^2 + 2bx_1x_2 + cx_2^2 = 1$. Clearly, if we rotated the axes of coordinates anticlockwise through an angle θ, then the ellipse would be in the standard position, shown in Fig. 7.1(b), and its equation would then be $y_1^2/\alpha^2 + y_2^2/\beta^2 = 1$ (the reader should confirm that in this case the matrix P is that given in (1.7)).

Quadratic forms (and their complex generalization—hermitian forms) arise in many applications (see Section 7.5), and their properties will be developed in this chapter.

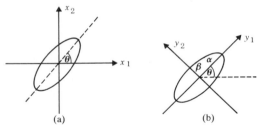

Fig. 7.1   (a) Ellipse $ax_1^2 + 2bx_1x_2 + cx_2^2 = 1$. (b) Ellipse $y_1^2/\alpha^2 + y_2^2/\beta^2 = 1$.

7.1 Definitions

The most general quadratic form in n real variables $x_1, \ldots, x_n$ is

$$q = a_{11}x_1^2 + 2a_{12}x_1x_2 + a_{22}x_2^2 + 2a_{13}x_1x_3 + a_{33}x_3^2 + 2a_{23}x_2x_3$$

$$+ \cdots + 2a_{n-1,n}x_{n-1}x_n + a_{nn}x_n^2 \tag{7.3}$$

$$= \sum_{i=1}^{n} \sum_{j=1}^{n} a_{ij}x_ix_j. \tag{7.4}$$

In (7.3) the total degree of each term in the x's is two, and the a's are real constants satisfying the condition

$$a_{ij} = a_{ji}, \qquad \text{all } i \neq j. \tag{7.5}$$

Generalizing the argument used to obtain (7.2), the form in (7.3) can be written

$$q = [x_1, \ldots, x_n] \begin{bmatrix} a_{11}x_1 + a_{12}x_2 + \cdots + a_{1n}x_n \\ \vdots \\ a_{1n}x_1 + a_{2n}x_2 + \cdots + a_{nn}x_n \end{bmatrix}$$

$$= x^{\mathsf{T}}Ax, \tag{7.6}$$

where $x = [x_1, x_2, \ldots, x_n]^{\mathsf{T}}$, and $A = [a_{ij}]$ is the $n \times n$ real symmetric matrix of the form. Given q, the matrix A can be written down on noticing that a_{ii} is the coefficient of x_i^2, and a_{ij} $(i \neq j)$ is *one-half* the coefficient of the x_ix_j term. In particular, A being diagonal corresponds to q containing only squared terms.

Example 7.1   For the quadratic form (with $n = 3$)

$$q = 2x_1^2 + 4x_1x_2 + 7x_2^2 + 5x_1x_3 + 6x_2x_3 - x_3^2 \qquad (7.7)$$

the matrix of the form is

$$A = \begin{bmatrix} 2 & 2 & \dfrac{5}{2} \\ 2 & 7 & 3 \\ \dfrac{5}{2} & 3 & -1 \end{bmatrix}. \qquad (7.8)$$

Exercise 7.1   Write down the $n \times n$ matrix of the form

$$q = 2x_1^2 - 4x_1x_2 + 3x_2^2 - 5x_1x_3 + 10x_3^2,$$

(a) when $n = 3$; (b) when $n = 4$.

It seems reasonable at this point to ask whether an expression

$$x^{\mathsf{T}}Cx \qquad (7.9)$$

in which C is an arbitrary (i.e. non-symmetric) real $n \times n$ matrix, represents a quadratic form more general than (7.6). The answer is 'No', which we demonstrate by splitting up C into its symmetric and skew symmetric parts M and S respectively. From (2.52), $C = M + S$, so that (7.9) becomes

$$x^{\mathsf{T}}Cx = x^{\mathsf{T}}Mx + x^{\mathsf{T}}Sx. \qquad (7.10)$$

However, since all quadratic forms are *scalar* quantities, in particular transposing $x^{\mathsf{T}}Sx$ does not alter its value. Therefore since $S^{\mathsf{T}} = -S$, we obtain

$$x^{\mathsf{T}}Sx = (x^{\mathsf{T}}Sx)^{\mathsf{T}} = x^{\mathsf{T}}S^{\mathsf{T}}(x^{\mathsf{T}})^{\mathsf{T}} = -x^{\mathsf{T}}Sx,$$

which shows that $x^{\mathsf{T}}Sx \equiv 0$. Hence in (7.10), $x^{\mathsf{T}}Cx \equiv x^{\mathsf{T}}Mx$, showing that the form (7.9) is identical to the form associated with the symmetric part $M = \frac{1}{2}(C + C^{\mathsf{T}})$ of C.

Exercise 7.2   Write out in full the form (7.9) when

$$C = \begin{bmatrix} 2 & 1 & 4 \\ 3 & -1 & 2 \\ 1 & -3 & 4 \end{bmatrix}.$$

Calculate the symmetric part M of C and verify that $x^{\mathsf{T}}Mx$ gives the same expression.

When the variables $x_1, \ldots, x_n$ are allowed to take complex values, the generalization is not to a complex quadratic form (i.e. letting the a's in (7.3) be complex), but instead to the *hermitian form* defined by

$$h = a_{11} |x_1|^2 + (a_{12}\bar{x}_1 x_2 + \bar{a}_{12}x_1\bar{x}_2) + a_{22} |x_2|^2$$
$$+ (a_{13}\bar{x}_1 x_3 + \bar{a}_{13}x_1\bar{x}_3) + \cdots + a_{nn} |x_n|^2 \qquad (7.11)$$

$$= \sum_{i=1}^{n} \sum_{j=1}^{n} a_{ij}\bar{x}_i x_j, \qquad (7.12)$$

where $|x_1|^2 = \bar{x}_1 x_1$, etc., and the a's are complex numbers satisfying

$$a_{ij} = \bar{a}_{ji}, \qquad \text{for all } i \neq j. \qquad (7.13)$$

The vector–matrix expression for (7.12) is

$$h = x^* A x \qquad (7.14)$$

in which, in view of (7.13), the matrix $A = [a_{ij}]$ is hermitian. The reason for the definition adopted is that h is real, whatever the values of the x's; this is shown by the following sequence of identities:

$$\bar{h} \equiv h^* \equiv (x^* A x)^* \equiv x^* A^* (x^*)^* \equiv x^* A x \equiv h.$$

For hermitian forms, (7.11) shows that a_{ii} is the coefficient of $|x_i|^2$, and a_{ij} $(i \neq j)$ is the coefficient of $\bar{x}_i x_j$.

Example 7.2 The matrix of the hermitian form

$$h = 3 |x_1|^2 + (1+i)\bar{x}_1 x_2 + (1-i)x_1\bar{x}_2 + 2 |x_2|^2$$

is

$$A = \begin{bmatrix} 3 & 1+i \\ 1-i & 2 \end{bmatrix}.$$

Also,

$$h = 3 |x_1|^2 + (1+i)\bar{x}_1 x_2 + \overline{[(1+i)\bar{x}_1 x_2]} + 2 |x_2|^2,$$

showing that h is real.

Exercise 7.3 Write down the hermitian form corresponding to the hermitian matrix in Example 2.14.

Exercise 7.4 If S is a real skew symmetric matrix prove that $x^\mathsf{T} S \bar{x}$ is either zero or purely imaginary.

It is worth mentioning here that any hermitian form in n complex variables can be expressed as a quadratic form in $2n$ real variables—see Problem 7.9.

We now consider the effect of the change of coordinates

$$x = Py \tag{7.15}$$

where P is a non-singular $n \times n$ matrix. For the quadratic form (7.6), substitution of (7.15) gives

$$q = y^\mathsf{T} B y \tag{7.16}$$

where

$$B = P^\mathsf{T} A P. \tag{7.17}$$

Any matrix B satisfying (7.17) with P non-singular is said to be *congruent* to A, and the relationship between A and B is called *congruence*. Comparison of (7.17) with (5.14) and (6.68) reveals that congruence is a special case of both equivalence and similarity. In particular, we know from Section 6.4.3 that for *any* real symmetric A, there exists an orthogonal matrix P in (7.17) such that B is the diagonal matrix of the eigenvalues $\lambda_1, \ldots, \lambda_n$ of A. In this case (7.16) becomes simply the sum of squares

$$\lambda_1 y_1^2 + \lambda_2 y_2^2 + \cdots + \lambda_n y_n^2, \tag{7.18}$$

where the λ's are all real.

Example 7.3 We can now reinterpret Example 6.11 in terms of quadratic forms. Corresponding to the matrix A in (6.86) we have

$$q = x_1^2 - 4x_1 x_2 - 2x_2^2, \tag{7.19}$$

which is transformed via

$$\begin{bmatrix} x_1 \\ x_2 \end{bmatrix} = \frac{1}{\sqrt{5}} \begin{bmatrix} 2 & 1 \\ -1 & 2 \end{bmatrix} \begin{bmatrix} y_1 \\ y_2 \end{bmatrix}$$

into $2y_1^2 - 3y_2^2$.

Exercise 7.5 Verify the preceding example by directly substituting into (7.19) for x_1 and x_2 in terms of y_1 and y_2.

Exercise 7.6 Interpret the result of Exercise 6.32 in terms of quadratic forms.

The reader who has attempted Exercise 7.5 will appreciate that, even in simple cases, matrices provide a convenient method for manipulating quadratic forms.

For hermitian forms the expression (7.18) is replaced by

$$h = y^*(P^*AP)y$$
$$= \lambda_1 |y_1|^2 + \lambda_2 |y_2|^2 + \cdots + \lambda_n |y_n|^2, \qquad (7.20)$$

with P in (7.15) now a unitary matrix, and $B = P^*AP$ is *conjunctive* to A.

We noted in Section 6.4.3 that the number of non-zero eigenvalues of a real symmetric or hermitian matrix A is equal to $R(A)$ ($= r$, say). Thus in both (7.18) and (7.20) the number of non-zero coefficients is equal to r, which is therefore also called the *rank* of the form.

Exercise 7.7 Determine the rank of the quadratic form in Exercise 7.1 in both cases (a) and (b).

7.2 Lagrange's reduction of quadratic forms

The preceding method of reducing a quadratic form by orthogonal similarity to a sum of squares requires calculation of eigenvalues and eigenvectors of the matrix of the form. This can be avoided by using Lagrange's procedure, which is a direct extension of the simple idea of 'completing the square'.

Example 7.4 Consider the following steps which reduce q to a sum of squares:

$$q = 2x_1^2 + 6x_1x_2 + x_2^2 = 2(x_1^2 + 3x_1x_2) + x_2^2$$

$$= 2\left[\left(x_1 + \frac{3}{2}x_2\right)^2 - \frac{9}{4}x_2^2\right] + x_2^2$$

$$= 2\left(x_1 + \frac{3}{2}x_2\right)^2 - \frac{7}{2}x_2^2$$

$$= 2y_1^2 - \frac{7}{2}y_2^2, \qquad (7.21)$$

where $y_1 = x_1 + \frac{3}{2}x_2$, $y_2 = x_2$. In matrix terms

$$\begin{bmatrix} y_1 \\ y_2 \end{bmatrix} = \underbrace{\begin{bmatrix} 1 & \frac{3}{2} \\ 0 & 1 \end{bmatrix}}_{P^{-1}} \begin{bmatrix} x_1 \\ x_2 \end{bmatrix}, \qquad \begin{bmatrix} x_1 \\ x_2 \end{bmatrix} = \underbrace{\begin{bmatrix} 1 & -\frac{3}{2} \\ 0 & 1 \end{bmatrix}}_{P} \begin{bmatrix} y_1 \\ y_2 \end{bmatrix}, \qquad (7.22)$$

which gives the transformation matrix P in (7.15), and

$$A = \begin{bmatrix} 2 & 3 \\ 3 & 1 \end{bmatrix}, \qquad B = P^{\mathrm{T}}AB = \begin{bmatrix} 2 & 0 \\ 0 & -\frac{7}{2} \end{bmatrix}.$$

The procedure in general is as follows:

(a) If $a_{11} \neq 0$, take out a factor a_{11} from all the terms involving x_1, giving $a_{11}(x_1^2 + x_1 k)$, where k depends only upon x_2, $x_3, \ldots, x_n$.

(b) Complete the square for these x_1 terms, giving $a_{11}[(x_1 + \frac{1}{2}k)^2 - \frac{1}{4}k^2]$, and let $y_1 = x_1 + \frac{1}{2}k$.

(c) Repeat with the x_2 terms, then the x_3 terms, etc.

As (7.22) shows, the transformation matrix P in (7.15) is not orthogonal in this case.

Example 7.5 We give a further illustration for the case of q in (7.7).

$$q = 2\left[x_1^2 + x_1\left(2x_2 + \frac{5}{2}x_3\right)\right] + 7x_2^2 + 6x_2x_3 - x_3^2$$

$$= 2\left[\left(x_1 + x_2 + \frac{5}{4}x_3\right)^2 - \frac{1}{4}\left(2x_2 + \frac{5}{2}x_3\right)^2 \right] + 7x_2^2 + 6x_2x_3 - x_3^2$$

$$= 2y_1^2 - \frac{1}{2}\left(2x_2 + \frac{5}{2}x_3\right)^2 + 7x_2^2 + 6x_2x_3 - x_3^2$$

$$= 2y_1^2 + 5x_2^2 + x_2x_3 - \frac{33}{8}x_3^2$$

$$= 2y_1^2 + 5\left(x_2^2 + \frac{1}{5}x_2x_3\right) - \frac{33}{8}x_3^2$$

$$= 2y_1^2 + 5\left[\left(x_2 + \frac{1}{10}x_3\right)^2 - \frac{1}{100}x_3^2 \right] - \frac{33}{8}x_3^2$$

$$= 2y_1^2 + 5y_2^2 - \frac{167}{40}y_3^2, \tag{7.23}$$

where $y_1 = x_1 + x_2 + \frac{5}{4}x_3$, $y_2 = x_2 + \frac{1}{10}x_3$, $y_3 = x_3$. It is very easy to solve for the x's in terms of the y's, and in matrix form we have

$$\begin{bmatrix} y_1 \\ y_2 \\ y_3 \end{bmatrix} = \begin{bmatrix} 1 & 1 & \frac{5}{4} \\ 0 & 1 & \frac{1}{10} \\ 0 & 0 & 1 \end{bmatrix}\begin{bmatrix} x_1 \\ x_2 \\ x_3 \end{bmatrix}, \qquad \begin{bmatrix} x_1 \\ x_2 \\ x_3 \end{bmatrix} = \begin{bmatrix} 1 & -1 & -\frac{23}{20} \\ 0 & 1 & -\frac{1}{10} \\ 0 & 0 & 1 \end{bmatrix}\begin{bmatrix} y_1 \\ y_2 \\ y_3 \end{bmatrix} \tag{7.24}$$

$$ P^{-1} P$$

Because of the nature of the procedure, it follows that the transformation matrix P^{-1} giving y in terms of x will be upper triangular, with all its diagonal elements unity. Thus (see Section 4.4) P will have the same form, as illustrated in (7.22) and (7.24).

Exercise 7.8 Reduce q in Exercise 7.1(a) to a sum of squares by Lagrange's method, stating the transformation obtained, and verify that $B = P^{T}AP$.

Example 7.6 If

$$q = x_1^2 + 4x_1x_2 + 4x_2^2, \qquad (7.25)$$

we can immediately see that

$$q = (x_1 + 2x_2)^2 = y_1^2 + 0y_2^2, \qquad (7.26)$$

where $y_1 = x_1 + 2x_2$, $y_2 = x_2$. The transformation is therefore

$$\begin{bmatrix} x_1 \\ x_2 \end{bmatrix} = \begin{bmatrix} 1 & 2 \\ 0 & 1 \end{bmatrix}^{-1} \begin{bmatrix} y_1 \\ y_2 \end{bmatrix} = \begin{bmatrix} 1 & -2 \\ 0 & 1 \end{bmatrix} \begin{bmatrix} y_1 \\ y_2 \end{bmatrix}.$$

Notice that it is necessary to define $y_2 = x_2$, so as to ensure that the transformation matrix is non-singular.

The Lagrange procedure detailed above breaks down if $a_{11} = 0$; in this case simply start with x_i, where a_{ii} is the first non-zero diagonal element of A. If all $a_{ii} = 0$ then a preliminary transformation is needed, as in the following example.

Example 7.7 For the form in three variables

$$q = x_1x_2 + 2x_1x_3 - x_2x_3 \qquad (7.27)$$

we first apply the transformation

$$x_1 = z_1, \qquad x_2 = z_1 + z_2, \qquad x_3 = z_3, \qquad (7.28)$$

which introduces a term in z_1^2, i.e.

$$q = z_1(z_1 + z_2) + 2z_1z_3 - (z_1 + z_2)z_3$$
$$= z_1^2 + z_1z_2 + z_1z_3 - z_2z_3. \qquad (7.29)$$

The transformation $x = Pz$ in (7.28) is easily confirmed to be non-singular, so the rank of q is unaltered. Lagrange's reduction can now be applied to (7.29), giving finally

$$q = y_1^2 - \frac{1}{4}y_2^2 + 2y_3^2, \qquad (7.30)$$

where $y_1 = z_1 + \frac{1}{2}z_2 + \frac{1}{2}z_3 = \frac{1}{2}(x_1 + x_2 + x_3)$, $y_2 = z_2 + 3z_3 = -x_1 + x_2 + 3x_3$, $y_3 = z_3 = x_3$. The choice of transformation in (7.28) is obviously not unique, for example $x_1 = z_1 + z_2$, $x_2 = z_1 - z_2$, $x_3 = z_3$ would produce terms in both z_1^2 and z_2^2.

The above modification of the procedure can be applied at any stage of the reduction when there are no x_i^2 terms.

Interestingly, we shall see in Section 7.4 that Lagrange's reduction can be effectively carried out via gaussian elimination. This avoids the rather tedious manipulations described above.

Exercise 7.9 Reduce to a sum of squares each of the following quadratic forms in three variables: (a) $3x_1x_2 + 5x_2x_3 + 2x_3^2$; (b) $x_1x_2 - x_1x_3 + x_2x_3$.

7.3 Sylvester's law of inertia

If a quadratic form $q = x^{\mathsf{T}}Ax$ in n variables is reduced to a sum of squares by *any* non-singular transformation $x = Py$, then since $B = P^{\mathsf{T}}AP$ in (7.17) is equivalent to A, we have $R(B) = R(A) = r$, so the number of non-zero coefficients in the sum is always equal to r. It is interesting, however, that not only the number of these coefficients remains fixed, but also the distribution of signs. Specifically, suppose q is reduced by two different non-singular transformations to two sums

$$\alpha_1 y_1^2 + \alpha_2 y_2^2 + \cdots + \alpha_r y_r^2, \qquad \beta_1 z_1^2 + \beta_2 z_2^2 + \cdots + \beta_r z_r^2$$

with all the α's and β's non-zero. Then Sylvester's result is that the number of *positive* α's is equal to the number of positive β's. That is, the numbers π and v of positive and negative terms in any sum of squares reduction of q remain *constant*, irrespective of the scheme of reduction used—hence the name: law of 'inertia'. The difference $\pi - v$ is called the *signature* of q (or of the matrix A).

The same result holds for a hermitian form, except that the reduction is to a sum of squares of moduli, as in (7.20).

Example 7.8

(a) Consider the form q in (7.19). The transformation $x_1 = 2z_1 + z_2$, $x_2 = z_1$ reduces it to $-6z_1^2 + z_2^2$ so $r = 2$, $\pi = 1$, agreeing with the expression obtained in Example 7.3 using $x_1 = (2y_1 + y_2)/\sqrt{5}$, $x_2 = (-y_1 + 2y_2)/\sqrt{5}$.

(b) For the form in (7.25), the transformation $x_1 = -2z_1 + z_2$, $x_2 = z_1 + 2z_2$ reduces it to $0z_1^2 + 25z_2^2$, so $r = 1$, $\pi = 1$, agreeing with (7.26).

(c) For the form in (7.27), the transformations $x_1 = z_1 + z_2$, $x_2 = z_1 - z_2$, $x_3 = z_3$, and then $z_1 = w_1 - \frac{1}{2}w_3$, $z_2 = w_2 + \frac{3}{2}w_3$, $z_3 = w_3$ produce $w_1^2 - w_2^2 + 2w_3^2$, so $r = 3$, $\pi = 2$, agreeing with (7.30).

Exercise 7.10 What are the rank and signature of the quadratic form in Exercise 7.9(b)? Reduce this form by starting with a different choice of z's from the one you used previously, and hence confirm Sylvester's law in this case.

7.4 Sign properties

Of particular importance in many applications are forms which do not change sign, whatever the values of the variables.

7.4.1 Definitions

The following definitions apply equally to both hermitian and quadratic forms.

A form $p(x)$ is called *positive definite* if it is positive everywhere except at the origin, i.e. $p(x) > 0$ for all $x \neq 0$, and $p(0) = 0$; $p(x)$ is *positive semidefinite* if $p(x) \geq 0$ for all x. Similarly, p is *negative* definite or semidefinite if $p(x) < 0$, or $p(x) \leq 0$, respectively. If $p(x)$ can take both negative and positive values then it is called *indefinite*. These various terms describe the *sign property* of a form.

Example 7.9

(a) If $n = 3$ the form $q = 2x_1^2 + x_2^2 + 3x_3^2$ is clearly positive definite; however, if $n = 4$ then we have $q = 2x_1^2 + x_2^2 + 3x_3^2 + 0x_4^2$, so $q = 0$ when $x_1 = x_2 = x_3 = 0$ but $x_4 \neq 0$, and hence q is positive semidefinite.

(b) The form $q = -x_1^2 - 2x_1x_2 - x_2^2 = -(x_1 + x_2)^2$ is negative semidefinite, since $q \leq 0$, and $q = 0$ when $x_1 = -x_2 \neq 0$.

(c) The form $q = 2x_1^2 + x_2^2 - 3x_3^2$ is indefinite, since for example, when $x_1 = 1$, $x_2 = x_3 = 0$ then $q > 0$, and when $x_3 = 1$, $x_1 = x_2 = 0$ then $q < 0$.

(d) The form $q = x_1x_2 + x_1x_3 + x_2x_3$ is indefinite, since, for example, taking $x_1 = \pm 1$, $x_2 = 1$, $x_3 = 0$ shows that q can take both positive and negative values.

The various terms are also applied to the matrix of the form. It is worth recalling two remarks made in previous chapters concerning a real symmetric matrix A being positive definite: in Section 3.3 we noted that it ensures that there exists a real non-singular triangular matrix U such that $A = U^T U$; and in Section 6.7.1 we stated that it provides a sufficient condition for convergence of the Gauss–Seidel method for solving $Ax = b$.

Exercise 7.11 What is the sign nature of the form $q = (x_1 + x_2 + x_3)^2$ (with $n = 3$)?

Exercise 7.12 If A and B are two positive definite hermitian $n \times n$ matrices, prove that $A + B$ is also positive definite.

Exercise 7.13 Generalize the result of Example 7.9(d) by showing that q in (7.4) is indefinite if all the coefficients a_{ii}, $i = 1, 2, \ldots, n$, are zero.

7.4.2 Tests

We now consider how to determine the sign property of a form. The first step is to notice that the transformation $x = Py$ applied to any form does not alter its sign property, provided P is non-singular. This is because, as the components of x vary through all possible values so do those of y, with $y = 0$ if and only if $x = 0$. In particular, suppose P is chosen so that a quadratic form $q = x^T A x$ is reduced to the sum of squares

$$\lambda_1 y_1^2 + \lambda_2 y_2^2 + \cdots + \lambda_r y_r^2, \qquad (7.31)$$

where as before $r = R(A)$ and the λ's are the eigenvalues of A. Then the sign properties of q are the same as those of (7.31). Since the λ's in (7.31) are all real, then it is obvious that (7.31) will be positive for all $y \neq 0$ only if $r = n$ and each λ is positive—otherwise it would be possible to make (7.31) zero or negative (compare with Example 7.9). Other cases can be argued similarly, to obtain the following: A quadratic or hermitian form is:

positive definite if $\lambda_i > 0$, $i = 1, 2, \ldots, n$;

negative definite if $\lambda_i < 0$, $i = 1, 2, \ldots, n$;

positive (negative) semidefinite if $\lambda_i \geq 0$ ($\lambda_i \leq 0$), $i = 1, 2, \ldots, n$;

indefinite if there is at least one positive and one negative λ.

Because of the law of inertia, the preceding conditions on the λ's also apply to the coefficients in *any* sum-of-squares reduction obtained by a non-singular transformation. We concentrate on quadratic forms, since these are most common in applications.

Example 7.10 We can now determine the sign properties of the forms which we earlier reduced to sums of squares by Lagrange's method. In Example 7.4 there is one positive and one negative coefficient in the reduced form (7.21), so q is indefinite—for example, when $y_1 = 1$, $y_2 = 0$ (corresponding from (7.22) to $x_1 = 1$, $x_2 = 0$) then $q > 0$; and when $y_1 = 0$, $y_2 = 1$ (corresponding to $x_1 = -\frac{3}{2}$, $x_2 = 1$) then $q < 0$.

Similarly, the form in Example 7.5 is indefinite, by inspection of (7.23). However, the form in Example 7.6 is positive semidefinite, since in (7.26) there is one positive and one zero coefficient.

Exercise 7.14 Determine the sign properties of the quadratic forms in Exercises 7.8 and 7.9.

Exercise 7.15 Prove that the trace and determinant of a positive definite hermitian (or real symmetric) matrix are both positive.

Exercise 7.16 Show that the condition for a real symmetric 2×2 matrix A to be positive definite is $a_{11} > 0$, $\det A > 0$.

Exercise 7.17 Deduce from the result in Exercise 7.12 that if two $n \times n$ hermitian matrices A and B have all their eigenvalues positive, then so does $A + B$.

Exercise 7.18 If A is a positive definite real symmetric matrix, prove that A^p has the same sign property for any positive integer p. Does this hold for any other values of p?

Example 7.10 shows that the sign property of a quadratic form can be determined immediately from the Lagrange sum of squares. It is now appropriate to show how Lagrange's method can be carried out via gaussian elimination. This is best done by a numerical example: consider the form q in (7.7) which has matrix A in (7.8). Applying gaussian elimination without partial pivoting (i.e. without row interchanges) to A gives

$$A \rightarrow \begin{bmatrix} 2 & \vdots & 2 & \frac{5}{2} \\ \hdashline 0 & \vdots & 5 & \frac{1}{2} \\ 0 & \vdots & \frac{1}{2} & -\frac{33}{8} \end{bmatrix} \rightarrow \begin{bmatrix} 2 & 2 & \frac{5}{2} \\ 0 & 5 & \frac{1}{2} \\ 0 & 0 & -\frac{167}{40} \end{bmatrix}. \tag{7.32}$$

Two interesting points are apparent in the final reduced form in (7.32). First, the pivots are identical to the coefficients in the Lagrange sum-of-squares reduction (7.23) of q. Second, if each row is divided by the pivot in that row, then the resulting matrix is precisely P^{-1} in (7.24), giving the transformation $y = P^{-1}x$. It can be shown that these results hold in general: Apply gaussian elimination without partial pivoting to the matrix A of the form q; provided there are no zero pivots, then the coefficients in the Lagrange reduction of q to a sum of squares are given by the pivots in the triangularized array, and if each row in this array is divided by its pivot then the resulting matrix is P^{-1} in the Lagrange transformation $y = P^{-1}x$.

It follows that for q to be positive (negative) definite all the pivots must be positive (negative). If pivots having opposite signs are encountered then q is indefinite and the elimination process can be stopped. If a zero pivot occurs then q is not definite.

It is worth noting that after each step of the elimination procedure the remaining non-triangular block in the bottom right-hand corner is still symmetric, as illustrated by the first array in (7.32) (see Problem 7.6).

Example 7.11 If $q = x_1^2 + 4x_1x_2 + 6x_2^2 + 2x_1x_3 + 2x_2x_3 + 5x_3^2$ then

$$A = \begin{bmatrix} 1 & 2 & 1 \\ 2 & 6 & 1 \\ 1 & 1 & 5 \end{bmatrix} \rightarrow \begin{bmatrix} 1 & 2 & 1 \\ 0 & 2 & -1 \\ 0 & 0 & \frac{7}{2} \end{bmatrix}$$

after elimination. The pivots are all positive, showing that q and A are positive definite. From the triangularized array, Lagrange's reduction of q would give $y_1^2 + 2y_2^2 + \frac{7}{2}y_3^2$, where $y_1 = x_1 + 2x_2 + x_3$, $y_2 = x_2 - \frac{1}{2}x_3$, $y_3 = x_3$ (the reader should confirm this).

Exercise 7.19 Test the form in Exercise 7.1(a) for definiteness using gaussian elimination. Compare your result with the Lagrange reduction obtained in Exercise 7.8.

If a zero pivot occurs, but it is required to determine the precise sign property of the form, then there are three different possibilities. The simplest is if $R(A) = r < n$, and all the first r pivots are non-zero. Then the correspondence with Lagrange's reduction still holds, so q is positive (negative) semidefinite if all the non-zero pivots are positive (negative). If, however, a zero pivot occurs at some stage (the ith, say, with $i \leqslant r$) then a little more care is needed: First, suppose that there is a *subsequent* non-zero diagonal element in some row k. The operations $(Ri) \leftrightarrow (Rk)$ *and* $(Ci) \leftrightarrow (Ck)$ bring this non-zero element into the ith diagonal position, and the triangularization can then be continued. Notice that it is necessary to carry out both row and column interchanges, so as to preserve the symmetry. This is equivalent to interchanging x_i and x_k in q, so does not affect the sign property. Finally, if no such k exists, then all the remaining diagonal elements are zero, and q in this case is indefinite, as the following argument shows. The reduced form of A at the ith stage is

$$i - 1 \begin{bmatrix} B & \vdots & C \\ \text{---} & & \text{---} \\ 0 & \vdots & E \end{bmatrix},$$
$$i - 1$$

where $B = [b_{st}]$ is upper triangular with all diagonal elements b_{ss} non-zero and $E = [e_{pq}]$ has all diagonal elements e_{pp} equal to zero. This corresponds to

$$q = b_{11} y_1^2 + b_{22} y_2^2 + \cdots + b_{i-1,i-1} y_{i-1}^2 + \sum_{\substack{p=i \\ p \neq q}}^{n} \sum_{q=i}^{n} e_{pq} x_p x_q, \quad (7.33)$$

where $y_s = (b_{ss} x_s + b_{s,s+1} x_{s+1} + \cdots + c_{sn} x_n)/b_{ss}$. Setting $y_1 = 0$, $y_2 = 0, \ldots, y_{i-1} = 0$ gives $i - 1$ homogeneous equations in the x's, which can always be solved for $x_1, \ldots, x_{i-1}$ in terms of $x_i, \ldots, x_n$. The form q then reduces to the last term in (7.33), which since it contains no squares shows that q is indefinite (see Exercise 7.13).

Example 7.12 We give an example of each of the three cases just discussed.

(a) For q in (7.25),

$$A = \begin{bmatrix} 1 & 2 \\ 2 & 4 \end{bmatrix} \rightarrow \begin{bmatrix} 1 & 2 \\ 0 & 0 \end{bmatrix}$$

showing that q and A are positive semidefinite (agreeing with the sum of squares (7.26)).

(b) If $q = x_1^2 + 4x_1x_2 + 4x_2^2 + 6x_1x_3 + 14x_2x_3 + 10x_3^2$, then

$$A = \begin{bmatrix} 1 & 2 & 3 \\ 2 & 3 & 7 \\ 3 & 7 & 10 \end{bmatrix} \rightarrow \begin{bmatrix} 1 & 2 & 3 \\ 0 & 0 & 1 \\ 0 & 1 & 1 \end{bmatrix} \quad \frac{(R2) \leftrightarrow (R3)}{(C2) \leftrightarrow (C3)} \rightarrow$$

$$\begin{bmatrix} 1 & 3 & 2 \\ 0 & 1 & 1 \\ 0 & 1 & 0 \end{bmatrix} \rightarrow \begin{bmatrix} 1 & 3 & 2 \\ 0 & 1 & 1 \\ 0 & 0 & -1 \end{bmatrix},$$

showing that q and A are indefinite.

(c) If $q = x_1^2 + 4x_1x_2 + 4x_2^2 - 2x_1x_3 + x_3^2$, then

$$A = \begin{bmatrix} 1 & 2 & -1 \\ 2 & 4 & 0 \\ -1 & 0 & 1 \end{bmatrix} \rightarrow \begin{bmatrix} 1 & 2 & -1 \\ 0 & 0 & 2 \\ 0 & 2 & 0 \end{bmatrix}.$$

No further nonzero pivots can be obtained, showing that q and A are indefinite (the reduced form of A corresponds to $q = (x_1 + 2x_2 - x_3)^2 + 4x_2x_3$, which becomes $4x_2x_3$ on taking $x_1 = -2x_2 + x_3$).

Exercise 7.20 Use gaussian elimination to test the following forms for definiteness: (a) $-x_1^2 + 2x_1x_2 - 2x_2^2 + 4x_2x_3 - 4x_3^2$; (b) the form of Exercise 7.9(a).

Exercise 7.21 Determine the sign properties of the following forms using gaussian elimination:

(a) $x_1^2 + 4x_1x_2 + 4x_2^2 + 6x_1x_3 - 4x_2x_3 + 9x_3^2$;

(b) $x_1^2 + 2x_1x_2 + x_2^2 + 4x_1x_3 + 4x_2x_3 + 5x_3^2$;

(c) $x_1^2 + 11x_2^2 + 13x_3^2 + 14x_4^2 + 2x_1x_2 - 6x_1x_3 + 4x_1x_4 + 10x_2x_4$
    $- 10x_2x_3 - 2x_3x_4$.

Exercise 7.22 Prove that if A and B are positive definite hermitian matrices, then so is $A \otimes B$ (hint: consider eigenvalues). If A and B have the same dimensions, does the result hold for the ordinary product AB?

It can also be shown that if A and B are positive definite then so is their Hadamard product $A \circ B$.

To close this section, we note that the leading principal minors of a positive definite matrix are all positive, since they are equal to successive products of pivots, which are themselves all positive. This is often stated as a theoretical test for definiteness.

Exercise 7.23 What are the sign conditions on leading principal minors of a negative definite matrix?

7.5 Some applications

7.5.1 Geometrical interpretation

As was seen at the beginning of this chapter, if A is a real symmetric 2×2 matrix then

$$x^{\mathsf{T}} A x = 1 \qquad (7.34)$$

is the equation of a curve in two dimensions. Applying an orthogonal transformation $x = Py$ to (7.34), which is equivalent to rotating the curve in the plane, reduces it to $\lambda_1 y_1^2 + \lambda_2 y_2^2 = 1$, where λ_1 and λ_2 are the eigenvalues of A. It is clear from this reduced form that if λ_1 and λ_2 are both positive (i.e. A is positive definite) then (7.34) represents an ellipse; if λ_1 and λ_2 have opposite signs (i.e. A is indefinite) then (7.34) represents a hyperbola; and if one eigenvalue is positive and the other zero, then A is positive semidefinite and (7.34) represents a pair of straight lines.

When A is 3×3 then (7.34) describes a surface in three dimensions. For example, if A is positive definite then the reduced form $\lambda_1 y_1^2 + \lambda_2 y_2^2 + \lambda_3 y_3^2 = 1$ shows that (7.34) represents an ellipsoid; in particular, if $\lambda_1 = \lambda_2 = \lambda_3 > 0$ this becomes a sphere. If $\lambda_1 = \lambda_2 > 0$, $\lambda_3 < 0$, the surface is called a 'rotational hyperboloid', and is the familiar shape of an electricity generating station's cooling tower.

Exercise 7.24 Prove that the plane curve $ax_1^2 + bx_1 x_2 + cx_2^2 = 1$, with $a > 0$, is an ellipse if $b^2 < 4ac$ and a hyperbola if $b^2 > 4ac$.

7.5.2 Optimization of functions

Let $\varphi(x) \equiv \varphi(x_1, x_2, \ldots, x_n)$ be a real scalar function of n real variables, having continuous first and second partial derivatives with respect to all the variables. The vector of partial derivatives $[\partial\varphi/\partial x_1, \partial\varphi/\partial x_2, \ldots, \partial\varphi/\partial x_n]^{\mathsf{T}}$ is called the *gradient* of φ, and is written $\nabla\varphi$ or grad φ. It is a standard result that for there to be a local maximum or minimum at some point $x = a$ then all the first partial derivatives must be zero there, i.e.

$$\nabla\varphi = 0, \qquad \text{at } x = a. \qquad (7.35)$$

Next, define the constant elements h_{ij} of the $n \times n$ *hessian matrix*

H (named after Hesse):

$$h_{ij} = \left(\frac{\partial^2 \varphi}{\partial x_i \, \partial x_j}\right)_{x=a}, \qquad i, j = 1, \ldots, n \qquad (7.36)$$

(notice that $h_{ij} = h_{ji}$, since by assumption of continuity $\partial^2 \varphi / \partial x_i \, \partial x_j = \partial^2 \varphi / \partial x_j \, \partial x_i$). Then $x = a$ is a local minimum if *H* is positive definite, and a local maximum if *H* is negative definite. The method of proving this result is to expand $\varphi(a + \delta x)$, where $\delta x = [\delta x_1, \ldots, \delta x_n]^T$ using Taylor's theorem. In view of (7.35), the terms which are linear in δx vanish, leaving

$$\varphi(a + \delta x) - \varphi(a) = \frac{1}{2}(\delta x)^T H(\delta x) + \left(\begin{array}{c}\text{cubic and higher-} \\ \text{degree terms in } \delta x\end{array}\right),$$

$$(7.37)$$

where *H* is the matrix defined in (7.36). For $x = a$ to be a local minimum we require $\varphi(a) < \varphi(a + \delta x)$ *for all* small variations $\delta x \neq 0$. Thus the right-hand side of (7.37) must be positive, and this will be the case for sufficiently small δx if *H* is positive definite.

Example 7.13 When $n = 2$ the condition for *H* to be positive definite is $h_{11} > 0$, $\det H > 0$ (see Exercise 7.16) which produces

$$\frac{\partial^2 \varphi}{\partial x_1^2} > 0, \qquad \left(\frac{\partial^2 \varphi}{\partial x_1^2}\right)\left(\frac{\partial^2 \varphi}{\partial x_2^2}\right) - \left(\frac{\partial^2 \varphi}{\partial x_1 \, \partial x_2}\right)^2 > 0$$

(all derivatives being evaluated at $x = a$) and this is the well-known condition for a point $x = a$, at which $\partial \varphi / \partial x_1 = \partial \varphi / \partial x_2 = 0$, to be a local minimum for $\varphi(x_1, x_2)$.

It is useful to note the case when φ is itself a quadratic form, i.e. $\varphi = x^T A x$. From the full expression in (7.3) it is obvious that

$$\partial \varphi_1 / \partial x_1 = 2a_{11}x_1 + 2a_{12}x_2 + 2a_{13}x_3 + \cdots + 2a_{1n}x_n$$
$$\vdots$$
$$\partial \varphi / \partial x_n = 2a_{n1}x_1 + 2a_{n2}x_2 + \cdots + 2a_{nn}x_n,$$

so in vector form we have

$$\nabla \varphi = 2Ax. \qquad (7.38)$$

Exercise 7.25 If $\varphi(x) = \frac{1}{2}x^T Q x - x^T b$, where *Q* is a positive definite real symmetric $n \times n$ matrix and *b* is a constant column *n*-vector, show that φ has a local minimum at $x = Q^{-1}b$.

Exercise 7.26 If u and v are two scalar differentiable functions of $x_1, \ldots, x_n$, prove that $\nabla(uv) = v(\nabla u) + u(\nabla v)$.

7.5.3 Rayleigh quotient

If A is a non-singular hermitian matrix, then from (7.20)

$$h = x^*Ax = \lambda_1 |y_1|^2 + \cdots + \lambda_n |y_n|^2,$$

where $x = Py$ with P unitary. Let the eigenvalues of A be numbered so that $\lambda_1 \geqslant \lambda_2 \geqslant \cdots \geqslant \lambda_n$. It then follows that

$$\lambda_n |y_1|^2 + \lambda_n |y_2|^2 + \cdots + \lambda_n |y_n|^2 \leqslant h \leqslant \lambda_1 |y_1|^2 + \cdots + \lambda_1 |y_n|^2,$$

i.e.

$$\lambda_n(y^*y) \leqslant h \leqslant \lambda_1(y^*y). \tag{7.39}$$

Because P is unitary, $x^*x = y^*P^*Py = y^*y$, so (7.39) implies that $\lambda_n(x^*x) \leqslant h \leqslant \lambda_1(x^*x)$, for arbitrary $y \neq 0$. Finally, since $x^*x \neq 0$ if $x \neq 0$, we can divide in (7.39) to obtain

$$\lambda_n \leqslant \frac{x^*Ax}{x^*x} \leqslant \lambda_1 \tag{7.40}$$

for arbitrary $x \neq 0$. The ratio $r = x^*Ax/x^*x$ ($= x^TAx/x^Tx$ when A is real symmetric) is called the *Rayleigh quotient*. From (7.40) we have $\lambda_1 \geqslant r$, $\lambda_n \leqslant r$, which provides an easy way of estimating bounds for the largest and smallest eigenvalues of A, as the following example demonstrates.

Example 7.14 For the quadratic form in (7.19) the Rayleigh quotient is

$$r = \frac{x_1^2 - 4x_1x_2 - 2x_2^2}{x_1^2 + x_2^2}. \tag{7.41}$$

Some choices for x_1 and x_2, and the corresponding values of r, are:

$$x^T = [1, 0], \quad r = 1; \qquad x^T = [0, 1], \quad r = -2;$$
$$x^T = [1, 1], \quad r = -2.5; \qquad x^T = [1, -1], \quad r = 1.5.$$

Since the largest and smallest of these values of r are 1.5 and -2.5 respectively, we have $\lambda_1 \geqslant 1.5$, $\lambda_2 \leqslant -2.5$. The exact values of λ_1 and λ_2 are 2 and -3 respectively (see Example 6.11).

Exercise 7.27 Obtain bounds for the largest and smallest eigenvalues of the matrix in (6.87) using the Rayleigh quotient.

The above procedure can be extended to obtain estimates for the eigenvalues of an *arbitrary* square matrix A as follows. The set of complex numbers x^*Ax obtained as x ranges over all vectors having unit length (i.e. $x^*x = 1$) is called the *numerical range* of A (the term *field of values* is also used). In particular, if A is hermitian then it follows from Section 7.1 that the numerical range is real. In general, if x is a normalized eigenvector corresponding to an eigenvalue λ of A then

$$x^*Ax = x^*(\lambda x) = \lambda x^*x = \lambda,$$

showing that the numerical range contains all the eigenvalues of A. It was seen in Section 2.3.2 (Exercise 2.27) that we can express any A as

$$A = A_1 + A_2,$$

where A_1 is hermitian and A_2 is skew hermitian. Let μ be any number in the numerical range of A, so that $\mu = x^*Ax$ for some x with $x^*x = 1$. Then we have

$$\bar{\mu} = \mu^* = (x^*Ax)^* = x^*A^*x$$

and hence

$$\mathrm{Re}(\mu) = \frac{1}{2}(\mu + \bar{\mu})$$

$$= \frac{1}{2}(x^*Ax + x^*A^*x)$$

$$= \frac{1}{2}x^*(A + A^*)x = x^*A_1x.$$

It follows from (7.40) that

$$\theta_n \leqslant \mathrm{Re}(\mu) \leqslant \theta_1,$$

where θ_1 and θ_n are the largest and smallest eigenvalues of the hermitian matrix A_1.

Exercise 7.28 Prove that

$$v_n \leqslant \mathrm{Im}(\mu) \leqslant v_1,$$

where v_1 and v_n are the largest and smallest eigenvalues of the hermitian matrix $-iA_2$.

The above results show that the numerical range of A (and hence all the eigenvalues of A) is contained in the rectangle in the complex plane with vertices $\theta_1 + iv_1$, $\theta_1 + iv_n$, $\theta_n + iv_1$, $\theta_n + iv_n$.

7.5.4 Liapunov stability

We now return to the system of linear differential equations discussed in Section 6.5, i.e.

$$\dot{x} = Ax, \tag{7.42}$$

where A is a real constant $n \times n$ matrix. As stated in Problem 6.23, the system is called asymptotically stable if the solution $x(t)$ of (7.42) tends to zero as $t \to \infty$, a necessary and sufficient condition for this being that all the eigenvalues of A have negative real parts. A method due to the Russian mathematician Liapunov avoids calculation of the eigenvalues. It can be shown that (7.42) is asymptotically stable if a positive definite quadratic form $V = x^{\mathrm{T}}Px$ can be found such that its derivative with respect to time is negative definite. This derivative is, by the product rule,

$$dV/dt = \dot{x}^{\mathrm{T}}Px + x^{\mathrm{T}}P\dot{x}$$

since P is a constant matrix, and substituting for $\dot{x}$ from (7.42) we have

$$dV/dt = (Ax)^{\mathrm{T}}Px + x^{\mathrm{T}}P(Ax)$$
$$= x^{\mathrm{T}}A^{\mathrm{T}}Px + x^{\mathrm{T}}PAx$$
$$= x^{\mathrm{T}}(A^{\mathrm{T}}P + PA)x = x^{\mathrm{T}}Qx,$$

showing that dV/dt is also a quadratic form, whose matrix

$$Q = A^{\mathrm{T}}P + PA \tag{7.43}$$

is required to be negative definite. In practice the method is to make a simple choice for Q, e.g. $-I_n$, and then solve (7.43) for P and test the sign property of this solution. We shall study (7.43) in more detail in Section 11.5.1, together with the corresponding result for linear difference equations.

Exercise 7.29 Verify that Q in (7.43) is symmetric.

Exercise 7.30 In (7.43) take $Q = -I_2$,

$$A = \begin{bmatrix} -1 & -1 \\ 2 & -4 \end{bmatrix}, \qquad P = \begin{bmatrix} p_1 & p_2 \\ p_2 & p_3 \end{bmatrix},$$

and solve for the elements of P. Hence determine whether (7.42) is asymptotically stable in this case.

Exercise 7.31 The equation (7.43) is a special case of (5.51). Use the criterion developed in Section 6.3.5 to determine the condition which A must satisfy in order that the solution P of (7.43) be unique.

The evaluation of a 'performance index' involving quadratic forms for a linear system of differential or difference equations is briefly explored in Problems 7.13 and 7.14. The use of such measures of performance to optimize the control of linear systems will be outlined in Section 11.5.2.

Problems

7.1 If A is real, symmetric, and non-singular, prove that $x^T A x$ and $x^T A^{-1} x$ have the same signature.

7.2 By considering the quadratic form $x^T A^T A x$, prove that if A is a real non-singular $n \times n$ matrix then $A^T A$ is positive definite. Prove also the converse, that any positive definite real symmetric matrix can be expressed as $B^T B$ with B non-singular.

What is the sign property of $A^T A$ if A is $m \times n$?

7.3 If $B = A^T A - A A^T$, where A is any real $n \times n$ matrix such that $B \neq 0$, show that B is symmetric. Use Problem 2.7 and (6.23) to prove that B is indefinite.

7.4 Determine the range of values of k for which each of the following forms is positive definite:

(a) $x_1^2 + 4x_1 x_2 + (7 - k)x_2^2 + 8x_1 x_3 + (46 - 10k)x_2 x_3 + (89 - 24k)x_3^2$;

(b) $x_1^2 + 6x_1 x_2 + 8x_1 x_3 + k(x_2^2 + x_3^2)$.

7.5 Use Gershgorin's theorem (given in Problem 6.4) to prove that if a real symmetric matrix $A = [a_{ij}]$ has $a_{ii} > 0$ for all i, and is diagonal dominant (defined in (6.131)), then A is positive definite.

7.6 If A is a real symmetric $n \times n$ matrix with $a_{11} \neq 0$, and the first column is reduced in the usual way by gaussian elimination, prove that the last $n - 1$ rows and columns of the resulting array also form a symmetric matrix.

7.7 If A is real and symmetric and $\det A < 0$, prove there exists a real vector x such that $x^{\mathrm{T}}Ax < 0$ (hint: use (6.1) and (6.21)).
Find such an x for the case when A is the matrix in (6.87).

7.8 If A is an arbitrary real symmetric matrix, prove that there exists a real number c such that $A + cI$ is positive definite.

7.9 In the hermitian form $h = x^*Ax$, let $A = A_1 + iA_2$, $x = u + iv$, with A_1, A_2, u, v purely real. Prove that $h = u^{\mathrm{T}}A_1u + v^{\mathrm{T}}A_1v - 2u^{\mathrm{T}}A_2v$, and hence show that $h = y^{\mathrm{T}}Dy$, where D is the real symmetric $2n \times 2n$ matrix in (2.77) and $y^{\mathrm{T}} = [u^{\mathrm{T}}, -v^{\mathrm{T}}]$.

7.10 Let A be an arbitrary positive definite real symmetric $n \times n$ matrix. Apply the definition of $A^{1/2}$ given in (6.77), with T orthogonal, to prove that $A^{1/2}$ is also real symmetric.
If B is any real symmetric $n \times n$ matrix, by considering the product $(A^{1/2})^{-1}ABA^{1/2}$, prove that all the eigenvalues of AB are real.

7.11 Let $C = A^*B - BA$, where A and B are $n \times n$ matrices.

(a) If B is hermitian, prove that the eigenvalues of A are all purely imaginary.

(b) If A is real, and all its eigenvalues are distinct, prove that it is always possible to choose a non-singular (but not necessarily hermitian) matrix B such that $C = 0$.

7.12 For the Rayleigh quotient r associated with a real symmetric matrix A, use (7.38) and the product rule in Exercise 7.26 to show that

$$\nabla r = 2Ax/(x^{\mathrm{T}}x) - 2(x^{\mathrm{T}}Ax)x/(x^{\mathrm{T}}x)^2.$$

Hence show that the critical points of r (where $\nabla r = 0$) are eigenvectors of A, and associated critical values of r are eigenvalues of A. Verify this for r in (7.41) by direct differentiation (see Example 6.11 for eigenvalues and eigenvectors in this case).

7.13 Integrate the identity

$$\frac{\mathrm{d}}{\mathrm{d}t}[x^{\mathrm{T}}(t)Px(t)] = x^{\mathrm{T}}(t)Qx(t)$$

from zero to infinity, where P and Q satisfy (7.43) and $x(t)$ is the solution of (7.42). Hence deduce that, provided the system (7.42) is asymptotically stable, then

$$\int_0^\infty x^{\mathrm{T}}(t)Qx(t)\,\mathrm{d}t = -x^{\mathrm{T}}(0)Px(0). \tag{7.44}$$

Similarly, by differentiating $tx^{\mathrm{T}}(t)Px(t)$ show that

$$\int_0^\infty tx^{\mathrm{T}}(t)Qx(t)\,\mathrm{d}t = x^{\mathrm{T}}(0)P_1x(0)$$

where $A^{\mathrm{T}}P_1 + P_1A = P$.

These integrals involving quadratic forms can be regarded as indices of performance for the system (7.42).

Apply (7.44) to Exercise 7.30 with $x(0) = [\alpha, \beta]^{\mathrm{T}}$, and hence evaluate

$$\int_0^\infty \|x(t)\|^2\,\mathrm{d}t$$

for this example.

7.14 Let $X(k)$ be the solution of the linear difference equations (6.99). Show that

$$X^{\mathrm{T}}(k)QX(k) = X^{\mathrm{T}}(k)PX(k) - X^{\mathrm{T}}(k+1)PX(k+1) \quad (7.45)$$

where $A^{\mathrm{T}}PA - P = -Q$. It can be shown (see Section 11.5.1) that if P and Q are positive definite then $X(k) \to 0$ as $k \to \infty$. By summing both sides in (7.45) from zero to infinity, show that in this case a performance index for (6.99) can be evaluated as

$$\sum_{k=0}^\infty X^{\mathrm{T}}(k)QX(k) = X^{\mathrm{T}}(0)PX(0).$$

References

Gantmacher (1959, Vol. 1), Mirsky (1963), Strang (1988).

8 Canonical forms

A *canonical form* is a standard form to which a certain class of matrices can be reduced by a defined set of operations. We have already encountered some simple canonical forms in earlier chapters, and it is useful to recall these:

(a) In Section 5.3.3 we noted that any $m \times n$ matrix A having rank r can be reduced by elementary row and column operations to the normal form N, where N is the $m \times n$ matrix having 1's in positions $(1, 1)$, $(2, 2)$, ..., (r, r) and zeros elsewhere. This is the canonical form of A under an equivalence transformation, as in (5.12).

(b) In Section 6.4.2 we saw that if all the eigenvalues of an $n \times n$ matrix A are distinct, then under the similarity transformation (6.74) the canonical form is the diagonal matrix consisting of these eigenvalues.

(c) Furthermore, we saw in Section 6.4.4 that an alternative canonical form, to which matrices with distinct eigenvalues are similar, is the companion matrix described in Section 6.3.4, together with the three alternative companion forms in Problem 6.12.

(d) When A is any real symmetric matrix then there exists a real orthogonal similarity transformation such that $T^{T}AT$ is diagonal (see (6.85)); and moreover A can also be reduced to diagonal form by a congruence transformation (Section 7.2). Indeed, any *normal* matrix is unitarily similar to a diagonal matrix (Section 6.4.3).

The importance of canonical forms is that they are easier to handle in applications than the original matrix, and they often reveal properties not readily apparent from the matrix itself. This was illustrated in the solution of sets of linear differential or difference equations in Section 6.5. This chapter is devoted to describing some other important canonical forms.

8.1 Jordan form

Example 8.1 A reader who attempted Problem 6.32 will have discovered that the matrix

$$A = \begin{bmatrix} 1 & a \\ 0 & 1 \end{bmatrix}, \qquad a \neq 0, \tag{8.1}$$

cannot be reduced to diagonal form by a similarity transformation $P^{-1}AP$. Clearly A in (8.1) has eigenvalues 1, 1, but this repeated eigenvalue is only part of the difficulty; if $a = 0$, then A has the same repeated eigenvalue, but of course is itself diagonal! The key to understanding what is going on lies in the fact that when $a \neq 0$, A in (8.1) has the single eigenvector $[1, 0]^T$, but when $a = 0$ there are two linearly independent eigenvectors $[1, 0]^T$ and $[0, 1]^T$.

Exercise 8.1 Show that for any value of the scalar a, the matrix

$$A = \begin{bmatrix} a & 1 & 0 \\ 0 & a & 1 \\ 0 & 0 & a \end{bmatrix}$$

cannot be transformed by similarity to a diagonal matrix, and that it has a single linearly independent eigenvector.

In general, we define a $k_i \times k_i$ upper triangular *Jordan block matrix*

$$J_i = \begin{bmatrix} \lambda_i & 1 & & & 0 \\ & \lambda_i & 1 & & \\ & & \cdot & \cdot & 1 \\ 0 & & & & \lambda_i \end{bmatrix} \tag{8.2}$$

which has the scalar λ_i along the principal diagonal, 1's along the *superdiagonal* (which is the line of elements immediately above the principal diagonal) and zero elsewhere. It is sometimes convenient to use J_i^T, which has the 1's on the *subdiagonal*. It is obvious that the characteristic polynomial of J_i is

$$(\lambda - \lambda_i)^{k_i}, \tag{8.3}$$

so that J_i has the eigenvalue λ_i repeated k_i times (λ_i is said to have *algebraic multiplicity* k_i—hereafter simply called *multiplicity*). As in the simple example of Exercise 8.1, consideration of the equation $J_i x = \lambda_i x$ shows that J_i cannot be diagonalized by similarity, and that J_i has only the single linearly independent eigenvector $[1, 0, \ldots, 0]^T$. The Jordan block J_i illustrates the general result, that an $n \times n$ matrix is similar to a diagonal matrix if and only if it has n linearly independent eigenvectors.

The important special case of this, which we used in Section 6.4.2, is that when all the eigenvalues of A are *simple* (i.e. each occurs only once in the characteristic polynomial) then A does indeed have n linearly independent eigenvectors. In general, for this to be true, each eigenvalue λ_i must have associated with it the same number k_i of linearly independent eigenvectors as its multiplicity. Otherwise the matrix is called *defective*.

An alternative characterization of when A can be diagonalized by similarity involves the concept of the minimum polynomial $m(\lambda)$, which (see Section 6.3.3) is the least-degree monic polynomial such that $m(A) \equiv 0$. The required condition is then that $m(\lambda)$ has no repeated factors. It is easy to verify (see Exercise 8.4) that J_i has minimum polynomial equal to its characteristic polynomial (8.3), and so is what we called non-derogatory (Section 6.3.3).

Example 8.2   The matrix A in (8.1) has minimum polynomial

$$m(\lambda) = \begin{cases} (\lambda - 1)^2, & a \neq 0 \\ (\lambda - 1), & a = 0. \end{cases}$$

Thus when $a \neq 0$, A is non-derogatory but not diagonalizable; when $a = 0$, $m(\lambda)$ has no repeated factors, and A can be (trivially) diagonalized.

Jordan canonical form theorem   Every $n \times n$ matrix is similar to the *Jordan form*

$$\mathcal{J} = \text{diag}[J_1, J_2, \ldots, J_q], \tag{8.4}$$

where J_i in (8.2) is $k_i \times k_i$, and $k_1 + k_2 + \cdots + k_q = n$.

Note that in (8.4) the eigenvalues $\lambda_1, \lambda_2, \ldots, \lambda_q$ need not necessarily be distinct. There can be several Jordan blocks associated with the same eigenvalue, so that if, for example, $J_1, J_2, \ldots, J_s$ are associated with λ_1 then the multiplicity of λ_1 is $k_1 + k_2 + \cdots k_s$. The number s is called the *geometric multiplicity* of λ_1, and clearly the geometric multiplicity of an eigenvalue is less than or equal to its (algebraic) multiplicity since each $k_i \geq 1$. For a non-derogatory matrix the geometric multiplicity of each eigenvalue is one. The number and dimensions of the blocks in (8.4) are unique, apart from their ordering, and $\mathcal{J}$ has q linearly independent eigenvectors (one for each block). The proof of the theorem is complicated, and can be found in Horn and Johnson (1985).

Example 8.3 Consider a 6×6 Jordan canonical form $\mathcal{J} = \text{diag}[J_1, J_2, J_3]$ with

$$J_1 = \begin{bmatrix} 2 & 1 \\ 0 & 2 \end{bmatrix}, \qquad J_2 = [2], \qquad J_3 = \begin{bmatrix} -4 & 1 & 0 \\ 0 & -4 & 1 \\ 0 & 0 & -4 \end{bmatrix}.$$

Then $\mathcal{J}$ has eigenvalues 2 and -4, each with multiplicity 3, and three linearly independent eigenvectors. By (8.3), the characteristic polynomials of J_1, J_2, J_3 are respectively $(\lambda - 2)^2$, $(\lambda - 2)$, $(\lambda + 4)^3$, so $\mathcal{J}$ has characteristic polynomial $(\lambda - 2)^3(\lambda + 4)^3$. Because the minimum polynomial of J_2 is contained in that for J_1, the minimum polynomial of $\mathcal{J}$ is $(\lambda - 2)^2(\lambda + 4)^3$, and $\mathcal{J}$ is derogatory.

However, if a second 6×6 Jordan form $\mathcal{J}'$ has blocks

$$J_1 = \begin{bmatrix} 2 & 1 & 0 \\ 0 & 2 & 1 \\ 0 & 0 & 2 \end{bmatrix}, \qquad J_2 = \begin{bmatrix} -4 & 1 & 0 \\ 0 & -4 & 1 \\ 0 & 0 & -4 \end{bmatrix},$$

then although $\mathcal{J}$ and $\mathcal{J}'$ have the same eigenvalues and characteristic polynomial they are not similar. The minimum polynomial of $\mathcal{J}'$ is $(\lambda - 2)^3(\lambda + 4)^3$, and $\mathcal{J}'$ is non-derogatory. This illustrates the fact that two matrices are similar if and only if their Jordan forms are *identical*.

Exercise 8.2 Write down all the possible Jordan canonical forms for a matrix of order 5 whose characteristic polynomial is $(\lambda - 2)^3(\lambda - 3)^2$.

The reader should be aware that the criteria we have referred to for testing whether a matrix is diagonalizable are mainly of theoretical interest, and that computation of the Jordan form of a defective matrix is in general very difficult. The classic reference here is Wilkinson (1965).

Exercise 8.3 Prove by induction on r that if $k_1 = 3$ then

$$J_1^r = \begin{bmatrix} \lambda_1^r & \binom{r}{1}\lambda_1^{r-1} & \binom{r}{2}\lambda_1^{r-2} \\ 0 & \lambda_1^r & \binom{r}{1}\lambda_1^{r-1} \\ 0 & 0 & \lambda_1^r \end{bmatrix}, \qquad r = 2, 3, \ldots,$$

where $\binom{r}{i}$ denotes the binomial coefficient $r!/i!(r - i)!$.

A similar result holds in general: if J_1 has order k_1, then

$$J_1^r = \begin{bmatrix} \lambda_1^r & \binom{r}{1}\lambda_1^{r-1} & \cdots & \binom{r}{k_1-1}\lambda_1^{r-k_1+1} \\ 0 & \lambda_1^r & \cdots & \binom{r}{k_1-2}\lambda_1^{r-k_1+2} \\ \cdot & \cdot & \cdots & \cdot \\ 0 & 0 & \cdots & \lambda_1^r \end{bmatrix}.$$

Exercise 8.4 By considering $(J_1 - \lambda_1 I)^r$, $r = 1, 2, \ldots, k_1$, deduce that the minimum polynomial of J_1 is $(\lambda - \lambda_1)^{k_1}$.

It is of some interest to extend the method in Section 6.4.2 for diagonalizing a matrix, if possible, using eigenvectors. We need the concept of a *generalized eigenvector* u of *grade* k corresponding to an eigenvalue λ of A. This is defined by

$$(A - \lambda I)^k u = 0, \qquad (A - \lambda I)^{k-1} u \neq 0. \qquad (8.5)$$

Thus k is the smallest positive integer such that $(A - \lambda I)^k u = 0$. When $k = 1$, (8.5) reduces to $(A - \lambda I)u = 0$, $u \neq 0$, which is the definition of an ordinary eigenvector of A in (6.2).

Exercise 8.5 Define a sequence of vectors

$$u_k = u, \qquad u_i = (A - \lambda I)u_{i+1}, \qquad i = k - 1, k - 2, \ldots, 2, 1, \qquad (8.6)$$

where u is the vector in (8.5). Show that u_i is a generalized eigenvector of A of grade i, and in particular u_1 is an eigenvector in the ordinary sense.

The set $u_k, u_{k-1}, \ldots, u_1$ is called a *chain* of generalized eigenvectors of length k belonging to λ.

Exercise 8.6 Prove that the set of k vectors defined in Exercise 8.5 is linearly independent.

Suppose that a $k_i \times k_i$ matrix A has a single eigenvalue λ_i with multiplicity k_i, and that there is a chain of generalized eigenvectors of length k_i. The definition (8.6) gives

$$Au_1 = \lambda_i u_1, \qquad Au_2 = u_1 + \lambda_i u_2, \ldots, Au_{k_i} = u_{k_i-1} + \lambda_i u_{k_i}.$$

Combining together these expressions produces

$$A[u_1, u_2, \ldots, u_{k_i}] = [u_1, u_2, \ldots, u_{k_i}]J_i,$$

where J_i is the Jordan block defined in (8.2). This shows that A is transformed by similarity into J_i by the matrix $[u_1, \ldots, u_{k_i}]$.

Example 8.4 Consider

$$A = \begin{bmatrix} 3 & 2 & 4 \\ 0 & 3 & -2 \\ 0 & 0 & 1 \end{bmatrix},$$

which we see by inspection (because of its upper triangular form) has $\lambda_1 = 3$ with multiplicity 2, and $\lambda_2 = 1$. Taking $\lambda = \lambda_1$ in (8.5) leads to

$$(A - 3I)^2 u = 0, \qquad (A - 3I)u \neq 0,$$

from which we obtain $u = [0, 1, 0]^T$ as a generalized eigenvector of grade 2. A chain of generalized eigenvectors of length 2 belonging to λ_1 is obtained from (8.6) as

$$u_2 = u, \qquad u_1 = (A - 3I)u_2 = [2, 0, 0]^T,$$

where u_1 is an eigenvector of A in the ordinary sense. Since λ_2 is not repeated, it has a single eigenvector $u_3 = [-3, 1, 1]^T$, obtained by solving $(A - I)u_3 = 0$. The required transformation matrix is

$$T = [u_1, u_2, u_3] = \begin{bmatrix} 2 & 0 & -3 \\ 0 & 1 & 1 \\ 0 & 0 & 1 \end{bmatrix} \qquad (8.7)$$

and the similarity transformation of A into its Jordan form is

$$T^{-1}AT = \begin{bmatrix} 3 & 1 & \vdots & 0 \\ 0 & 3 & \vdots & 0 \\ \hdashline 0 & 0 & \vdots & 1 \end{bmatrix}.$$

The preceding example is relatively simple, because the length of the chain of generalized eigenvectors belonging to λ_1 is equal to the multiplicity of λ_1. When this does not hold, the situation is more complicated, since there may be several blocks associated with each eigenvalue. Nevertheless, it is still possible in general to construct m linearly independent eigenvectors associated with an eigenvalue of multiplicity m, so a desired transformation matrix T can be constructed. For full details, including worked examples, see Chen (1984). The way in which the Jordan form depends upon the rank of $(A - \lambda_i I)^k$ is discussed by Horn and Johnson (1985).

Exercise 8.7 Determine the transformation matrix T and Jordan form when

$$A = \begin{bmatrix} 0 & 1 & 0 \\ 0 & 0 & 1 \\ 1 & -3 & 3 \end{bmatrix}.$$

When A is a real matrix then in general its Jordan form will be complex. However, this Jordan form can be modified as follows to give a *real* canonical form: The eigenvalues of A are either real or occur in complex conjugate pairs. Jordan blocks (8.2) corresponding to the former are also real, but for a complex eigenvalue λ_j we have a pair of blocks in the form

$$\text{diag}[J_j, \bar{J}_j] \tag{8.8}$$

occurring in (8.4). Permuting rows and columns in (8.8) produces the similar matrix

$$L = \begin{bmatrix} L_j & I & 0 & 0 \\ 0 & L_j & I & \\ & & \ddots & \ddots & I \\ 0 & & & & L_j \end{bmatrix}$$

where there are k_j block rows and columns, I is the 2×2 unit matrix and $L_j = \text{diag}[\lambda_j, \bar{\lambda}_j]$. However, we have seen in Exercise 6.3 that if $\lambda_j = a_j + ib_j$ then

$$A_j = \begin{bmatrix} a_j & b_j \\ -b_j & a_j \end{bmatrix}$$

has eigenvalues λ_j and $\bar{\lambda}_j$. It follows that L, and hence also the matrix (8.8), are similar to the $2k_j \times 2k_j$ matrix

$$K_j = \begin{bmatrix} A_j & I & 0 & 0 \\ 0 & A_j & I & \\ & & \ddots & \ddots & I \\ 0 & & & & A_j \end{bmatrix}. \tag{8.9}$$

In fact, the promised modification of the Jordan form theorem is that every real matrix A is real similar to the real matrix

$$\text{diag}[K_1, K_2, \ldots, K_p, J_{p+1}, \ldots, J_q], \tag{8.10}$$

where $K_1, \ldots, K_p$ correspond to the complex eigenvalues and $J_{p+1}, \ldots, J_q$ to the real eigenvalues of A.

In Section 6.5 we applied the diagonalization result to the solution of a set of linear differential (or difference) equations. This can now be extended to the general case when $T^{-1}AT = \mathcal{J}$, so that as in (6.92), the transformation $y = T^{-1}x$ turns the differential equations

$$\dot{x} = Ax \tag{8.11}$$

into

$$\dot{y} = \mathcal{J}y, \tag{8.12}$$

where $\mathcal{J}$ is the Jordan form in (8.4). On writing $y = [Y_1, Y_2, \ldots, Y_q]^T$, where Y_i is a column vector with k_i components, we see from the block diagonal form of $\mathcal{J}$ that (8.12) reduces to the sets of equations

$$\dot{Y}_i = J_i Y_i, \qquad i = 1, 2, \ldots, q, \tag{8.13}$$

where J_i is the Jordan block in (8.2). We need therefore only consider the solution of one of the sets of triangular equations in (8.13), which for convenience we write as

$$\dot{z} = Jz, \tag{8.14}$$

where $z = [z_1, z_2, \ldots, z_k]^T$ and J is a $k \times k$ block in the form (8.2) with λ's along the diagonal. Writing out in full the equations in (8.14), starting with the last, gives

$$\dot{z}_k = \lambda z_k, \quad \dot{z}_{k-1} = \lambda z_{k-1} + z_k,$$
$$\dot{z}_{k-2} = \lambda z_{k-2} + z_{k-1}, \ldots, \dot{z}_1 = \lambda z_1 + z_2. \tag{8.15}$$

The solutions of the equations in (8.15) are successively

$$z_k = \alpha_k e^{\lambda t}$$
$$z_{k-1} = \alpha_{k-1} e^{\lambda t} + \alpha_k t e^{\lambda t}$$
$$z_{k-2} = \alpha_{k-2} e^{\lambda t} + \alpha_{k-1} t e^{\lambda t} + \tfrac{1}{2}\alpha_k t^2 e^{\lambda t}$$

and so on, where the α's are arbitrary constants determined by the initial values $z_i(0)$, $i = 1, 2, \ldots, k$. In general,

$$z_i(t) = p_i(t) e^{\lambda t},$$

where $p_i(t)$ is a polynomial in t of degree $k - i$.

Exercise 8.8 Determine the general solution of the differential equations (8.11) where A is the matrix in Example 8.4, using the transformation into Jordan form.

Exercise 8.9 Determine the form of the general solution of the system of difference equations

$$\begin{bmatrix} X_1(n+1) \\ X_2(n+1) \\ X_3(n+1) \end{bmatrix} = \begin{bmatrix} \lambda & 1 & 0 \\ 0 & \lambda & 1 \\ 0 & 0 & \lambda \end{bmatrix} \begin{bmatrix} X_1(n) \\ X_2(n) \\ X_3(n) \end{bmatrix}, \qquad n = 0, 1, 2, \ldots.$$

8.2   Normal forms

We remarked in paragraph (c) at the beginning of this chapter when all the eigenvalues of A are distinct, then A is similar to the companion matrix C associated with its characteristic polynomial (defined in Section 6.3.4). This is a special case of the fact that A is similar to C if and only if A is non-derogatory. The similarity transformation between C and the Jordan form of A was given in (6.89) for the case of distinct eigenvalues. More generally, when the Jordan form of A is (8.4) with a *single* Jordan block J_i associated with each distinct λ_i, $i = 1, 2, \ldots, q$, then $\mathscr{J}$ can be transformed into companion form by

$$H\mathscr{J}H^{-1} = C,$$

where $H = [H_1, H_2, \ldots, H_q]$ and H_j is the $n \times k_j$ matrix

$$
H_j =
\begin{bmatrix}
1 & 0 & \cdots & 0 \\
\lambda_j & 1 & \cdots & 0 \\
\vdots & \vdots & & \vdots \\
\lambda_j^{n-2} & \binom{n-2}{1}\lambda_j^{n-3} & \cdots & \binom{n-2}{k_j-1}\lambda_j^{n-k_j-1} \\
\lambda_j^{n-1} & \binom{n-1}{1}\lambda_j^{n-2} & \cdots & \binom{n-1}{k_j-1}\lambda_j^{n-k_j}
\end{bmatrix}.
$$

When each $k_j = 1$, then H reduces to the Vandermonde matrix in (6.88).

Exercise 8.10   For the companion form matrix A in Exercise 8.7, use the expression for H given above to transform A into Jordan form.

Exercise 8.11   Show that when $a = 0$ in (8.1), so that A is derogatory, then it is not similar to the companion matrix associated with its characteristic polynomial.

A derogatory matrix A is similar to its *first normal canonical form* (also sometimes called the 'Frobenius' or 'rational' form)

$$\text{diag}[C_1, C_2, \ldots, C_t], \qquad (8.16)$$

where each C_i is a companion form matrix. As for the Jordan form, the order in which the blocks are arranged is irrelevant. To describe the dimensions of the blocks in (8.16) we need to introduce a new idea. Consider the *characteristic matrix* $\lambda I - A$: its minors will be polynomials in λ. For a given value of k, let $d_k(\lambda)$ denote the monic greatest common divisor of all minors of order k of $\lambda I - A$. Then the *invariant factors* of $\lambda I - A$ are the polynomials

$$i_k(\lambda) = \frac{d_k(\lambda)}{d_{k-1}(\lambda)}, \qquad k = 1, 2, \ldots, n, \qquad (8.17)$$

where $d_0 = 1$ and $d_n(\lambda) = \det(\lambda I - A)$. Notice that it follows from the determinantal expansion formulae in Section 4.1.2 that $d_{k-1}(\lambda)$ is a factor of $d_k(\lambda)$, so the expressions defined by (8.17) are indeed polynomials. Furthermore, i_k is a factor of i_{k+1} for $k = 1, 2, \ldots, n - 1$. These invariant factors are sometimes called the *similarity invariants* of A, because two matrices are similar if and only if their characteristic matrices have the same invariant factors. Furthermore, the minimum polynomial of A is equal to the invariant factor of highest degree, and the characteristic polynomial is equal to the product of the invariant factors.

Exercise 8.12 Show that the similarity invariants of the $n \times n$ companion matrix C are $1, 1, \ldots, 1, k(\lambda) = \det(\lambda I - C)$. Show also that the polynomials in (8.17) satisfy

$$i_1 i_2 \cdots i_n = \det(\lambda I - A).$$

Hence deduce that if $k(\lambda)$ is equal to the characteristic polynomial of A, then A and C are similar if and only if A is non-derogatory.

Exercise 8.13 Show that the invariant factors of $\lambda I - J_i$, where J_i is the Jordan block in (8.2), are $1, 1, \ldots, 1, (\lambda - \lambda_i)^{k_i}$.

It follows that the matrices C_i in (8.16) are the companion matrices associated with the *non-constant* invariant factors i_l, $i_{l+1}, \ldots, i_n$ of $\lambda I - A$. This is because (see Exercise 8.12) $\lambda I - C_1$ has invariant factors $1, 1, \ldots, 1, i_l$, $\lambda I - C_2$ has factors $1, 1, \ldots, 1, i_{l+1}$, and so on; overall the invariant factors for (8.16) are therefore identical to those of $\lambda I - A$. Notice that when A is itself non-derogatory, its invariant factors are $1, 1, \ldots, 1, \det(\lambda I - A)$, so (8.16) reduces to just a single companion matrix.

Example 8.5 The invariant factors of a certain 7×7 matrix A are found to be $1, 1, 1, \lambda^3 - 2\lambda^2 + 1, \lambda^4 + 3\lambda^3 - 10\lambda^2 + \lambda + 5$. Hence the first normal canonical form (8.16) of A has two blocks, which are the companion forms for the two non-constant invariant factors, namely

$$C_1 = \begin{bmatrix} 0 & 1 & 0 \\ 0 & 0 & 1 \\ -1 & 0 & 2 \end{bmatrix}, \qquad C_2 = \begin{bmatrix} 0 & 1 & 0 & 0 \\ 0 & 0 & 1 & 0 \\ 0 & 0 & 0 & 1 \\ -5 & -1 & 10 & -3 \end{bmatrix}.$$

A second canonical form of this type requires the invariant factors of $\lambda I - A$ to be decomposed into irreducible factors: over the complex field these will be powers of linear factors, in the form $(\lambda - \lambda_r)^s$; over the real field, pairs of complex conjugate eigenvalues of A will lead to corresponding quadratic factors. These factors are called the *elementary divisors* of $\lambda I - A$, and depend upon the field of numbers being used. In particular, if A has the Jordan form (8.4), where say J_{m_1}, J_{m_2}, ..., J_{m_u} are associated with λ_1, J_{n_1}, ..., J_{n_v} with λ_2, and so on, down to J_{t_1}, ..., J_{t_w} associated with λ_q, then the elementary divisors of $\lambda I - A$ over the complex field are

$$(\lambda - \lambda_1)^{m_1}, (\lambda - \lambda_1)^{m_2}, \ldots, (\lambda - \lambda_1)^{m_u}, (\lambda - \lambda_2)^{n_1}, \ldots,$$
$$(\lambda - \lambda_2)^{n_v}, \ldots, (\lambda - \lambda_q)^{t_1}, \ldots, (\lambda - \lambda_q)^{t_w}.$$

In the *second normal canonical form* (confusingly, also sometimes called the Frobenius form) the companion matrices C_i in (8.16) are associated with the elementary divisors of $\lambda I - A$, and so have orders $m_1, m_2, \ldots, m_u, n_1, \ldots, n_v, \ldots, t_1, \ldots, t_w$.

Exercise 8.6 Consider a real matrix A of order 6 which has a Jordan form (8.4) with blocks

$$J_1 = \begin{bmatrix} 2 & 1 \\ 0 & 2 \end{bmatrix}, \qquad J_2 = \begin{bmatrix} i & 1 \\ 0 & i \end{bmatrix}, \qquad J_3 = \begin{bmatrix} -i & 1 \\ 0 & -i \end{bmatrix}.$$

By the result of Exercise 8.13, the non-trivial similarity invariants of J_1, J_2 and J_3 are respectively

$$(\lambda - 2)^2, (\lambda - i)^2, (\lambda + i)^2,$$

so the characteristic polynomial of A is

$$(\lambda - 2)^2(\lambda - i)^2(\lambda + i)^2 = (\lambda - 2)^2(\lambda^2 + 1)^2.$$

Over the field of real numbers the elementary divisors of A are

$$(\lambda - 2)^2 = \lambda^2 - 4\lambda + 4, \qquad (\lambda^2 + 1)^2 = \lambda^4 + 2\lambda^2 + 1.$$

The second normal form of A is (8.16) where the blocks are the companion matrices for these polynomials, namely

$$C_1 = \begin{bmatrix} 0 & 1 \\ -4 & 4 \end{bmatrix}, \qquad C_2 = \begin{bmatrix} 0 & 1 & 0 & 0 \\ 0 & 0 & 1 & 0 \\ 0 & 0 & 0 & 1 \\ -1 & 0 & -2 & 0 \end{bmatrix}. \qquad (8.18)$$

Over the field of complex numbers the elementary divisors of A are

$$(\lambda - 2)^2, \ (\lambda - i)^2 = \lambda^2 - 2i\lambda - 1, \ (\lambda + i)^2 = \lambda^2 + 2i - 1.$$

In this case the blocks in the second normal form (8.16) are the companion matrices C_1 in (8.18) and

$$C_2 = \begin{bmatrix} 0 & 1 \\ 1 & 2i \end{bmatrix}, \qquad C_3 = \begin{bmatrix} 0 & 1 \\ 1 & -2i \end{bmatrix}.$$

The real Jordan form of A is given by (8.10) with

$$K_1 = \begin{bmatrix} 0 & 1 & 1 & 0 \\ -1 & 0 & 0 & 1 \\ 0 & 0 & 0 & 1 \\ 0 & 0 & -1 & 0 \end{bmatrix}, \qquad J_2 = \begin{bmatrix} 2 & 1 \\ 0 & 2 \end{bmatrix}.$$

Exercise 8.14 A certain 6×6 matrix has invariant factors $\lambda + 3$, $\lambda^2 - 9$, $\lambda^3 + 5\lambda^2 - 9\lambda - 45$. Write down the first normal canonical form. Determine the elementary divisors over the real field, and hence write down the second normal canonical form.

8.3 Schur form

We have seen that not every matrix is similar to a diagonal matrix, which is unfortunate since the latter are especially easy to handle. However, it is remarkable that every $n \times n$ complex matrix A can be reduced by a similarity transformation to *triangular* form S (the *Schur canonical form*); the latter retains the nice property of diagonal matrices that the eigenvalues are displayed along the principal diagonal. Moreover, the transformation matrix U in

$$U^{-1}AU = S \tag{8.19}$$

is *unitary*, i.e. $U^*U = I_n$ (see Problem 4.11), so that (8.19) is equivalent to $U^*AU = S$, or $A = USU^*$. The result holds whether S in (8.19) is upper or lower triangular, and we shall keep to the former in our derivation, which uses induction: the result is obviously true for $n = 1$; suppose it is true for $n - 1$, $n \geqslant 2$, so that for any $(n - 1) \times (n - 1)$ matrix A_1 there exists a unitary matrix U_1 such that

$$U_1^*A_1U_1 = S_1, \tag{8.20}$$

where the suffix 1 in (8.20) indicates that all the matrices have orders $n - 1$. Let λ_1 be an eigenvalue of A with corresponding normalized eigenvector x_1, and consider a unitary matrix V with first column x_1. It follows that

$$
\begin{aligned}
V^*AV &= V^*[Ax_1, \ldots] \\
&= V^*[\lambda_1 x_1, \ldots] \\
&= \begin{bmatrix} \lambda_1 & y \\ 0 & A_1 \end{bmatrix} \begin{matrix} 1 \\ n-1 \end{matrix}, \\
& \quad\; \begin{matrix} 1 & n-1 \end{matrix}
\end{aligned}
\tag{8.21}
$$

where details of the elements of y and A_1 are unimportant. It is then easy to check, using (8.20), that

$$
U^* \begin{bmatrix} \lambda_1 & y \\ 0 & A_1 \end{bmatrix} U = \begin{bmatrix} \lambda_1 & yU_1 \\ 0 & S_1 \end{bmatrix},
\tag{8.22}
$$

where $U = \text{diag}[1, U_1]$. The matrix on the right side in (8.22) (Δ, say) has the required triangular form. Thus the result holds for $n \times n$ matrices, thereby completing the induction argument.

Exercise 8.15 Verify that (8.22) is correct, and that the matrix VU which transforms A into Δ is indeed unitary.

One way of constructing V in the above derivation is to use the so-called *Gram–Schmidt orthogonalization procedure,* which is of importance in its own right. This states that if $x_1, x_2, \ldots, x_n$ are a set of n linearly independent n-vectors, then the vectors $y_1, y_2, \ldots, y_n$, defined by

$$
y_1 = x_1, \qquad y_2 = x_2 - \frac{\langle y_1, x_2 \rangle}{\langle y_1, y_1 \rangle} y_1
$$

and generally

$$
y_j = x_j - \sum_{i=1}^{j-1} \frac{\langle y_i, x_j \rangle}{\langle y_i, y_i \rangle} y_i, \qquad j = 2, 3, \ldots, n,
\tag{8.23}
$$

are linearly independent and mutually orthogonal. The notation $\langle a, b \rangle$ denotes *scalar* (or *inner*) product $\sum \bar{a}_i b_i$, defined in (2.44) for the real case, and is useful because it avoids the need to specify whether we are dealing with row or column vectors.

Exercise 8.16 Confirm that the vectors $[1, 0, -1]$, $[4, 1, 4]$, $[-3, 24, -3]$ are mutually orthogonal, and normalize them.

If it is required that all the y_j be normalized, so that the complete set forms an *orthonormal basis*, then the procedure described by (8.23) is conveniently rewritten as follows: set $y_1 = x_1/\|x_1\|$, where it can be recalled (see (6.8)) that

$$\|x_1\| = (\langle x_1, x_1 \rangle)^{1/2}.$$

Then the remaining orthonormal vectors are given by $y_j = z_j/\|z_j\|$ with

$$z_j = x_j - \sum_{i=1}^{j-1} \langle y_i, x_j \rangle y_i, \qquad j \geqslant 2. \tag{8.24}$$

The application of this scheme to determination of the Schur form of a given matrix A is now straightforward. The vector x_1 is taken to be a normalized eigenvector for λ_1, and using *any* other $n - 1$ vectors to give a linearly independent set, the orthonormal basis obtained using (8.24) gives the matrix $V = [y_1, y_2, \ldots, y_n]$ in (8.21). The procedure is then repeated for A_1, using an eigenvector corresponding to λ_2, to produce a unitary matrix U_2, say, such that $U_2^* A_1 U_2$ is upper triangular with λ_2 in the $(1, 1)$ position. The required transformation matrix for A is $V_2 = \text{diag}(1, U_2)$. This is continued for the remaining eigenvalues $\lambda_3, \ldots, \lambda_n$ of A, and the overall unitary transformation matrix is $U = VV_2V_3 \cdots V_{n-1}$.

Example 8.7 Consider the matrix A in (6.24). It was found in Example 6.6 that its eigenvalues are $\lambda_1 = 0$, $\lambda_2 = 3$, $\lambda_3 = -3$, and that a normalized eigenvector corresponding to λ_1 is $x_1 = \frac{1}{3}[2, -2, 1]^\mathsf{T}$. A convenient basis is $x_1, x_2 = [0, 1, 0]^\mathsf{T}$, $x_3 = [0, 0, 1]^\mathsf{T}$. The Gram–Schmidt orthonormalization (8.24) gives

$$y_1 = x_1$$

$$z_2 = x_2 - \langle y_1, x_2 \rangle y_1$$

$$= [0, 1, 0]^\mathsf{T} - \left(-\frac{2}{3}\right)\left(\frac{2}{3}, -\frac{2}{3}, \frac{1}{3}\right)^\mathsf{T}$$

$$= \frac{1}{9}[4, 5, 2]^\mathsf{T}$$

$$y_2 = \frac{z_2}{\|z_2\|} = \frac{1}{3\sqrt{5}}[4, 5, 2]^\mathsf{T}$$

$$z_3 = x_3 - \langle y_1, x_3 \rangle y_1 - \langle y_2, x_3 \rangle y_2$$

$$= [0, 0, 1]^T - \frac{1}{3}\left[\frac{2}{3}, -\frac{2}{3}, \frac{1}{3}\right]^T - \frac{2}{3\sqrt5}\left[\frac{4}{3\sqrt5}, \frac{5}{3\sqrt5}, \frac{2}{3\sqrt5}\right]^T$$

$$= \frac{1}{5}[-2, 0, 4]^T$$

$$y_3 = \frac{z_3}{\|z_3\|} = \frac{1}{\sqrt5}[-1, 0, 2]^T.$$

The matrix V in (8.21) is therefore

$$V = [y_1, y_2, y_3]$$

$$= \begin{bmatrix} \dfrac{2}{3} & \dfrac{4}{3\sqrt5} & -\dfrac{1}{\sqrt5} \\[2mm] -\dfrac{2}{3} & \dfrac{5}{3\sqrt5} & 0 \\[2mm] \dfrac{1}{3} & \dfrac{2}{3\sqrt5} & \dfrac{2}{\sqrt5} \end{bmatrix}$$

and (8.21) becomes

$$V^*AV = \begin{bmatrix} 0 & 0 & 0 \\ 0 & \dfrac{9}{5} & \dfrac{12}{5} \\ 0 & \dfrac{12}{5} & -\dfrac{9}{5} \end{bmatrix}. \qquad (8.25)$$

The reader should verify that the 2×2 matrix A_1 in the lower right corner in (8.25) has a normalized eigenvector $u_1 = (1/\sqrt5)[2, 1]^T$ corresponding to $\lambda_2 = 3$. By inspection, a suitable orthonormal basis is u_1, $(1/\sqrt5)[1, -2]^T$, so $U_2 = [u_1, u_2]$ and

$$V_2 = \begin{bmatrix} 1 & 0 \\ 0 & U_2 \end{bmatrix} = \frac{1}{\sqrt5}\begin{bmatrix} \sqrt5 & 0 & 0 \\ 0 & 2 & 1 \\ 0 & 1 & -2 \end{bmatrix}.$$

It is left as a further exercise for the reader to check that $U^*AU = (VV_2)^*AVV_2$ is upper triangular.

In the above example the transformation matrix is real, and hence orthogonal. In general, a real matrix is orthogonally similar to a triangular matrix with real diagonal elements if and only if all its eigenvalues are real.

Exercise 8.17 Use the Gram–Schmidt orthonormalization procedure to obtain the Schur canonical form of

$$\begin{bmatrix} 2 & 2 & 1 \\ 1 & 3 & 1 \\ 1 & 2 & 2 \end{bmatrix},$$

together with a corresponding orthogonal transformation matrix.

It is interesting to note that if the Gram–Schmidt orthonormalization (8.23) is applied to the set of columns of a non-singular matrix A, then this is equivalent to the factorization $A = QR$, where $Q = [y_1, y_2, \ldots, y_n]$ is unitary and R is upper triangular. This follows because each column x_j of A is expressed via (8.23) only in terms of $y_1, y_2, \ldots, y_j$.

Exercise 8.18 Apply the Gram–Schmidt procedure to the columns of

$$A = \begin{bmatrix} 1 & 1 & 1 \\ 1 & 1 & 0 \\ 1 & 0 & 0 \end{bmatrix}.$$

Express each of these columns in terms of the orthonormal vectors, and hence obtain the factorization $A = QR$.

The QR factorization is important because it forms the core of the powerful QR algorithm for computing eigenvalues and eigenvectors, mentioned in Section 6.6.2.

Notice that in the transformation (8.19) to Schur form, neither the transformation matrix U, nor the off-diagonal elements in S, are unique. The diagonal elements of S are the eigenvalues of A, but the order in which they occur is not unique.

Exercise 8.19 Repeat Exercise 8.17, using different choices for the basis vectors at the first step, so as to obtain different transformation and triangular matrices.

If A is real but has complex eigenvalues, then a real orthogonal matrix can be found such that $U^{\mathrm{T}}AU$ has the real block triangular form

$$\begin{bmatrix} \Delta_1 & & \\ & \Delta_2 & X \\ & & \ddots & \\ 0 & & & \Delta_k \end{bmatrix}, \tag{8.26}$$

where each block Δ_i on the principal diagonal is either a real scalar, or a real 2×2 block corresponding to a pair of complex conjugate eigenvalues, as in (8.10) for the real Jordan form. When A is real and symmetric, its eigenvalues are all real, so that it is in fact orthogonally similar to a *diagonal* matrix (see Section 6.4.3). The most general result of this type is quoted in part (c) of Problem 8.5.

8.4 Hessenberg form

Both (8.9) and (8.26) are upper triangular matrices with additional elements along the subdiagonal (the line immediately below the principal diagonal). Such a matrix is said to have *upper Hessenberg form,* and its transpose has *lower* Hessenberg form. For example, the companion matrix defined in Section 6.3.4 has lower Hessenberg form. Specifically, $H = [h_{ij}]$ is upper Hessenberg if $h_{ij} = 0$ for $i > j + 1$. In order to show how to transform a real matrix A into Hessenberg form, it is first convenient to introduce the *Householder transformation*

$$P = I_n - \frac{2vv^\mathsf{T}}{v^\mathsf{T}v} \tag{8.27}$$

associated with a real non-zero column n-vector v.

Exercise 8.20 Prove that P in (8.27) is real symmetric, orthogonal and *involutory* (i.e. $P^{-1} = P$).

The crucial property of the Householder transformation is that given any non-zero real column n-vector x, it is possible to choose v such that $Px = \pm \|x\| e_1$, where e_1 is the first column of I_n.

Exercise 8.21 Prove that, if we take in (8.27)

$$v = x \pm \|x\| e_1, \tag{8.28}$$

then $Px = \mp \|x\| e_1$.

Exercise 8.22 Use (8.27) and (8.28) to obtain a matrix P such that $P[1, 1, 2]^\mathsf{T} = -\sqrt{6}[1, 0, 0]^\mathsf{T}$.

A real orthogonal matrix U such that $U^\mathsf{T}AU$ is in upper Hessenberg form is now found by expressing U as a product of

Householder transformations (by Problem 4.10(e) the product of orthogonal matrices is also orthogonal). This is best understood through a specific example.

Example 8.8 Suppose we require to reduce a general 4×4 real matrix $A = [a_{ij}]$ to upper Hessenberg form. First determine, using (8.27) and (8.28), a 3×3 Householder matrix P_1, such that the first column of A *below* the principal diagonal is reduced to a multiple of e_1, i.e.

$$P_1 \begin{bmatrix} a_{21} \\ a_{31} \\ a_{41} \end{bmatrix} = -k_1 \begin{bmatrix} 1 \\ 0 \\ 0 \end{bmatrix}, \qquad k_1 = \left\| \begin{bmatrix} a_{21} \\ a_{31} \\ a_{41} \end{bmatrix} \right\| = (a_{21}^2 + a_{31}^2 + a_{41}^2)^{1/2}.$$

Since P_1 is symmetric and orthogonal, so is $U_1 = \mathrm{diag}[1, P_1]$, and it is left for the reader to check that

$$U_1^T A U_1 = \begin{bmatrix} a_{11} & x & x & x \\ -k_1 & x & x & x \\ 0 & b_{32} & x & x \\ 0 & b_{42} & x & x \end{bmatrix} = B, \text{ say,} \qquad (8.29)$$

where the x's denote elements which are non-zero in general. Next, reduce the *second* column of (8.29) below the principal diagonal in a similar fashion: compute a 2×2 Householder matrix P_2 such that

$$P_2 \begin{bmatrix} b_{32} \\ b_{42} \end{bmatrix} = -k_2 \begin{bmatrix} 1 \\ 0 \end{bmatrix}, \qquad k_2 = \left\| \begin{bmatrix} b_{32} \\ b_{42} \end{bmatrix} \right\| = (b_{32}^2 + b_{42}^2)^{1/2}.$$

Then $U_2 = \mathrm{diag}[I_2, P_2]$ gives

$$U_2^T B U_2 = \begin{bmatrix} x & x & x & x \\ x & x & x & x \\ 0 & -k_2 & x & x \\ 0 & 0 & x & x \end{bmatrix},$$

which has the required upper Hessenberg form. Overall, the orthogonal transformation matrix to be applied to A is $U = U_1 U_2$.

Exercise 8.23 Obtain an orthogonal transformation of

$$A = \begin{bmatrix} 1 & -1 & 0 \\ 1 & 2 & 4 \\ 2 & 3 & -5 \end{bmatrix}$$

into Hessenberg form.

In general, P_i is an $(n-i) \times (n-i)$ Householder transformation which reduces the last $n-i$ elements of column i of the matrix under consideration. If $U_i = \mathrm{diag}[I_i, P_i]$ then the overall orthogonal transformation $U = U_1 U_2 \cdots U_{n-2}$ makes $U^T A U$ have upper Hessenberg form (notice however that the decomposition is not unique). The algorithm outlined above is known to be very satisfactory numerically. In practice, before using the QR algorithm for eigenvalue computation, a given non-symmetric matrix is first put into Hessenberg form, and this speeds up the process since the Gram–Schmidt decomposition into QR (see Section 8.3) of a Hessenberg matrix requires only $O(n^2)$ operations, compared to $O(n^3)$ for an arbitrary matrix. For further discussion on the numerical complexities, consult Wilkinson (1965) or Golub and Van Loan (1989).

Exercise 8.24 The Householder transformation associated with a complex column n-vector v is defined by

$$P = I_n - \frac{2vv^*}{v^*v}.$$

Prove that P is hermitian, unitary and involutory.

Exercise 8.25 Show that if A is symmetric and $U^T A U = H$ is upper Hessenberg, then H must be tridiagonal (defined in Problem 3.2) and symmetric.

Exercise 8.26 Reduce the matrix

$$A = \begin{bmatrix} 1 & 3 & 4 \\ 3 & 2 & 4 \\ 4 & 4 & -1 \end{bmatrix}$$

to tridiagonal form by an orthogonal transformation using the Householder method.

Transformation of an arbitrary matrix A to Hessenberg form $H = [h_{ij}]$ also provides a satisfactory numerical procedure for determining the characteristic polynomial of A, since the characteristic polynomial of H can be computed using the following simple scheme. Firstly, for a vector $\alpha_i = [\alpha_{i1}, \alpha_{i2}, \ldots, \alpha_{i,n+1}]$ define

$$\tilde{\alpha}_i = [\alpha_{i2}, \alpha_{i3}, \ldots, \alpha_{i,n+1}, \alpha_{i1}],$$

which is obtained from α_i by shifting each element one position to the left, with the 'overflow' element α_{i1} being moved to the last position. Then

$$\det(\lambda I - H) = \lambda^n + \alpha_{n2}\lambda^{n-1} + \alpha_{n3}\lambda^{n-2} + \cdots + \alpha_{n,n+1}, \quad (8.30)$$

where

$$\alpha_i = \tilde{\alpha}_{i-1} - h_{ii}\alpha_{i-1} - \sum_{j=1}^{i-1} h_{i-j,i}h_{i,i-1}h_{i-1,i-2} \cdots h_{i-j+1,i-j}\alpha_{i-j-1},$$

$$i = 1, 2, \ldots, n \quad (8.31)$$

with $\alpha_0 = [0, 0, \ldots, 0, 1]$ (note that the sum in (8.31) is defined to be zero when $i = 1$).

Example 8.9 Consider the Hessenberg matrix

$$H = \begin{bmatrix} 2 & 1 & 4 \\ -3 & 4 & 5 \\ 0 & 7 & 1 \end{bmatrix}.$$

Since $n = 3$, we have $\alpha_0 = [0, 0, 0, 1]$, and from (8.31)

$$\alpha_1 = \tilde{\alpha}_0 - h_{11}\alpha_0$$
$$= [0, 0, 1, 0] - 2[0, 0, 0, 1] = [0, 0, 1, -2].$$

Repeating this procedure gives

$$\alpha_2 = \tilde{\alpha}_1 - h_{22}\alpha_1 - h_{12}h_{21}\alpha_0$$
$$= [0, 1, -2, 0] - 4[0, 0, 1, -2] - 1 \cdot (-3)[0, 0, 0, 1]$$
$$= [0, 1, -6, 11]$$
$$\alpha_3 = \tilde{\alpha}_2 - h_{33}\alpha_2 - h_{23}h_{32}\alpha_1 - h_{13}h_{32}h_{21}\alpha_0$$
$$= [1, -6, 11, 0] - 1[0, 1, -6, 11] - 5 \cdot 7[0, 0, 1, -2]$$
$$\quad - 4 \cdot 7 \cdot (-3)[0, 0, 0, 1]$$
$$= [1, -7, -18, 143].$$

Hence from (8.30)

$$\det(\lambda I - H) = \lambda^3 + \alpha_{32}\lambda^2 + \alpha_{33}\lambda + \alpha_{34}$$
$$= \lambda^3 - 7\lambda^2 - 18\lambda + 143.$$

Notice that it can be assumed when using the above scheme for determining $\det(\lambda I - H)$ that all the elements on the subdiagonal are non-zero, i.e. $h_{i+1,i} \neq 0$, $i = 1, 2, \ldots, n - 1$. If this is not the case, then H can be partitioned into block triangular form, and (see Problem 4.5(b)) its characteristic polynomial is equal to the product of those of the diagonal blocks, each of which is a smaller Hessenberg matrix.

Exercise 8.27 Use (8.30) and (8.31) to obtain the characteristic polynomial of

$$H = \begin{bmatrix} 4 & -5 & -1 \\ 3 & 5 & 6 \\ 0 & -1 & 2 \end{bmatrix}.$$

Hence determine the characteristic polynomial of the upper Hessenberg matrix

$$\begin{bmatrix} 1 & 2 & 14 & -11 & 12 \\ -1 & 0 & -9 & 0 & 8 \\ 0 & 0 & 4 & -5 & -1 \\ 0 & 0 & 3 & 5 & 6 \\ 0 & 0 & 0 & -1 & 2 \end{bmatrix}.$$

8.5 Singular value and polar decompositions

Let us recall that only certain square matrices are similar to a diagonal matrix, but that we have seen in Section 8.3 that all complex square matrices are unitarily similar to a triangular matrix. An important result, which can be thought of informally as a compromise between these two extremes, is the *singular value decomposition,* for square or rectangular matrices, which can be stated as follows:

For any $m \times n$ complex matrix A with rank r there exist unitary matrices $U(m \times m)$ and $V(n \times n)$ such that

$$A = U\Sigma V. \tag{8.32}$$

In (8.32) $\Sigma = [\sigma_{ij}]$ is an $m \times n$ matrix all of whose entries is zero except for the diagonal elements $\sigma_{ii} = \sigma_i$, called the *singular values* of A, which satisfy

$$\sigma_1 \geqslant \sigma_2 \geqslant \cdots \geqslant \sigma_r > 0, \quad \sigma_{r+1} = \cdots = \sigma_p = 0, \quad p = \min(m, n).$$

If $m = n$ then Σ is a diagonal matrix; otherwise the form of Σ depends upon which of m and n is the greater. For example, if $m = 2$, $n = 3$ then

$$\Sigma = \begin{bmatrix} \sigma_1 & 0 & 0 \\ 0 & \sigma_2 & 0 \end{bmatrix}.$$

When A is real then U and V in (8.32) are real and orthogonal.

Exercise 8.28 Prove using (8.32) that A^*A is hermitian and positive semidefinite or definite (this generalizes Problem 7.2). Show that the non-zero singular values of A are equal to the positive square roots of the non-zero eigenvalues of A^*A.

Exercise 8.29 Prove using (8.32) that when $m = n$,

$$|\det A| = \sigma_1\sigma_2\cdots\sigma_n.$$

To see how the singular value decomposition arises, let us simplify the discussion by considering the case when A is square and non-singular. The argument is easily extended to the general case. We begin by recalling from Section 6.4.3 that since A^*A is hermitian and positive definite, then it is always possible to find an orthonormal set of eigenvectors $x_1, \ldots, x_n$ such that

$$A^*Ax_i = \sigma_i^2 x_i, \qquad i = 1, 2, \ldots, n, \qquad (8.33)$$

since by Exercise 8.28, A^*A has eigenvalues $\sigma_1^2, \ldots, \sigma_n^2$. Define vectors

$$y_i = \frac{1}{\sigma_i} Ax_i \qquad (8.34)$$

and construct matrices

$$V^* = [x_1, x_2, \ldots, x_n], \qquad U = [y_1, y_2, \ldots, y_n]. \qquad (8.35)$$

Then

$$\begin{aligned}U^*AV^* &= U^*[Ax_1, \ldots, Ax_n]\\ &= U^*[\sigma_1 y_1, \ldots, \sigma_n y_n]\\ &= \text{diag}[\sigma_1, \ldots, \sigma_n], \qquad (8.36)\end{aligned}$$

the last step following from the fact that the y_i also form an orthonormal set of vectors. Since by construction U and V are unitary, premultiplying (8.36) by U and postmultiplying by V gives the required result (8.32).

Exercise 8.30 Show that the vectors $y_1, \ldots, y_n$ in (8.34) form an orthonormal set.

The procedure described above can be used to construct the singular value decomposition of a non-singular square matrix, by finding eigenvalues and eigenvectors of A^*A as in (8.33) and (8.34). A general algorithm for the singular value decomposition can be found in Golub and Van Loan (1989).

Exercise 8.31 Determine the singular value decomposition of

$$A = \begin{bmatrix} 6 & 2 \\ 0 & 5 \end{bmatrix}.$$

An important feature of the singular value decomposition (8.32) is the light it sheds on the difficult numerical problem of determining the rank $R(A) = r$ of the $m \times n$ matrix A. Because of rounding errors and inaccuracy of data, it is easy to make a discrete error when computing $R(A)$. Suppose B is any matrix having the same dimensions as A but with rank $k < r$. Define the *spectral norm* (or *2-norm*) $\|A\|_2$ of A as its largest singular value (i.e. the square root of the largest eigenvalue of A^*A). Then it can be shown that the spectral norm distance $\|A - B\|_2$ between A and all such matrices B is at least σ_{k+1}. In particular, the smallest non-zero singular value σ_r of A is the spectral norm distance between A and matrices B having rank one less than A. Thus σ_r provides a measure of how 'close' a given matrix A is to being rank deficient. Further material on the spectral and other matrix norms will be given in Section 14.3.

Exercise 8.32 As for square matrices (see (6.117)), the euclidean norm of the $m \times n$ matrix A is defined by

$$\|A\| = [\mathrm{tr}(A^*A)]^{1/2} = \left[\sum_{i=1}^{m} \sum_{j=1}^{n} |a_{ij}|^2 \right]^{1/2}.$$

Prove that $\|UA\| = \|A\|$ for any unitary matrix U. Hence deduce that $\|UAV\| = \|A\|$ for unitary matrices U, V.

Exercise 8.33 Use the preceding exercise to prove that for A in (8.32) $\|A\|^2 = \sigma_1^2 + \sigma_2^2 + \cdots + \sigma_p^2$.

Return to (8.33), in which A is assumed square and non-singular, and rewrite it as

$$A^*AV^* = V^*D^2$$

where $D = \mathrm{diag}[\sigma_1, \ldots, \sigma_n]$ and V^* is the unitary matrix in (8.35). This represents the unitary similarity of A^*A to the diagonal matrix of its eigenvalues, i.e.

$$VA^*AV^* = D^2. \tag{8.37}$$

We can define a *square root H* of A^*A by

$$H = V^*DV, \tag{8.38}$$

since it is trivial to confirm that $H^2 = A^*A$ (see also Exercise 6.30). It is customary to write formally $H = (A^*A)^{1/2}$. We see from (8.38) that H is hermitian and is similar to the positive diagonal matrix D, and so is positive definite.

Exercise 8.34 Use (8.37) and (8.38) to prove that the matrix defined by $Q = AH^{-1}$ is unitary

It follows from (8.38) and Exercise 8.34 that when A is non-singular we can express it in the form

$$A = QH, \tag{8.39}$$

where Q is unitary and H is positive definite hermitian. This is called the *polar decomposition* of A, since Q can be expressed as $\exp(iR)$ for some hermitian matrix R (see Chapter 9, Problem 9.14), so (8.39) can be regarded as a matrix generalization of the polar form $\exp(i\theta) |a|$ of a complex number a. Comparing (8.39) with (8.32) shows that since $\Sigma = D$ we have $Q = UV$. In fact the expression (8.39) still holds when A is singular or rectangular, in which cases H is only positive semidefinite and $Q^*Q = I$. For example, if $m \geq n$ and in (8.32)

$$\Sigma = \begin{bmatrix} D \\ 0 \end{bmatrix},$$

then H is still given by (8.38), and $Q = U_1 V$ where $U = [U_1, U_2]$ with U_1 $m \times n$.

Example 8.10 It is easy to verify that for the matrix

$$A = \begin{bmatrix} 8 & 6 \\ 16 & 12 \\ 16 & 12 \end{bmatrix} \tag{8.40}$$

the singular value decomposition (8.32) is

$$A = \underbrace{\begin{bmatrix} \dfrac{1}{3} & \dfrac{2}{3} & \dfrac{2}{3} \\ \dfrac{2}{3} & -\dfrac{2}{3} & \dfrac{1}{3} \\ \dfrac{2}{3} & \dfrac{1}{3} & -\dfrac{2}{3} \end{bmatrix}}_{U} \underbrace{\begin{bmatrix} 30 & 0 \\ 0 & 0 \\ 0 & 0 \end{bmatrix}}_{\Sigma} \underbrace{\begin{bmatrix} \dfrac{4}{5} & \dfrac{3}{5} \\ \dfrac{3}{5} & -\dfrac{4}{5} \end{bmatrix}}_{V}. \tag{8.41}$$

Hence the polar decomposition (8.39) has

$$
Q = \begin{bmatrix} \dfrac{1}{3} & \dfrac{2}{3} \\[2mm] \dfrac{2}{3} & -\dfrac{2}{3} \\[2mm] \dfrac{2}{3} & \dfrac{1}{3} \end{bmatrix} \qquad V = \begin{bmatrix} \dfrac{2}{3} & -\dfrac{1}{3} \\[2mm] \dfrac{2}{15} & \dfrac{14}{15} \\[2mm] \dfrac{11}{15} & \dfrac{2}{15} \end{bmatrix}
$$
$$
\underset{U_1}{}
$$

$$
H = V^* \underset{D}{\begin{bmatrix} 30 & 0 \\ 0 & 0 \end{bmatrix}} V = \begin{bmatrix} \dfrac{96}{5} & \dfrac{72}{5} \\[2mm] \dfrac{72}{5} & \dfrac{54}{5} \end{bmatrix}
$$

and it is left as an exercise for the reader to check that $Q^*Q = I$.

Exercise 8.35 Determine the polar decomposition for A in Exercise 8.31.

Exercise 8.36 Use (8.33) and (8.34) to show that $AA^*y_i = \sigma_i^2 y_i$, $i = 1, 2, \ldots, n$. This shows that AA^* has eigenvalues σ_i^2 and eigenvectors y_i. Hence deduce that $AA^* = UD^2U^*$.

Exercise 8.37 Show that $H_1 = UDU^*$ is a square root of AA^*. Hence show that an alternative polar decomposition is $A = H_1 Q_1$, where $Q_1 Q_1^* = I$.

Exercise 8.38 Show that if Q and H in (8.39) commute, then A is normal.
   The converse of this result also holds.

Problems

8.1 Suppose that A is similar to a diagonal matrix and that all its eigenvalues are roots of unity. Show that there exists a positive integer N such that $A^N = I$.

8.2 (a) Let $f(\lambda)$ be any polynomial such that $f(A) = 0$, for an $n \times n$ matrix A. Prove that the minimum polynomial $m(\lambda)$ of A is a factor of $f(\lambda)$.

    (b) Let A be such that $A^M = I$ for some positive integer $M \geq 1$. Deduce that $m(\lambda)$ has no repeated factors, so that A is similar to a diagonal matrix.

8.3 Write down seven matrices which have characteristic polynomial $(\lambda - 2)^5$, but no two of which are similar.

8.4 Consider the system of differential equations

$$\dot{x}(t) = J_i x(t) + b u(t),$$

where $x(t) = [x_1(t), \ldots, x_{k_i}(t)]^\mathsf{T}$, $b = [0, 0, \ldots, 0, 1]^\mathsf{T}$ and J_i is the Jordan block (8.2). Assume that $x(0) = 0$ and $u(0) = 0$, and apply the Laplace transform, denoted by an overbar, to show that

$$\bar{x}_1(s) = \frac{\bar{u}(s)}{(s - \lambda_i)^{k_i}}.$$

8.5 Let A be an $n \times n$ complex matrix.

   (a) Show that if A is normal (defined in Problem 2.10) then its Schur form is normal.

   (b) Show that any normal, upper triangular matrix is diagonal.

   (c) Hence show that A is unitarily similar to a diagonal matrix if and only if it is normal.

   (d) Deduce that if A is normal then its singular values are the moduli of the eigenvalues.

8.6 Let $\mathscr{C}(A, b)$ be the controllability matrix defined in (4.37). If $U^\mathsf{T} A U = H$ is a Hessenberg decomposition of A, and $b = U e_1$, where e_1 is the first column of I_n and U is orthogonal, show that

$$\mathscr{C}(A, b) = U \mathscr{C}(H, e_1).$$

Deduce that this is a QR factorization (see Section 6.6.2) of $\mathscr{C}(A, b)$.

8.7 Prove that a square matrix is unitary if and only if all its singular values are equal to one.

8.8 Show that if an $n \times n$ non-singular matrix A is involutory (i.e. $A = A^{-1}$) then its Jordan form has the same property. Use the result in Exercise 8.3 to show that A is involutory if and only if A is similar to

$$\begin{bmatrix} I_m & 0 \\ 0 & -I_{m-n} \end{bmatrix}$$

for some $0 \leqslant m \leqslant n$.

8.9 Consider the orthogonal transformation $x = Uy$ applied to the differential equations $\dot{x} = Ax + bu$ describing a linear control system. Suppose b is an eigenvector of A, and let U be constructed as in Section 8.4 with first column equal to $b/\|b\|$. Show that the resulting system has the form

$$\dot{y} = Hy + fu,$$

where $f = [\|b\|, 0, \ldots, 0]^{\mathsf{T}}$ and H is in upper Hessenberg form. Compare this with Problem 6.26.

8.10 Every real symmetric positive definite matrix A can be expressed in the *Choleski decomposition* $A = LL^{\mathsf{T}}$, where L is a lower triangular matrix (see Problems 3.6 and 7.2). If L^{T} has the polar decomposition QH as in (8.39), show that $H = A^{1/2}$.

8.11 Consider two real $m \times n$ matrices A and B, and let $U\Sigma V$ be the singular value decomposition of $B^{\mathsf{T}}A$. Define the matrix $Z = VQ^{\mathsf{T}}U$, where Q is any $n \times n$ orthogonal matrix. Show that

(a) Z is orthogonal;

(b) $\operatorname{tr}(Q^{\mathsf{T}}B^{\mathsf{T}}A) \leq \sum_{i=1}^{n} \sigma_i$, where $\sigma_1, \ldots, \sigma_n$ are the singular values of $B^{\mathsf{T}}A$.
  (Hint: Use Problems 2.7(b) and 4.10(g).)

8.12 If A, B and Q are as given in the preceding problem, show that

$$\operatorname{tr}(A - BQ)^{\mathsf{T}}(A - BQ) = \operatorname{tr}(A^{\mathsf{T}}A) + \operatorname{tr}(B^{\mathsf{T}}B) - 2\operatorname{tr}(Q^{\mathsf{T}}B^{\mathsf{T}}A)$$
$$(= r, \text{ say}).$$

Hence deduce that r is minimized by taking $Z = I$, so that $Q = UV$. This choice of the orthogonal matrix Q is the one which transforms B by postmultiplication (equivalent to rotation of coordinates) 'most nearly' into A, since $r = \|A - BQ\|^2$ is minimized (Golub and Van Loan, 1983).

8.13 It can be shown that if A is square and non-singular then the sequence defined by

$$X_0 = A, \qquad X_{k+1} = \frac{1}{2}[X_k + (X_k^*)^{-1}], \qquad k = 0, 1, 2, \ldots$$

converges quadratically to the unitary factor Q in the polar decomposition (8.39) of A. Apply this method to the matrix A in Exercise 8.31.

8.14 It is known (Wilkinson 1965) that if H is an $n \times n$ hermitian matrix in the form

$$\begin{matrix} & n-1 & 1 \\ H = & \begin{bmatrix} H' & a \\ a^* & b \end{bmatrix} & \begin{matrix} n-1 \\ 1 \end{matrix} \end{matrix},$$

then the eigenvalues $\lambda_1 \geq \lambda_2 \geq \lambda_3 \geq \cdots \geq \lambda_n$ of H *interlace* the eigenvalues $\mu_1 \geq \mu_2 \geq \cdots \geq \mu_{n-1}$ of H', that is

$$\lambda_1 \geq \mu_1 \geq \lambda_2 \geq \mu_2 \geq \cdots \geq \lambda_{n-1} \geq \mu_{n-1} \geq \lambda_n.$$

Suppose the companion matrix

$$C = \begin{bmatrix} 0 & 1 & & & \\ 0 & 0 & 1 & & \\ & & & \ddots & \\ & & & & 1 \\ a_1 & a_2 & \cdots & & a_n \end{bmatrix}$$

has singular values $\sigma_1 \geq \sigma_2 \geq \cdots \geq \sigma_n$. Recall (Exercise 8.36) that CC^* has eigenvalues $\sigma_1^2, \sigma_2^2, \ldots, \sigma_n^2$.

(a) By applying the above result to CC^*, show that

$$\sigma_2 = \sigma_3 = \cdots = \sigma_{n-1} = 1.$$

(b) By considering the eigenvalue–eigenvector equation for CC^* in the form

$$CC^* \begin{bmatrix} x \\ x_n \end{bmatrix} = \lambda \begin{bmatrix} x \\ x_n \end{bmatrix},$$

where x is a column $(n-1)$-vector, show that

$$(\lambda^2 - \lambda s + |a_1|^2)x_n = 0.$$

Hence deduce that

$$\sigma_1^2, \sigma_n^2 = \frac{1}{2}[s \pm (s^2 - 4|a_1|^2)^{1/2}]$$

where $s = 1 + |a_1|^2 + |a_2|^2 + \cdots + |a_n|^2$.

References

Chen (1984), Gantmacher (1959, Vol. 1), Golub and Van Loan (1989), Horn and Johnson (1985), Mirsky (1963), Wilkinson (1965).

9 Matrix functions

So far in this book we have mainly been concerned with algebraic manipulations of matrices, and have encountered limiting processes only in relation to iterative methods in Section 6.7. We now give an introduction to what can be called 'matrix analysis', where the idea of convergence plays a fundamental role.

9.1 Definition and properties

A scalar infinite power series

$$f(\lambda) = f_0 + f_1\lambda + f_2\lambda^2 + f_3\lambda^3 + \cdots \tag{9.1}$$

has *radius of convergence* R if for every $|\lambda| < R$ the series on the right in (9.1) has a finite limit as the number of terms tends to infinity. Consider the matrix series obtained by replacing λ in (9.1) by an $n \times n$ matrix A:

$$f(A) = f_0 I_n + f_1 A + f_2 A^2 + f_3 A^3 + \cdots. \tag{9.2}$$

The matrix series on the right in (9.2) is *convergent* if the sequence of partial sums $\sum_{i=0}^{k} f_i A^i$ tends to a matrix with finite elements as $k \to \infty$.

Assume first that all the eigenvalues $\lambda_1, \ldots, \lambda_n$ of A are distinct. Then we can use the diagonalization formula (6.74), i.e. $A = T\Lambda T^{-1}$, which together with (6.76), namely $A^k = T\Lambda^k T^{-1}$, reduces (9.2) to

$$
\begin{aligned}
f(A) &= f_0 I + f_1 T\Lambda T^{-1} + f_2 T\Lambda^2 T^{-1} + f_3 T\Lambda^3 T^{-1} + \cdots \\
&= T(f_0 I + f_1\Lambda + f_2\Lambda^2 + \cdots)T^{-1}.
\end{aligned}
\tag{9.3}
$$

Since $\Lambda = \mathrm{diag}[\lambda_1, \lambda_2, \ldots, \lambda_n]$ it follows that the matrix within brackets in (9.3) is also diagonal, and the i, i element is $f_0 + f_1\lambda_i + f_2\lambda_i^2 + \cdots$, which will converge to $f(\lambda_i)$ if $|\lambda_i| < R$, and diverge if $|\lambda_i| > R$. Thus, provided *all* the eigenvalues of A lie within the radius of convergence of the series (9.1), we can write (9.3) as

$$f(A) = T \, \text{diag}[f(\lambda_1), f(\lambda_2), \ldots, f(\lambda_n)]T^{-1}. \qquad (9.4)$$

In this case it therefore makes sense to use (9.2) as the definition of the *matrix function $f(A)$*. We shall see shortly that this remains valid even if A does not have a diagonal Jordan form. Many authors use the term 'function of a matrix', but the more concise expression 'matrix function' is preferred in this book.

Example 9.1 It is well known that the logarithmic series

$$\ln(1 + \lambda) = \lambda - \frac{1}{2}\lambda^2 + \frac{1}{3}\lambda^3 - \cdots$$

is convergent provided $|\lambda| < 1$. Thus we can write

$$\ln(I_n + A) = A - \frac{1}{2}A^2 + \frac{1}{3}A^3 - \cdots, \qquad (9.5)$$

provided all the eigenvalues of A have modulus less than unity. For example, if

$$A = \begin{bmatrix} 0 & 1 \\ -\frac{1}{8} & \frac{3}{4} \end{bmatrix}, \qquad (9.6)$$

then it is easily checked that $\lambda_1 = \frac{1}{4}$, $\lambda_2 = \frac{1}{2}$ and the convergence criterion is satisfied, so (9.5) holds. In fact (9.6) is in companion form, so from (6.88) the matrix transforming A to diagonal form is given by

$$T = \begin{bmatrix} 1 & 1 \\ \lambda_1 & \lambda_2 \end{bmatrix} = \begin{bmatrix} 1 & 1 \\ \frac{1}{4} & \frac{1}{2} \end{bmatrix}, \qquad T^{-1} = \begin{bmatrix} 2 & -4 \\ -1 & 4 \end{bmatrix}. \qquad (9.7)$$

Hence from (9.4) we have

$$\ln(I_2 + A) = T \, \text{diag}\left[\ln\left(1 + \frac{1}{4}\right), \quad \ln\left(1 + \frac{1}{2}\right)\right]T^{-1},$$

which becomes, on using (9.6) and (9.7),

$$\ln\begin{bmatrix} 1 & 1 \\ -\frac{1}{8} & \frac{7}{4} \end{bmatrix} = \begin{bmatrix} \ln\frac{25}{24} & 4\ln\frac{6}{5} \\ \frac{1}{2}\ln\frac{5}{6} & \ln\frac{9}{5} \end{bmatrix}. \qquad (9.8)$$

If the scalar series (9.1) converges for all finite values of λ, then the matrix series (9.2) converges for *all* $n \times n$ matrices having finite elements. For example,

$$\sin A = A - \frac{1}{3!} A^3 + \frac{1}{5!} A^5 - \cdots \qquad (9.9)$$

is valid for all such matrices A.

Exercise 9.1 Use (9.2) to prove: (a) if A is symmetric then so is $f(A)$; (b) if A is triangular then so is $f(A)$ (see Problem 2.8); (c) if u_i is an eigenvector of A corresponding to λ_i, then $f(\lambda_i)$ and u_i are the corresponding eigenvalue and eigenvector of $f(A)$.

Exercise 9.2 If

$$A = \begin{bmatrix} 0 & 2 & 1 \\ 0 & 0 & 3 \\ 0 & 0 & 0 \end{bmatrix} \qquad (9.10)$$

verify that $A^3 = 0$ and hence calculate $\cos A$ directly from (9.2).

Exercise 9.3 Use (9.4) to calculate $\sin A$ for the matrix in (9.6).

We now turn to the situation when A is not similar to a diagonal matrix, so that (9.3) does not hold. Instead, suppose $A = T \mathcal{J} T^{-1}$ where $\mathcal{J}$ is the Jordan form (8.4) of A, so that (9.3) is replaced by

$$f(A) = T(f_0 I + f_1 \mathcal{J} + f_2 \mathcal{J}^2 + \cdots) T^{-1}. \qquad (9.11)$$

Since $\mathcal{J}$ is block diagonal, the matrix within brackets in (9.11) is also block diagonal (see Problem 9.1), so to investigate convergence we only need to consider a typical block J_i in $\mathcal{J}$ in the form (8.2) associated with an eigenvalue λ_i. For convenience take J_i to be $k \times k$, and consider the series

$$P_i = f_0 I_k + f_1 J_i + f_2 J_i^2 + \cdots. \qquad (9.12)$$

From Exercise 8.3 we see that each term in (9.12) is upper triangular, so P_i is also upper triangular. The diagonal elements of P_i are

$$f_0 + f_1 \lambda_i + f_2 \lambda_i^2 + \cdots, \qquad i = 1, 2, 3, \ldots,$$

which as before converges to $f(\lambda_i)$ provided $|\lambda_i| < R$. Using the form of J_i^r in Exercise 8.3, (9.12) shows that the elements of P_i

along the superdiagonal are

$$f_1 + 2\lambda_i f_2 + 3\lambda_i^2 f_3 + \cdots + r\lambda_i^{r-1} f_r + \cdots,$$

which converges to the derivative $f^{(1)}(\lambda_i)$, again provided $|\lambda_i| < R$. Continuing this argument, we end up with

$$P_i = \begin{bmatrix} f(\lambda_i) & \dfrac{1}{1!}f^{(1)}(\lambda_i) & \cdots & \dfrac{1}{(k-1)!}f^{(k-1)}(\lambda_i) \\ 0 & f(\lambda_i) & \cdots & \dfrac{1}{(k-2)!}f^{(k-2)}(\lambda_i) \\ & 0 & & \vdots \\ & & & f(\lambda_i) \end{bmatrix}, \quad (9.13)$$

where $f^{(p)}(\lambda)$ denotes the pth derivative. We can therefore define the matrix (9.13) to be $f(J_i)$.

Exercise 9.4 Use the result in Exercise 8.3 to verify (9.13).

We therefore conclude that provided all the eigenvalues of A satisfy $|\lambda_i| < R$, then the power series definition (9.2) gives

$$f(A) = T \text{ diag}[f(J_1), f(J_2), \ldots, f(J_q)]T^{-1}, \quad (9.14)$$

where $J_1, \ldots, J_q$ are the blocks in the Jordan form (8.4) of A.

An interesting alternative approach to establishing the convergence of (9.2) in the general case begins with the Schur canonical form $S = [s_{ij}]$ of Section 8.3. For *any* matrix A we have $A = TST^{-1}$, $A^k = TS^kT^{-1}$ where S is upper triangular, so (9.3) is now replaced by

$$T(f_0 I + f_1 S + f_2 S^2 + \cdots)T^{-1}. \quad (9.15)$$

The diagonal elements of S are the eigenvalues $\lambda_1, \ldots, \lambda_n$ of A and these may not all be distinct, but we can always choose a set of real positive numbers $\mu_1, \ldots, \mu_n$ which *are* distinct and satisfy

$$|\lambda_i| \leq \mu_i \leq r, \quad i = 1, 2, \ldots, n,$$

where $r < R$. Let $X = [x_{ij}]$ be any $n \times n$ upper triangular matrix such that $x_{ij} \geq |s_{ij}|$, for all i, j, with $x_{ii} = \mu_i$, $i = 1, 2, \ldots, n$ (X is said to *majorize* S, and we write $X \gg S$; notice that X is *non-negative*, i.e. $x_{ij} \geq 0$ for all $i, j = 1, 2, \ldots, n$). Since X has the *distinct* eigenvalues $\mu_1, \ldots, \mu_n$ it can be diagonalized as in

(6.74), i.e.

$$T_1^{-1} X T_1 = \text{diag}[\mu_1, \ldots, \mu_n] = M. \qquad (9.16)$$

Exercise 9.5 Show that $X^k \gg S^k$ for any positive integer k.

Exercise 9.6 A *norm* of an $n \times n$ matrix A can be defined by

$$\|A\| = n \max |a_{ij}|, \qquad i, j = 1, \ldots, n$$

(see Section 14.3.2 and Problem 14.11). Prove

(a) if $B \gg A$ then $\|B\| \geq \|A\|$;
(b) $\|A_1 A_2 A_3\| \leq \|A_1\| \, \|A_2\| \, \|A_3\|$ for any $n \times n$ matrices A_1, A_2, A_3;
(c) $\|M^k\| \leq nr^k$, where M is the diagonal matrix defined in (9.16).

Exercise 9.7 Show that the series $\sum\limits_{k=0}^{\infty} A_k$ converges if the series $\sum\limits_{k=0}^{\infty} \|A_k\|$ converges, for any $n \times n$ matrices $A_1, A_2, A_3, \ldots$ (you can assume the well known scalar result that $\sum\limits_{0}^{\infty} a_k$ converges if $\sum\limits_{0}^{\infty} |a_k|$ converges).

In order to establish the convergence of (9.2) using this approach, we only need to show that the series within brackets in (9.15) converges. From (9.16) $X = T_1 M T_1^{-1}$, and by Exercise 9.5 we have

$$S^k \ll T_1 M^k T_1^{-1}.$$

Hence, using Exercise 9.6(a) and (b), we obtain

$$\|S^k\| \leq \|T_1 M^k T_1^{-1}\|$$
$$\leq \|T_1\| \, \|M^k\| \, \|T_1^{-1}\|. \qquad (9.17)$$

Furthermore, the norms of both T_1 and T_1^{-1} are independent of k, and using Exercise 9.6(c) reduces (9.17) to

$$\|S^k\| \leq \alpha r^k,$$

where α is a constant, independent of k. Hence

$$\sum_{k=0}^{\infty} f_k \, \|S^k\| \leq \alpha \sum_{k=0}^{\infty} f_k r^k \qquad (9.18)$$

and, since by definition $r < R$, the series on the right in (9.18) is convergent. Therefore, so is the series on the left in (9.18), and the desired convergence of $\sum\limits_{0}^{\infty} f_k S^k$ then follows from Exercise 9.7.

Exercise 9.8 Use the preceding method of proof to show directly that if all the eigenvalues of A have modulus less than 1 then $A^k \to 0$ as $k \to \infty$.

In addition, use the Schur canonical form to prove the converse: a necessary condition for $A^k \to 0$ as $k \to \infty$ is that $|\lambda_i| < 1$ for all i.

It is important to realize that just because a matrix function is defined by an infinite series like that for a scalar function, it does *not* follow that properties of the scalar function still apply. For example, the matrix A in (9.10) when substituted into (9.9) gives $\sin A = A$ (since $A^3 = A^5 = \cdots = 0$), but for scalar functions $\sin z = z$ only when $z = 0$.

To illustrate this point further, consider the important *exponential matrix* (or *matrix exponential*) $\exp A$ or e^A, defined by

$$e^A = I_n + A + \frac{1}{2!}A^2 + \frac{1}{3!}A^3 + \cdots, \tag{9.19}$$

which converges for all $n \times n$ matrices A having finite elements, since the series for e^λ converges for all finite λ. If B is a second $n \times n$ matrix then

$$e^{A+B} = I + (A + B) + \frac{1}{2!}(A + B)^2 + \cdots$$

$$= I + (A + B) + \frac{1}{2!}(A^2 + AB + BA + B^2) + \cdots \tag{9.20}$$

and

$$(e^A)(e^B) = \left[I + A + \frac{1}{2!}A^2 + \cdots\right]\left[I + B + \frac{1}{2!}B^2 + \cdots\right]$$

$$= I + A + B + \frac{1}{2!}(A^2 + 2AB + B^2) + \cdots. \tag{9.21}$$

Comparison of (9.20) and (9.21) reveals that the relation

$$e^{A+B} = e^A e^B \tag{9.22}$$

holds *only if* $AB = BA$. In general, formulae based on scalar identities and involving two or more matrices hold only if all the matrices commute with each other. For example

$$\sin(A + B) = \sin A \cos B + \cos A \sin B \tag{9.23}$$

only if A and B commute with each other. Formulae involving only a single matrix usually do carry over from the scalar case; for example, for any finite A

$$\sin 2A = 2 \sin A \cos A.$$

Exercise 9.9 If

$$A = \begin{bmatrix} 1 & 1 \\ 0 & 0 \end{bmatrix}, \qquad B = \begin{bmatrix} 1 & -1 \\ 0 & 0 \end{bmatrix}$$

show that $A^2 = A$ and hence obtain e^A using (9.19). Similarly, obtain e^B. Determine also e^{A+B}, $e^A e^B$, and $e^B e^A$ (notice that they are all different from each other).

Exercise 9.10 Using (9.20) with $B = -A$, prove that e^A is non-singular for any $n \times n$ matrix A. Hence deduce that

$$(e^A)^{-1} = I - A + \frac{1}{2} A^2 - \cdots = e^{-A}.$$

Exercise 9.11 Use (9.19) to prove that $(\exp A)^* = \exp(A^*)$. Hence show that if A is skew symmetric then e^A is orthogonal, and if A is skew hermitian then e^A is unitary.

Students are often disappointed that even simple results like (9.22) do not hold for matrix functions. However, as on previous occasions in this book, it is possible to use Kronecker products to overcome some of the snags arising from non-commutativity. To begin this development, we first establish using the basic definition (9.2) that if A is $n \times n$, then

$$\begin{aligned}
f(A \otimes I_m) &= f_0 I_{nm} + f_1 (A \otimes I_m) + f_2 (A \otimes I_m)^2 + \cdots \\
&= f_0 I_n \otimes I_m + f_1 (A \otimes I_m) + f_2 (A^2 \otimes I_m) + \cdots \\
&= (f_0 I_n + f_1 A + f_2 A^2 + \cdots) \otimes I_m \\
&= f(A) \otimes I_m
\end{aligned} \qquad (9.24)$$

(notice that we have used $(A \otimes I)^2 = A^2 \otimes I$, etc., easily proved using (2.76)). Similarly, it is left as a simple task for the reader to verify that if B is $m \times m$ then

$$f(I_n \otimes B) = I_n \otimes f(B). \qquad (9.25)$$

Now consider the matrix D introduced in (5.53) and studied further in Section 6.3.5, namely

$$D = A \otimes I_m + I_n \otimes B \qquad (9.26)$$

(the transpose on B has been dropped here for convenience, without affecting the argument; D is often called the *Kronecker sum* of A and B). From the rule in (2.76) we have

$$(A \otimes I_m)(I_n \otimes B) = (AI_n) \otimes (I_m B) = (I_n \otimes B)(A \otimes I_m),$$

showing that $A \otimes I_m$ and $I_n \otimes B$ commute with each other. Thus scalar formulae do carry over for these matrices. For example,

$$\sin D = \sin(A \otimes I_m + I_n \otimes B)$$
$$= \sin(A \otimes I_m)\cos(I_n \otimes B) + \cos(A \otimes I_m)\sin(I_n \otimes B) \quad (9.27)$$

and, using (9.24) and (9.25), the expression (9.27) can be simplified as follows:

$$\sin D = [(\sin A) \otimes I_m][I_n \otimes (\cos B)] + [(\cos A) \otimes I_m][I_n \otimes (\sin B)]$$
$$= (\sin A) \otimes (\cos B) + (\cos A) \otimes (\sin B) \quad (9.28)$$

again using (2.76). The identity (9.28) can be regarded as a generalization of the usual scalar result for $\sin(a + b)$, obtained by setting $n = m = 1$.

Exercise 9.12 For the matrix D in (9.26) prove that

$$e^D = e^A \otimes e^B.$$

9.2 Sylvester's formula

A disadvantage of using (9.4) or (9.11) to evaluate $f(A)$ is the need to calculate the eigenvectors of A in order to obtain T. We now show how this can be avoided. As in (6.47), we can divide $f(\lambda)$ by the characteristic polynomial $k(\lambda)$ of A to give

$$f(\lambda) = q(\lambda)k(\lambda) + \alpha(\lambda), \quad (9.29)$$

where the remainder polynomial $\alpha(\lambda)$ has degree at most $n - 1$. Since by definition $k(\lambda_r) = 0$ for each eigenvalue λ_r, substitution of $\lambda = \lambda_r$ into (9.29) gives

$$f(\lambda_r) = \alpha(\lambda_r)$$
$$= \alpha_0 + \alpha_1 \lambda_r + \alpha_2 \lambda_r^2 + \cdots + \alpha_{n-1} \lambda_r^{n-1} \quad (9.30)$$

for $r = 1, 2, \ldots, n$. Consider first the case when all the eigenvalues of A are distinct. Equation (9.30) then represents n equations for the n unknown constants $\alpha_0, \ldots, \alpha_{n-1}$, and these equations can be written as

$$[\alpha_0, \alpha_1, \ldots, \alpha_{n-1}]V_n = [f(\lambda_1), \ldots, f(\lambda_n)],$$

where V_n is the Vandermonde matrix defined in (4.80). In fact (9.30) can be regarded as determining the $(n-1)$th degree polynomial $\alpha(\lambda)$ which takes the value $f(\lambda)$ when $\lambda = \lambda_r$, for each value of r (see Problem 4.17). This is a standard problem of interpolation, and the desired (unique) polynomial is given by Lagrange's formula (Barnett and Cronin 1986, Formula (9.5.6)):

$$\alpha(\lambda) = \sum_{r=1}^{n} f(\lambda_r)z_r, \qquad z_r = \prod_{\substack{i=1 \\ i \neq r}}^{n} [(\lambda - \lambda_i)/(\lambda_r - \lambda_i)]. \quad (9.31)$$

To obtain $f(A)$ we use the fact that by the Cayley–Hamilton theorem (6.37), $k(A) \equiv 0$, so on replacing λ by A in the identity (9.29) we obtain simply $f(A) \equiv \alpha(A)$. Hence, replacing λ by A in (9.31) gives *Sylvester's formula* (also known as the *Lagrange–Sylvester* formula)

$$f(A) = \sum_{r=1}^{n} f(\lambda_r)Z_r, \quad (9.32)$$

where the Z_r are now $n \times n$ matrices given by

$$Z_r = \prod_{\substack{i=1 \\ i \neq r}}^{n} [(A - \lambda_i I)/(\lambda_r - \lambda_i)]. \quad (9.33)$$

Notice that (9.32) and (9.33) require a knowledge only of the eigenvalues of A, and that the matrices Z_r in (9.33) are determined entirely by A, i.e. the Z_r are independent of the function f.

Example 9.2 Let A be a 3×3 matrix having eigenvalues $\lambda_1 = 1$, $\lambda_2 = 2$, $\lambda_3 = -1$. Applying (9.33) with $n = 3$ produces

$$Z_1 = (A - \lambda_2 I)(A - \lambda_3 I)/(\lambda_1 - \lambda_2)(\lambda_1 - \lambda_3)$$

$$= -\frac{1}{2}(A - 2I)(A + I)$$

$$Z_2 = (A - \lambda_1 I)(A - \lambda_3 I)/(\lambda_2 - \lambda_1)(\lambda_2 - \lambda_3)$$

$$= \frac{1}{3}(A - I)(A + I)$$

$$Z_3 = (A - \lambda_1 I)(A - \lambda_2 I)/(\lambda_3 - \lambda_1)(\lambda_3 - \lambda_2)$$

$$= \frac{1}{6}(A - I)(A - 2I).$$

Then for *any* function $f(\lambda)$ for which the convergence condition is satisfied we have in this case

$$f(A) = f(1)Z_1 + f(2)Z_2 + f(-1)Z_3$$

with the matrices Z_r given above.

Example 9.3 Return to the matrix A in (9.6). We have

$$Z_1 = (A - \lambda_2 I)/(\lambda_1 - \lambda_2) = \left(A - \frac{1}{2}I\right)\Big/\left(\frac{1}{4} - \frac{1}{2}\right)$$

$$= \begin{bmatrix} 2 & -4 \\ \frac{1}{2} & -1 \end{bmatrix}$$

$$Z_2 = (A - \lambda_1 I)/(\lambda_2 - \lambda_1) = \left(A - \frac{1}{4}I\right)\Big/\left(\frac{1}{2} - \frac{1}{4}\right)$$

$$= \begin{bmatrix} -1 & 4 \\ -\frac{1}{2} & 2 \end{bmatrix}.$$

Again, if $f(A)$ exists, then $f(A) = f(\frac{1}{4})Z_1 + f(\frac{1}{2})Z_2$. For example, if $f(\lambda) = \ln(1 + \lambda)$, then

$$\ln(I_2 + A) = \left(\ln\frac{5}{4}\right)Z_1 + \left(\ln\frac{3}{2}\right)Z_2 \tag{9.34}$$

and on substituting for Z_1 and Z_2 it can be confirmed that (9.34) agrees with (9.8).

Notice that it is easily proved (see Problem 9.5) that the matrices Z_r in (9.33) satisfy

$$\sum_{r=1}^{n} Z_r = I_n, \qquad Z_r Z_s = 0 \quad (r \neq s) \tag{9.35}$$

and these provide a simple numerical check on the calculations. The reader can easily confirm that in the preceding example $Z_1 + Z_2 = I_2$, $Z_1 Z_2 = 0$.

Exercise 9.13 Verify that (9.35) holds for the matrices Z_1, Z_2, Z_3 in Example 9.2.

Exercise 9.14 Repeat Exercise 9.3 using Z_1 and Z_2 in Example 9.3.

Exercise 9.15   If

$$A = \begin{bmatrix} 3 & 2 \\ 2 & 3 \end{bmatrix}$$

use Sylvester's formula to determine $\sin A$. Similarly, write down an expression for A^{100}.

Exercise 9.16   For the matrix A in Exercise 6.8(b) calculate e^A: (a) using Sylvester's formula; (b) using (9.4) and the similarity transformation obtained in the solution of Exercise 6.28.

We now turn to the case when A has repeated eigenvalues. Suppose that the minimum polynomial of A (defined in Section 6.3.3) is

$$m(\lambda) = (\lambda - \lambda_1)^{j_1}(\lambda - \lambda_2)^{j_2} \cdots (\lambda - \lambda_s)^{j_s}, \qquad (9.36)$$

where $\lambda_1, \lambda_2, \ldots, \lambda_s$ are the distinct eigenvalues of A, and $p = j_1 + j_2 + \cdots + j_s$ is the degree of $m(\lambda)$. The p numbers

$$f(\lambda_i), \quad f^{(1)}(\lambda_i), \ldots, f^{(j_i-1)}(\lambda_i), \qquad i = 1, 2, \ldots, s \quad (9.37)$$

(where as before the superscripts denote the derivatives of $f(\lambda)$ with respect to λ) are called the *values of $f(\lambda)$ on the spectrum of A*.

Exercise 9.17   Show that the values of $m(\lambda)$ on the spectrum of A are all zero.

Exercise 9.18   If $q(\lambda)$ is any polynomial which *annihilates* A, i.e. $q(A) \equiv 0$, and $m(\lambda)$ is the minimum polynomial of A, show that $m(\lambda)$ is a factor of $q(\lambda)$.

In particular, it follows from the Cayley–Hamilton theorem (Section 6.3.3) that $m(\lambda)$ is a factor of the characteristic polynomial of A.

Exercise 9.19   Let $f_1(\lambda)$ be a polynomial such that $f_1(A) = f(A)$. By considering $f_2(\lambda) = f_1(\lambda) - f(\lambda)$, and using Exercises 9.17 and 9.18, show that $f_1(\lambda)$ and $f(\lambda)$ have the same values on the spectrum of A.

The converse of the result in Exercise 9.19 also holds, showing that two polynomials give rise to the *same* matrix on substituting A for λ if and only if the polynomials have the same values on the spectrum of A. This can be used to derive the following general version of Sylvester's formula:

$$f(A) = \sum_{r=1}^{s} [f(\lambda_r)Z_{r1} + f^{(1)}(\lambda_r)Z_{r2} + \cdots + f^{(j_r-1)}(\lambda_r)Z_{rj_r}], \quad (9.38)$$

where the j_i are the indices in (9.36). The *component* (or *constituent*) matrices Z_{rl} are all determined entirely by A, and are independent of the function f. Two particular cases are worthy of special mention. Firstly, if all the eigenvalues of A are distinct, so that $s = n$ in (9.36), then (9.38) reduces to the original Sylvester formula (9.32) with $Z_{r1} = Z_r$. Secondly, if the minimum polynomial has only simple roots, so that $j_1 = j_2 = \cdots = j_s = 1$ in (9.36), then the formula (9.38) simplifies to

$$f(A) = \sum_{r=1}^{s} f(\lambda_r)Z_r,$$

where the $Z_r(= Z_{r1})$ are defined exactly as in (9.33) except that the upper index on the product is s instead of n. In general, an explicit expression can be derived for the matrices Z_{rl} in (9.38), but it is preferable to find them in simple problems by selecting appropriate expressions for the function $f(\lambda)$. The procedure is best illustrated by an example.

Example 9.4 We wish to determine the component matrices for

$$A = \begin{bmatrix} 2 & -1 & 1 \\ 1 & 0 & 2 \\ 1 & -1 & 3 \end{bmatrix}. \tag{9.39}$$

First recall from Section 8.2 that the minimum polynomial of A is equal to the similarity invariant of A of highest degree. To determine this we compute $d_3(\lambda)$, the determinant of

$$\lambda I - A = \begin{bmatrix} \lambda - 2 & 1 & -1 \\ -1 & \lambda & -2 \\ -1 & 1 & \lambda - 3 \end{bmatrix},$$

to obtain

$$d_3(\lambda) = (\lambda - 1)(\lambda - 2)^2.$$

The greatest common divisor $d_2(\lambda)$ of all minors of order 2 of $\lambda I - A$ is obviously 1. Hence by (8.17) the required similarity invariant is $i_3(\lambda) = d_3(\lambda)/d_2(\lambda)$, showing that $m(\lambda) = d_3(\lambda)$. The distinct eigenvalues of A are $\lambda_1 = 1$, $\lambda_2 = 2$, and in (9.36) we have $j_1 = 1$, $j_2 = 2$, so (9.38) gives

$$f(A) = f(1)Z_{11} + f(2)Z_{21} + f^{(1)}(2)Z_{22}. \tag{9.40}$$

Since (9.40) holds for *any* function $f(\lambda)$ for which $f(A)$ is defined, we make three simple choices for $f(\lambda)$ which produce three easily solved equations for the component matrices. Firstly, set $f(\lambda) = 1$ in (9.40) to give

$$I = Z_{11} + Z_{21}. \tag{9.41}$$

A natural second choice would be $f(\lambda) = \lambda$, but this would produce three terms arising from (9.40); a better choice is $f(\lambda) = \lambda - 2$, which makes $f(2) = 0$, $f^{(1)}(2) = 1$, and (9.40) then gives

$$A - 2I = -Z_{11} + Z_{22}. \tag{9.42}$$

Similarly, $f(\lambda) = (\lambda - 2)^2$ is preferable to λ^2, since $f(2) = 0$ and $f^{(1)}(2) = 0$, so (9.40) becomes

$$(A - 2I)^2 = Z_{11}. \tag{9.43}$$

On substituting for A from (9.39), we therefore obtain from (9.43)

$$Z_{11} = \begin{bmatrix} 0 & 1 & -1 \\ 0 & 1 & -1 \\ 0 & 0 & 0 \end{bmatrix},$$

so from (9.41)

$$Z_{21} = I - Z_{11} = \begin{bmatrix} 1 & -1 & 1 \\ 0 & 0 & 1 \\ 0 & 0 & 1 \end{bmatrix},$$

and finally from (9.42)

$$Z_{22} = A - 2I + Z_{11} = \begin{bmatrix} 0 & 0 & 0 \\ 1 & -1 & 1 \\ 1 & -1 & 1 \end{bmatrix}.$$

Exercise 9.20 Write down e^{At} for the matrix in (9.39).

Exercise 9.21 For the matrix A in Exercise 8.7 determine A^{50} (a) using (9.38); (b) using the transformation to Jordan form obtained previously. (Hint: write J as $I + X$ and expand $(I + X)^{50}$ by the binomial theorem.)

Exercise 9.22 Determine the minimum polynomial of

$$A = \begin{bmatrix} 2 & 0 & 0 \\ -11 & 0 & 1 \\ -22 & -4 & 4 \end{bmatrix}.$$

Use (9.38) to obtain the expression for $\sin(At)$.

Numerical procedures for evaluating matrix functions are discussed by Golub and Van Loan (1989).

9.3 Linear differential and difference equations

We return to the systems of equations studied in Section 6.5, and show how they can be solved using matrix functions. First consider the differential equations (6.91), i.e.

$$\dot{x} = Ax, \tag{9.44}$$

where A is a constant $n \times n$ matrix. If a is a scalar, then the solution of $\dot{x} = ax$ is $x(t) = e^{at}x(0)$. This suggests that we use

$$e^{At} = I_n + tA + \frac{t^2}{2!}A^2 + \frac{t^3}{3!}A^3 + \cdots. \tag{9.45}$$

From our earlier discussion, it follows that the series (9.45) converges for all finite values of t. Differentiating (9.45) gives

$$\frac{d}{dt}(e^{At}) = 0 + A + \frac{2t}{2!}A^2 + \frac{3t^2}{3!}A^3 + \cdots$$

$$= A\left(I + tA + \frac{t^2}{2!}A^2 + \cdots\right) = Ae^{At},$$

which is the same result as for scalar exponentials. It follows by analogy with the scalar case that the solution of (9.44) can be written

$$x(t) = e^{At}x_0, \tag{9.46}$$

where $x_0 = x(0)$.

Exercise 9.23 Prove that $\exp(At_1)\exp(At_2) = \exp[A(t_1 + t_2)]$ for any finite values of t_1 and t_2.

Exercise 9.24 Determine e^{At} for the matrix A in (9.10).

An advantage of using (9.46) instead of (6.96) is that there is no need to calculate eigenvectors of A, since e^{At} can be obtained using Sylvester's formula.

Example 5.9 Return to the problem in Example 6.13 where A is the matrix in (6.5) and $\lambda_1 = -1$, $\lambda_2 = 4$. A direct application of (9.32) with $n = 2$ produces

$$Z_1 = -\frac{1}{5}\begin{bmatrix} -3 & 3 \\ 2 & -2 \end{bmatrix}, \quad Z_2 = \frac{1}{5}\begin{bmatrix} 2 & 3 \\ 2 & 3 \end{bmatrix}. \tag{9.47}$$

From (9.32) with $f(\lambda) = e^{\lambda t}$, we have $e^{At} = e^{-t}Z_1 + e^{4t}Z_2$. Then (9.46) gives, with $x_0 = [x_{10}, x_{20}]^T$, the solution of (9.44) in this case:

$$x(t) = \frac{1}{5}\begin{bmatrix} (2e^{4t} + 3e^{-t}) & 3(e^{4t} - e^{-t}) \\ 2(e^{4t} - e^{-t}) & (3e^{4t} + 2e^{-t}) \end{bmatrix}\begin{bmatrix} x_{10} \\ x_{20} \end{bmatrix},$$

It is readily confirmed that this agrees with the solution previously obtained in (6.98), with appropriate expressions for the constants α_1 and α_2 in terms of x_{10} and x_{20}.

Exercise 9.25 Solve (9.44) using (9.46) when A is the matrix in Exercise 6.8(b) (compare your answer with that for Exercise 6.36).

The question of *stability* of the equations (9.44) is an important one for applications. Roughly speaking, the system (9.44) is said to be *asymptotically stable* if the effect of any initial perturbation x_0 from equilibrium eventually dies away, i.e. $x(t) \to 0$ as $t \to \infty$. Using the solution (9.46) together with Sylvester's formula (9.38) for e^{At}, it can be shown that the necessary and sufficient condition for asymptotic stability is that all the eigenvalues of A have negative real parts. This fact was noted in Problem 6.23, and investigated further in Section 7.5.4. A more detailed treatment will be given in Chapter 11 (Sections 11.4.1, 11.5.1).

It is interesting to consider the Laplace transformation of (9.44). As in Problem 4.8, this gives

$$s\bar{x} - x_0 = A\bar{x},$$

where $\bar{x} = \mathcal{L}\{x(t)\}$, and on rearrangement this becomes

$$\bar{x} = (sI_n - A)^{-1}x_0. \tag{9.48}$$

However, from (9.46)

$$\bar{x} = \mathcal{L}\{e^{At}x_0\} \tag{9.49}$$

and since x_0 is an arbitrary constant vector, it follows by comparing (9.48) and (9.49) that

$$\mathcal{L}\{e^{At}\} = (sI - A)^{-1}, \tag{9.50}$$

which is a generalization of the well-known result for scalars: $\mathcal{L}\{e^{at}\} = 1/(s - a)$ (Barnett and Cronin 1986, Formula (6.4.7)).

Turning to the linear difference equations in (6.99), i.e.

$$X(k + 1) = AX(k), \qquad k = 0, 1, 2, \ldots, \tag{9.51}$$

the solution can in this case be written

$$X(k) = A^k X(0). \tag{9.52}$$

This is readily verified, since (9.52) gives $X(k + 1) = A^{k+1}X(0) = AA^kX(0) = AX(k)$ as required. Evaluation of A^k can again be carried out using Sylvester's formula by setting $f(\lambda) = \lambda^k$, thus avoiding the calculation of eigenvectors of A needed in (6.101).

Exercise 9.26 Solve the difference equations in Example 6.14 using (9.52) and Sylvester's formula (Z_1 and Z_2 are given in (9.47)).

In the case of (9.51), the necessary and sufficient condition for asymptotic stability is that all the eigenvalues of A have modulus less than one (see Exercise 9.8). Again, this will be discussed in more detail in Chapter 11 (Sections 11.4.2, 11.5.1).

9.4 Matrix sign function

We assume in this section that A has no purely imaginary or zero eigenvalues. Let $\mathscr{J}$ be the Jordan form of A, so that

$$A = T\mathscr{J}T^{-1} \tag{9.53}$$

where T is non-singular. We now define a matrix version of the scalar *sign function*:

$$\mathrm{sgn}(x) = \begin{cases} +1, & \mathrm{Re}(x) > 0 \\ -1, & \mathrm{Re}(x) < 0, \end{cases}$$

where x is an arbitrary complex number ($\mathrm{sgn}(x)$ is undefined if x is purely imaginary or zero). The *matrix sign function* $\mathrm{sgn}(A)$ is defined by

$$\mathrm{sgn}(A) = T\,\mathrm{sgn}(\mathscr{J})T^{-1}, \tag{9.54}$$

where

$$\mathrm{sgn}(\mathscr{J}) = \mathrm{diag}[\mathrm{sgn}(\lambda_1), \mathrm{sgn}(\lambda_2), \ldots, \mathrm{sgn}(\lambda_n)] \tag{9.55}$$

and the λ_i are the eigenvalues of A, not necessarily all distinct.

Example 9.6 Return to the matrix A in (6.5), which was found in Example 6.1 to have eigenvalues $-1, 4$. The matrix T such that

$$A = T\,\mathrm{diag}[-1, 4]T^{-1}$$

is given in (6.75). Since the eigenvalues are both real (9.55) gives

$$\text{sgn}(\mathcal{J}) = \text{diag}[-1, 1]$$

and from (9.54)

$$\text{sgn}(A) = T \begin{bmatrix} -1 & 0 \\ 0 & 1 \end{bmatrix} T^{-1}$$

$$= \frac{1}{5} \begin{bmatrix} -1 & 6 \\ 4 & 1 \end{bmatrix}. \tag{9.56}$$

Exercise 9.27 Show that when all the eigenvalues of A have either positive or negative real parts, then $\text{sgn}(A) = I$ or $-I$, respectively.

Exercise 9.28 If A has a diagonal Jordan form and all its eigenvalues are ± 1, show that $\text{sgn}(A) = A$.

Exercise 9.29 Prove that $[\text{sgn}(A)]^2 = I$.

Exercise 9.30 Determine $\text{sgn}(A)$ when A is the matrix in (6.87).

The definition (9.54) is impractical as a means of computing the matrix sign function, since it requires a knowledge of the real parts of the eigenvalues and the transformation matrix T. A much better computational procedure is based on the Newton–Raphson method for determining the roots of scalar equations (see Section 6.7.2). To develop this, we first consider the matrix

$$B_1 = (A - S)(A + S)^{-1}, \tag{9.57}$$

where for convenience we have written S to stand for $\text{sgn}(A)$. Substituting for A and S from (9.53) and (9.54) respectively produces

$$B_1 = T(\mathcal{J} - S_J)(\mathcal{J} + S_J)^{-1} T^{-1} \tag{9.58}$$

where S_J stands for $\text{sgn}(\mathcal{J})$. Now $\mathcal{J}$ is upper triangular and S_J is diagonal, so both $\mathcal{J} - S_J$ and $\mathcal{J} + S_J$ are upper triangular. Furthermore (see Section 4.4) the inverse of $\mathcal{J} + S_J$ is upper triangular, and therefore so is the product $(\mathcal{J} - S_J)(\mathcal{J} + S_J)^{-1}$. It is easy to check that the diagonal elements in this upper triangular product are

$$\mu_j = \frac{\lambda_j - \text{sgn}(\lambda_j)}{\lambda_j + \text{sgn}(\lambda_j)}, \qquad j = 1, 2, \ldots, n \tag{9.59}$$

and these are the eigenvalues of $(\mathcal{J} - S_J)(\mathcal{J} + S_J)^{-1}$. Because of the similarity relationship (9.58), the μ_j in (9.59) are also the eigenvalues of B_1.

Exercise 9.31 Prove that the numbers μ_j defined by (9.59) satisfy $|\mu_j| < 1$.

We now extend (9.57) to define a sequence of matrices

$$B_r = (A_r - S)(A_r + S)^{-1} \qquad r = 1, 2, 3, \ldots, \qquad (9.60)$$

where $A_1 = A$ and

$$A_{r+1} = \frac{1}{2}(A_r + A_r^{-1}), \qquad r = 1, 2, 3, \ldots. \qquad (9.61)$$

Our objective is to show that as $r \to \infty$ then $A_r \to S$, the required sign matrix. This relies on the following two exercises which the reader is asked to solve.

Exercise 9.32 Prove that $AS = SA$, and hence show that $A_r S = SA_r$, $r \geqslant 1$.

Exercise 9.33 Use (9.60) and (9.61) to show that

$$B_{r+1} = (A_r^2 + I - 2SA_r)(A_r^2 + I + 2SA_r)^{-1}.$$

Hence use the results in Exercises 9.29 and 9.32 to show that $B_{r+1} = B_r^2$.

It follows from Exercise 9.33 that

$$B_r = (B_1)^{2^{r-1}}, \qquad r \geqslant 2. \qquad (9.62)$$

Since (see Exercise 9.31) all the eigenvalues of B_1 have modulus less than 1, we have $B_r \to 0$ as $r \to \infty$ (see Exercise 9.8). Finally, it is left to the reader to rearrange (9.60) to give

$$A_r = (I - B_r)^{-1}(I + B_r)S,$$

showing that $A_r \to S$ as $r \to \infty$. In fact, (9.62) implies that the convergence is quite rapid.

Exercise 9.34 By considering its eigenvalues, show that $I - B_r$ is non-singular for all r.

When A is a (complex) scalar, then (9.61) is equivalent to the well-known Newton–Raphson iterative method for obtaining the roots of scalar equations (see Problem 9.17).

Exercise 9.35 Repeat Example 9.6 using (9.61), going as far as $r = 3$.

The matrix sign function will be applied in Section 11.5.4 to the solution of certain matrix equations. However, it is interesting here to consider the rank of the matrix

$$I + \text{sgn}(A) = I + T \,\text{sgn}(\mathscr{J})T^{-1}, \quad \text{by (9.54)}$$
$$= T[I + \text{sgn}(\mathscr{J})]T^{-1}.$$

Exercise 9.36 Prove that

$$\text{rank}[I + \text{sgn}(\mathscr{J})] = p,$$

where p is the number of eigenvalues of A with positive real parts.

Since T in (9.63) is non-singular, and since by assumption A has no eigenvalues on the imaginary axis in the complex plane, it follows from the result in Exercise 9.36 that A has p and $n - p$ eigenvalues with positive and negative real parts respectively, where p is the rank of $I + \text{sgn}(A)$, and $\text{sgn}(A)$ can be computed using (9.61).

Example 9.6 (continued) Using the expression for $\text{sgn}(A)$ in (9.56) gives

$$I + \text{sgn}(A) = \frac{1}{5}\begin{bmatrix} 4 & 6 \\ 4 & 6 \end{bmatrix},$$

which clearly has rank 1, confirming that A has one eigenvalue on either side of the imaginary axis.

Exercise 9.37 Using $\text{sgn}(A)$ determined in Exercise 9.30, confirm that A in (6.87) has two eigenvalues with positive real parts.

Problems

9.1 If A is a block diagonal matrix, i.e.

$$A = \text{diag}[A_1, A_2, \ldots, A_k],$$

where each A_i is a square matrix (e.g. see (2.64)) prove that $f(A)$ is also block diagonal, and

$$f(A) = \text{diag}[f(A_1), f(A_2), \ldots, f(A_k)].$$

9.2 Let A be a 4×4 matrix having eigenvalues $\pm k, 0, 0$. Use the Cayley–Hamilton theorem to show that $A^4 = k^2 A^2$. Hence, using the infinite series definition for $\cos A$, prove that $\cos A = I + (\cos k - 1)A^2/k^2$.

9.3 Prove using (9.4) that when A has a diagonal Jordan form then

$$\det f(A) = f(\lambda_1)f(\lambda_2)\cdots f(\lambda_n).$$

Hence show that $\det(\exp A) = \exp(\text{tr} A)$. Deduce that when A is real and skew symmetric then $\det(\exp A) = 1$.

These results still hold when the Jordan form of A is not diagonal.

9.4 Write the equation of simple harmonic motion

$$\ddot{z} + \omega^2 z = 0 \tag{9.64}$$

in the form (9.44) by taking $x_1 = z$, $x_2 = \dot{z}/\omega$. By calculating A^2 and using (9.45), show that

$$e^{At} = \begin{bmatrix} \cos \omega t & \sin \omega t \\ -\sin \omega t & \cos \omega t \end{bmatrix}$$

and hence determine the solution $z(t)$ of (9.64).

9.5 By making suitable choices for $f(\lambda)$ in (9.32) prove:

(a) $\sum\limits_{r=1}^{n} Z_r = I_n$;

(b) $\sum\limits_{r=1}^{n} \lambda_r^k Z_r = A^k$, $k = 1, 2, 3, \ldots$.

9.6 For the matrix A in (2.25) use the result of Exercise 2.9 to prove by induction that

$$A^k = (2^k - 1)A - (2^k - 2)I_3$$

for any positive integer k. Hence show that

$$e^A = (e^2 - e)A - (e^2 - 2e)I_3.$$

9.7 If A is a real $n \times n$ matrix prove that

$$e^{iA} = \cos A + i \sin A$$

and hence show that

$$\cos A = \frac{1}{2}(e^{iA} + e^{-iA}), \qquad \sin A = \frac{1}{2i}(e^{iA} - e^{-iA}).$$

9.8 The eigenvalues of an $n \times n$ matrix A are the nth roots of unity. Expand $(I - cA)^{-1}$ using the binomial theorem. What is the condition on the scalar c for this expression to be valid?

Hence show that in this case

$$(I - cA)^{-1} = \frac{1}{1 - c^n}(I + cA + c^2A^2 + \cdots + c^{n-1}A^{n-1}).$$

9.9 Let $m(\lambda)$ and $m_1(\lambda)$ be two polynomials of least possible degree which annihilate A. Prove that $m_1(\lambda)$ is a scalar multiple of $m(\lambda)$. This shows that the *monic* minimum polynomial of A is unique.

9.10 By making suitable choices for $f(\lambda)$ in (9.38) prove:

(a) $\sum\limits_{r=1}^{s} Z_{r1} = I_n$;   (b) $\sum\limits_{r=1}^{s} (\lambda_r Z_{r1} + Z_{r2}) = A$.

9.11 Consider again the control system described by $\dot{x} = Ax + bu$. Use the transformation $z(t) = e^{-At}x(t)$ and the product rule (2.70) to obtain a differential equation for $z(t)$. Hence show that the general solution of the original equation, subject to $x(0) = x_0$, can be written

$$x(t) = e^{At}\left[x_0 + \int_0^t e^{-A\tau}bu\, d\tau\right].$$

9.12 Verify that the solution of the differential equation

$$\frac{dX(t)}{dt} = AX(t) + X(t)B,$$

where A, X, B are all $n \times n$, can be written in the form

$$X(t) = e^{At}X(0)e^{Bt}.$$

9.13 The *z-transform* $\bar{X}(z)$ or $\mathcal{Z}\{X(k)\}$ of a discrete function $X(k)$, $k = 0, 1, 2, 3, \ldots$, is defined by

$$\bar{X}(z) = \sum_{k=0}^{\infty} X(k)z^{-k}.$$

(a) Show that

$$\mathcal{Z}\{a^k\} = \frac{z}{z - a},$$

where a is a constant.

(b) Show that

$$\mathcal{Z}\{X(k + 1)\} = z\bar{X}(z) - zX(0).$$

(c) By considering the z-transformation of (9.51), deduce using (9.52) that

$$\mathcal{Z}\{A^k\} = z(zI - A)^{-1},$$

which is the generalization of the result in part (a), and the analogue of (9.50).

9.14 Show that a square matrix A is unitary if and only if it can be expressed as e^{iR}, where R is hermitian. (Hint: If A is unitary, recall that it is normal and so is unitarily similar to a diagonal matrix.)

9.15 The *spectral radius* of an $n \times n$ (complex) matrix A having eigenvalues $\lambda_1, \ldots, \lambda_n$ is defined by

$$\rho(A) = \max_{1 \leqslant i \leqslant n} |\lambda_i|.$$

Let $= [b_{ij}]$ be an $n \times n$ non-negative matrix (i.e. all $b_{ij} \geq 0$).

(a) Prove that B^k is non-negative for all positive integers k.

(b) Suppose $B \gg A$ (i.e. $b_{ij} \geq |a_{ij}|$, all i and j) so that as in Exercise 9.5 $B^k \gg A^k$. Hence deduce that if $\rho(B) < 1$ then $A^k \to 0$ as $k \to \infty$.

This result can be used to show that if $B \gg A$ then $\rho(B) \gg \rho(A)$ (Mirsky 1963).

9.16 Determine e^{At} where A is the matrix in Exercise 6.8(b) by using (9.50) (compare with Exercise 9.16).

9.17 Consider the model of a beetle population in Problem 1.6 in the form $X_{n+1} = AX_n$, where $X_n = [x_n, y_n, z_n]^T$. Determine the characteristic polynomial of A, and hence deduce that $A^3 = I$. Show that this implies that the distribution of age-groups in the population goes through a three-year cycle.

9.18 Suppose that there is a different birth pattern for the beetle population model in the previous problem, namely that each beetle produces on average one offspring in the second year of life and three more in the third year of life. The other assumptions in Problem 1.6 remain unchanged. Obtain the new matrix A.

Suppose also that the initial population X_0 is equally distributed amongst the three age groups. By obtaining an expression for A^n using Sylvester's formula (9.32), deduce that as $n \to \infty$ the distribution amongst the age groups approaches the ratios $1:3:6$.

9.19 By applying the Newton–Raphson formula (see (6.132))

$$a_{r+1} = a_r - \frac{f(a_r)}{f'(a_r)}, \qquad r \geq 1,$$

to the scalar function $f(x) = x^2 - 1$ with $a_1 = a$ (where $\mathrm{Re}(a) \neq 0$), show that

$$a_{r+1} = \frac{1}{2}\left(a_r + \frac{1}{a_r}\right).$$

Prove that $a_r \to \mathrm{sgn}(a)$ as $r \to \infty$.

9.20 Prove that if A and B are similar so that $B = P^{-1}AP$ with P non-singular, then $\mathrm{sgn}(B) = P^{-1}\,\mathrm{sgn}(A)P$.

References

Barnett and Cronin (1986), Chen (1984), Gantmacher (1959, Vol. 1), Golub and Van Loan (1989), Mirsky (1963).

10 Generalized inverses

Consider again the set of linear equations

$$Ax = b \qquad (10.1)$$

in n unknowns. We saw in Chapter 4 that when A is $n \times n$ and has a non-zero determinant, then the unique solution of (10.1) is $x = A^{-1}b$, where A^{-1} is the inverse of A. However, in Chapter 5 we studied the important problem of solving (10.1) when A is singular or rectangular. We now investigate this topic further, and develop several different generalizations of the concept of inverse.

10.1 The Moore–Penrose inverse

10.1.1 Definition

It is assumed throughout that A has dimensions $m \times n$. We consider initially the case when A is real, and generalize the discussion given in Section 5.5 for determining the *least-squares* solution of a set of inconsistent linear equations. Suppose that when some vector x is substituted into the left-hand side of (10.1) it gives a vector $b' = [b'_1, b'_2, \ldots, b'_m]^\mathsf{T}$. The objective is to choose x so that the sum of the squares of the differences between the elements of b' and those of the required right-hand side $b = [b_1, \ldots, b_m]^\mathsf{T}$ is as small as possible. In other words, the sum of the squares of the *residuals*, namely

$$\sum_{i=1}^{m} (b'_i - b_i)^2$$

is minimized. Using the definition of euclidean norm in (6.8), this is equivalent to minimizing

$$S = \|Ax - b\|^2 = (Ax - b)^\mathsf{T}(Ax - b)$$
$$= x^\mathsf{T}A^\mathsf{T}Ax - x^\mathsf{T}A^\mathsf{T}b - b^\mathsf{T}Ax + b^\mathsf{T}b. \qquad (10.2)$$

To determine the appropriate vector x we set the gradient vector ∇S (defined in Section 7.5.2) equal to zero, and use the property established in (7.38) that

$$\nabla(x^\mathsf{T}A^\mathsf{T}Ax) = 2A^\mathsf{T}Ax. \tag{10.3}$$

Exercise 10.1 Show that

$$\nabla(x^\mathsf{T}A^\mathsf{T}b) = A^\mathsf{T}b = \nabla(b^\mathsf{T}Ax). \tag{10.4}$$

Using (10.3) and (10.4) we obtain the gradient of (10.2) as

$$\nabla S = 2A^\mathsf{T}Ax - 2A^\mathsf{T}b,$$

which on equating to zero gives

$$A^\mathsf{T}Ax = A^\mathsf{T}b. \tag{10.5}$$

It was noted in Exercise 5.6 that $R(A^\mathsf{T}A) = R(A)$, so in particular if $R(A) = n \le m$ then the $n \times n$ matrix $A^\mathsf{T}A$ is non-singular. Thus the unique least-squares solution of (10.5) is

$$x = (A^\mathsf{T}A)^{-1}A^\mathsf{T}b. \tag{10.6}$$

This generalizes the result we obtained in (5.43) for $R(A) = 2$. We recall (see Exercise 5.19) that when A is square and non-singular then (10.6) reduces to the unique solution $x = A^{-1}b$ of (10.1). When A is complex, the transpose in (10.6) is replaced by conjugate transpose.

Exercise 10.2 Assuming that A^*A is non-singular, verify that

$$C = (A^*A)^{-1}A^* \tag{10.7}$$

satisfies the condition $CA = I_n$.

Since multiplication of A on the left by C in (10.7) gives the unit matrix, it is called a *left inverse* of A.

Exercise 10.3 Show that

$$C' = A^*(AA^*)^{-1} \tag{10.8}$$

is a *right* inverse of A, ie. $AC' = I_m$ (assuming AA^* is non-singular).

Exercise 10.4 Calculate C using (10.7) when

$$A = \begin{bmatrix} 1 & 1 \\ 1 & 2 \\ 1 & -4 \end{bmatrix}$$

and verify that $CA = I_2$.

We now consider a more general set of equations whose right-hand side is an $m \times m$ matrix B, namely

$$AX = B \qquad (10.9)$$

where X is $n \times m$, and A is $m \times n$ and is allowed to be complex. We seek the *minimal least-squares solution*: this means the solution X which, as before, minimizes the sum of squares $\|AX - B\|^2$ and also has itself the smallest euclidean norm $\|X\|$ amongst all least-squares solutions. It can be shown that the solution to this problem is uniquely given by

$$X = A^+ B \qquad (10.10)$$

where A^+ is the unique $n \times m$ *Moore–Penrose (MP) generalized inverse* of A, defined by the equations

$$AA^+A = A \qquad (10.11)$$

$$A^+AA^+ = A^+ \qquad (10.12)$$

$$(AA^+)^* = AA^+ \qquad (10.13)$$

$$(A^+A)^* = A^+A. \qquad (10.14)$$

Note that the defining equations apply for any values of m and n. In particular, when $m \geqslant n$ and A has maximum possible rank n, then A^*A is non-singular and A^+ is equal to the expression in (10.7); and similarly when $m \leqslant n$ and $R(A) = m$, A^+ is equal to the expression in (10.8).

Exercise 10.5 Verify that the expressions for A^+ in (10.7) and (10.8), when they are valid, satisfy the conditions (10.11) to (10.14).

Example 10.1 When A is square and non-singular then $A^+ = A^{-1}$, but it is interesting to realize that the elements of A^+ may not be continuous functions of those of A. This is easily seen from the matrix

$$A = \begin{bmatrix} a & 1 \\ 1 & 1 \end{bmatrix}. \qquad (10.15)$$

When $a \neq 1$, A is non-singular and from (4.7)

$$A^{-1} = \frac{1}{a-1} \begin{bmatrix} 1 & -1 \\ -1 & a \end{bmatrix}.$$

However, when $a = 1$ it is left as an exercise for the reader to verify that

$$A^+ = \frac{1}{4} \begin{bmatrix} 1 & 1 \\ 1 & 1 \end{bmatrix} \qquad (10.16)$$

satisfies the defining equations (10.11) to (10.14). Clearly $A^{-1} \not\rightarrow A^+$ as $a \rightarrow 1$. In general, such a discontinuity only arises when the rank of A changes; here $R(A) = 2$, $a \neq 1$ but $R(A) = 1$ when $a = 1$.

When A is a zero matrix, then A^+ is defined to be zero also, so in particular when A is a scalar a then $a^+ = 1/a$ if $a \neq 0$, and $a^+ = 0$ if $a = 0$.

Exercise 10.6 If $A = \mathrm{diag}[a_1, a_2, \ldots, a_n]$ verify that
$$A^+ = \mathrm{diag}[a_1^+, a_2^+, \ldots, a_n^+] \quad \text{(compare with Exercise 4.20)}.$$

10.1.2 Properties

The following properties are analogous to those of the ordinary inverse, developed in Section 4.3.1.

$$(A^+)^+ = A \tag{10.17}$$
$$(A^*)^+ = (A^+)^* \tag{10.18}$$
$$(kA)^+ = k^+A^+ \text{ for any scalar } k \tag{10.19}$$
$$(A^*A)^+ = A^+(A^*)^+. \tag{10.20}$$

Exercise 10.7 Verify that the properties (10.17) to (10.20) hold by using the defining equations (10.11) to (10.14).

Exercise 10.8 Prove that
$$(AA^*)^+ = (A^*)^+A^+. \tag{10.21}$$

The reader should recognise the similarity between (10.20), (10.21) and the familiar formula $(BA)^{-1} = A^{-1}B^{-1}$ (see (4.51)) when $B = A^*$. However, this expression for the inverse of a product does not hold in general for the MP-inverse, i.e. $(AB)^+ \neq B^+A^+$, as the following example illustrates.

Example 10.2 If
$$A = [1, 1], \qquad B = \begin{bmatrix} -1 \\ 3 \end{bmatrix},$$

then (as we shall see in the next section) it is easy to compute
$$A^+ = \begin{bmatrix} \dfrac{1}{2} \\ \dfrac{1}{2} \end{bmatrix}, \qquad B^+ = \begin{bmatrix} \dfrac{-1}{10}, \dfrac{3}{10} \end{bmatrix},$$

so that $B^+A^+ = 1/10$. However, $AB = 2$ so $(AB)^+ = \frac{1}{2} \neq B^+A^+$.

In the special case when the explicit expressions (10.7) and (10.8) are valid, then it is left to the reader to verify the following result.

Exercise 10.9 If A is $m \times n$ with $R(A) = n \leqslant m$, and B is $n \times m$ with $R(B) = n$, show that

$$(AB)^+ = B^*(BB^*)^{-1}(A^*A)^{-1}A^* \tag{10.22}$$

by verifying that (10.11) to (10.14) hold. Hence deduce, using (10.7) and (10.8), that in this case $(AB)^+ = B^+A^+$.

Other than the simple case in (10.22), in general for $(AB)^+$ to equal B^+A^+ necessary and sufficient conditions are somewhat complicated, typical ones being that

$$A^+ABB^*A^* = BB^*A^* \quad \text{and} \quad BB^+A^*AB = A^*AB.$$

The result (10.22) is interesting from another point of view, since it can be used to compute A^+, as we shall see in the next section.

A further way in which the MP-inverse differs from the usual inverse of non-singular square matrices is that the eigenvalues of A^+ are not in general the reciprocals of those of A (see Exercise 6.13 for the non-singular case).

Example 10.3 Consider a triangular matrix

$$B = \begin{bmatrix} 0 & 3 \\ 0 & 2 \end{bmatrix} \tag{10.23}$$

which has eigenvalues $\lambda_1 = 0$, $\lambda_2 = 2$. It is easy to compute directly from the defining equations that

$$B^+ = \frac{1}{13} \begin{bmatrix} 0 & 0 \\ 3 & 2 \end{bmatrix},$$

which has eigenvalues 0, $2/13$, whereas $\lambda_1^+ = 0$ and $\lambda_2^+ = 1/2$. Furthermore, the reader can check that

$$u_1 = \begin{bmatrix} 1 \\ 0 \end{bmatrix}, \qquad u_2 = \begin{bmatrix} 3 \\ 2 \end{bmatrix}$$

are eigenvectors of B, whereas B^+ has eigenvectors

$$\begin{bmatrix} 2 \\ -3 \end{bmatrix}, \qquad \begin{bmatrix} 0 \\ 1 \end{bmatrix}.$$

Hence u_1 and u_2 are *not* eigenvectors of B, which again differs from the non-singular case: any eigenvector of B is also an eigenvector of B^{-1} (see Exercise 6.13).

An $n \times n$ matrix A is called an *EP-matrix* if $A^+A = AA^+$. A necessary and sufficient condition for A to be an EP-matrix is that

$$Au = \lambda u \quad \text{if and only if} \quad A^+u = \lambda^+u.$$

Example 10.3 (continued) We have seen that the matrix in (10.23) does not satisfy the above property. In fact

$$BB^+ = \frac{1}{13}\begin{bmatrix} 9 & 6 \\ 6 & 4 \end{bmatrix}, \qquad B^+B = \begin{bmatrix} 0 & 0 \\ 0 & 1 \end{bmatrix},$$

showing that B is not an EP-matrix.

Another characterization of an EP-matrix A is that $AX = 0$ if and only if $A^*X = 0$, where X is also $n \times n$, so in particular any hermitian matrix is EP. An alternative condition will be given in Section 14.1.2. It is worth noting that any normal matrix (defined in Problem 2.10) is EP.

Even when two square matrices A and B are similar, so that they have the same eigenvalues and Jordan form, these properties do not carry over to A^+ and B^+, and the only fact which holds in general is that $R(A^+) = R(B^+)$.

Example 10.3 (continued) In fact B in (10.23) is similar to A in (10.15) when $a = 1$, with $B = PAP^{-1}$ where

$$P = \begin{bmatrix} 2 & 1 \\ 1 & 1 \end{bmatrix}, \qquad P^{-1} = \begin{bmatrix} 1 & -1 \\ -1 & 2 \end{bmatrix}.$$

The eigenvalues of A and B are $0, 2$, but those of A^+ in (10.16) are $0, \frac{1}{2}$, whereas B^+ has eigenvalues $0, 2/13$.

Exercise 10.10 Show that A^+A is an idempotent matrix (defined in Problem 4.14). Hence deduce that $R(A^+A) = \text{tr}(A^+A)$ (see Problem 5.15).

Exercise 10.11 Prove that if H is hermitian and idempotent then $H^+ = H$.

Exercise 10.12 Prove that

$$A^+ = (A^*A)^+A^*.$$

This shows that the Moore–Penrose generalized inverse of an *arbitrary* rectangular matrix A can be expressed in terms of that of the *hermitian* matrix A^*A (which is positive definite or semidefinite, compare with Problem 7.2). When A^*A is positive definite then (10.7) applies.

Exercise 10.13 Prove that

$$(UAV)^+ = V^*A^+U^*, \qquad (10.24)$$

where $U(m \times m)$ and $V(n \times n)$ are unitary matrices.

Exercise 10.14 Verify that

$$\begin{array}{c} r \\ s \end{array}\!\!\left[\begin{array}{c} X \\ 0 \end{array}\right]^+ = [\,X^+ \quad 0\,]\, t \qquad (10.25)$$
$$\quad t \qquad\qquad r \quad s$$

for any dimensions r, s, t.

10.1.3 Computation

Unfortunately there is no straightforward explicit formula for A^+ like that for A^{-1} in terms of $\mathrm{adj}\,A$ and $\det A$ in (4.46). Computing A^+ directly using the defining equations (10.11) to (10.14) is only a feasible proposition when the dimensions of A are small. We therefore outline some procedures which can be used to determine A^+.

For the first of these, it is necessary to begin by noting that if A is $m \times n$ and has rank r, then it is always possible to find matrices $C(m \times r)$ and $D(r \times n)$, each having rank r, such that

$$A = CD. \qquad (10.26)$$

A scheme for performing this full-rank factorization of A will be described shortly. The reason for requiring (10.26) is that since C and D have maximum possible rank, the expressions (10.7) and (10.8) can be used for their generalized inverses, giving

$$A^+ = D^*(DD^*)^{-1}(C^*C)^{-1}C^*. \qquad (10.27)$$

Notice that (10.27) is the same as (10.22) with a slight change of notation (A, B in (10.22) are replaced by C, D respectively). When $r = 1$, C and D are respectively column and row vectors, so that both DD^* and C^*C are scalars, and (10.27) reduces to the particularly simple formula,

$$A^+ = \frac{D^*C^*}{(DD^*)(C^*C)}$$

$$= \frac{1}{\alpha}A^*, \tag{10.28}$$

where

$$\alpha = \|A\|^2 = \sum_{i,j} |a_{ij}|^2. \tag{10.29}$$

The formula (10.28) is particularly useful when A is itself a row or column vector.

Exercise 10.15 Write down the MP-inverse of

$$A = \begin{bmatrix} 1 & 3 \\ 4 & 12 \\ 1 & 3 \end{bmatrix}$$

using (10.28), and verify that the defining equations (10.11) to (10.14) are satisfied.

Exercise 10.16 Verify the expression for α in (10.29) by showing that $\alpha = \mathrm{tr}(A^*A)$ (use property (a) in Problem 2.7).

Exercise 10.17 When A is the general 2×2 matrix in (4.4) but having $\det A = 0$, write down the expression for A^+ using (10.28).
  Notice how it differs from the expression (4.7) for A^{-1} when A is non-singular.

The factorization CD of A in (10.26) can be accomplished using elementary row operations, defined in Section 5.3.1. To obtain D, we reduce A to the *reduced row echelon form*

$$\begin{bmatrix} D \\ 0 \end{bmatrix} \begin{matrix} r \\ m-r \end{matrix}, \tag{10.30}$$

$$n$$

where the elements d_{ij} of D satisfy the conditions:

(i) For each row i ($i = 1, 2, \ldots, r$) there is a column j_i such that $d_{ij_i} = 1$, and $d_{ij} = 0$, $j < j_i$, with $j_1 < j_2 < \cdots < j_r$.

(ii) In each column j_i ($i = 1, 2, \ldots, r$) there is only a single non-zero entry, i.e. $d_{kj_i} = 0$ when $k \neq i$.

Note that (10.30) can be regarded as a step on the route to the normal form of A described in Section 5.3.3. To obtain the normal form, further elementary column operations must be applied to (10.30).

Example 10.4 Consider the 3×4 matrix A in (5.4) having rank $r = 2$. The elementary row operations listed in (5.5) reduce it to the form given in (5.6). The additional operations

$$(R2) \leftrightarrow (R3), \qquad (R1) - 4(R2)$$

applied to (5.6) produce

$$\begin{bmatrix} 1 & 2 & 0 & -\dfrac{1}{5} \\ 0 & 0 & 1 & \dfrac{3}{10} \\ 0 & 0 & 0 & 0 \end{bmatrix}, \qquad (10.31)$$

which is the reduced row echelon form of A.

Having found D in (10.30), the matrix C in the required factorization (10.26) of A simply consists of columns $j_1, j_2, \ldots, j_r$ of A.

Example 10.4 *(continued)* In (10.31) we have $j_1 = 1$, $j_2 = 3$, so for the matrix A in (5.4)

$$C = \begin{bmatrix} 1 & 4 \\ 2 & 8 \\ 3 & 2 \end{bmatrix}, \qquad D = \begin{bmatrix} 1 & 2 & 0 & -\dfrac{1}{5} \\ 0 & 0 & 1 & \dfrac{3}{10} \end{bmatrix}. \qquad (10.32)$$

In order to compute A^+ from (10.27) the inverses of DD^* and C^*C can then be determined by a standard method, for example by again applying elementary row operations, as described in Section 4.4.

Exercise 10.18 Compute A^+ for A in (5.4) using (10.27) and the factors in (10.32).

Another route to finding A^+ from a preliminary factorization of A uses the singular value decomposition developed in Section 8.5. Recall from (8.32) that any $m \times n$ matrix A can be written as $A = U\Sigma V$, where U and V are unitary. From (10.24) we have

$$A^+ = V^* \Sigma^+ U^* \qquad (10.33)$$

and the specific form of Σ^+ depends on the relative values of m and n. For example, if $m = n$ then by Exercise 10.6 $\Sigma^+ = \text{diag}[\sigma_1^+, \sigma_2^+, \ldots, \sigma_n^+]$; and if $m > n$ then by Exercise 10.14

$$\Sigma^+ = [\text{diag}(\sigma_1^+, \sigma_2^+, \ldots, \sigma_n^+), 0]. \tag{10.34}$$

Example 10.5 Consider the 3×2 matrix A in (8.40). The singular value decomposition of A is given in (8.41), with $\sigma_1 = 30$, $\sigma_2 = 0$. Hence from (10.34)

$$\Sigma^+ = \begin{bmatrix} \dfrac{1}{30} & 0 & 0 \\ 0 & 0 & 0 \end{bmatrix}$$

and substituting this into (10.33) with the matrices U, V given in (8.41) produces

$$A^+ = \begin{bmatrix} \dfrac{2}{225} & \dfrac{4}{225} & \dfrac{4}{225} \\ \dfrac{1}{150} & \dfrac{2}{150} & \dfrac{2}{150} \end{bmatrix}.$$

Of course, in this example it is much simpler to determine A^+ using (10.28), since $R(A) = 1$.

The next procedure which we describe involves partitioning. Let A_k be an $m \times k$ matrix having columns $a_1, a_2, \ldots, a_k$ and MP-inverse A_k^+. Then the MP-inverse of the matrix obtained by appending an additional column a_{k+1} to A_k is given by

$$A_{k+1}^+ = [A_k, a_{k+1}]^+ = \begin{bmatrix} A_k^+ - d_k b_k \\ b_k \end{bmatrix}, \tag{10.35}$$

where we define

$$d_k = A_k^+ a_{k+1}, \qquad c_k = a_{k+1} - A_k d_k \tag{10.36}$$

and

$$b_k = \begin{cases} c_k^+, & \text{if } c_k \neq 0 \\ (1 + d_k^* d_k)^{-1} d_k^* A_k^+, & \text{if } c_k = 0. \end{cases} \tag{10.37}$$

For a given $m \times n$ matrix A having columns $a_1, a_2, \ldots, a_n$ the matrices $A_1^+, A_2^+, \ldots, A_n^+ = A^+$ can be found successively using (10.35), (10.36) and (10.37).

Example 10.6 Return again to the 3×4 matrix A in (5.4), namely

$$A = \begin{bmatrix} 1 & 2 & 4 & 1 \\ 2 & 4 & 8 & 2 \\ 3 & 6 & 2 & 0 \end{bmatrix}.$$
$$\quad a_1 \quad a_2 \quad a_3 \quad a_4$$

From (10.28) we have

$$A_1^+ = [a_1]^+ = \frac{1}{14}[1, 2, 3].$$

Setting $k = 1$ in (10.35) gives

$$A_2^+ = [a_1, a_2]^+ = \begin{bmatrix} A_1^+ - d_1 b_1 \\ b_1 \end{bmatrix}, \tag{10.38}$$

where from (10.36)

$$d_1 = A_1^+ a_2 = 2$$

$$c_1 = a_2 - A_1 d_1 = \begin{bmatrix} 0 \\ 0 \\ 0 \end{bmatrix}.$$

Hence from (10.37)

$$b_1 = (1 + d_1^* d_1)^{-1} d_1^* A_1^+$$
$$= (5)^{-1} 2 A_1^+$$
$$= \frac{1}{35}[1, 2, 3]$$

and substitution into (10.38) produces

$$A_2^+ = \frac{1}{70} \begin{bmatrix} 1 & 2 & 3 \\ 2 & 4 & 6 \end{bmatrix}$$

(notice that this expression could in fact have been obtained from (10.28) since $R(A_2) = 1$). Proceeding in the same fashion produces

$$d_2 = A_2^+ a_3 = \frac{13}{35} \begin{bmatrix} 1 \\ 2 \end{bmatrix}$$

$$c_2 = a_3 - A_2 d_2 = \frac{5}{7} \begin{bmatrix} 3 \\ 6 \\ -5 \end{bmatrix}$$

$$b_2 = c_2^+ = \frac{1}{50}[3, 6, -5], \quad \text{using (10.28),}$$

$$A_3^+ = \begin{bmatrix} A_2^+ - d_2 b_2 \\ b_2 \end{bmatrix}$$

$$= \frac{1}{1750} \begin{bmatrix} -14 & -28 & 140 \\ -28 & -56 & 280 \\ 105 & 210 & -175 \end{bmatrix}.$$

The final step is left to the reader:

Exercise 10.19 Use (10.35) to (10.37) to find A_4^+ ($= A^+$) and check your answer with that obtained in Exercise 10.18.

Exercise 10.20 Use the iterative procedure to find A^+ when

$$A = \begin{bmatrix} 1 & 0 & 1 \\ 2 & 1 & 0 \\ 3 & 1 & 1 \\ 1 & 1 & -1 \end{bmatrix}.$$

Another, quite different, iterative procedure is interesting because it is a direct extension of the Newton–Raphson type method given in Section 6.7.2 for computing the ordinary inverse. It can be shown that $X_k \to A^+$ as $k \to \infty$, where $X_0 = pA^*$,

$$X_{k+1} = X_k(2I - AX_k), \qquad k = 0, 1, 2, \ldots \qquad (10.39)$$

and p is any real scalar satisfying $0 < p < 2/\mu$, where μ is the largest eigenvalue of AA^*. The formula (10.39) has precisely the same form as that in (6.134), although of course A now has dimensions $m \times n$. The condition on p ensures convergence, and the eigenvalue calculation can be avoided by appealing to Gershgorin's theorem (see Problem 6.4) as follows:

Exercise 10.21 Deduce that if $AA^* = [\alpha_{ij}]$ then

$$\mu \le \max_i \sum_{j=1}^m |a_{ij}|. \qquad (10.40)$$

Example 10.7 Consider

$$A = \begin{bmatrix} 2 & 1 \\ 4 & 2 \end{bmatrix},$$

which by using the result in Exercise 10.17 is found to have the exact

MP-inverse

$$A^+ = \begin{bmatrix} 0.08 & 0.16 \\ 0.04 & 0.08 \end{bmatrix}. \tag{10.41}$$

Since

$$AA^* = \begin{bmatrix} 5 & 10 \\ 10 & 20 \end{bmatrix}$$

has maximum row sum equal to 30, we infer from (10.40) that we can take $0 < p < 2/30$, and a convenient value is $p = 0.05$. This gives

$$X_0 = \begin{bmatrix} 0.1 & 0.2 \\ 0.05 & 0.1 \end{bmatrix}$$

and from (10.39)

$$X_1 = X_0(2I - AX_0) = \begin{bmatrix} 0.075 & 0.15 \\ 0.0375 & 0.075 \end{bmatrix}$$

$$X_2 = X_1(2I - AX_1) = \begin{bmatrix} 0.0797 & 0.1594 \\ 0.0398 & 0.0797 \end{bmatrix},$$

which is a good approximation to (10.41).

Exercise 10.22   Compute X_0, X_1 and X_2 when

$$A = \begin{bmatrix} 1 & 2 \\ 2 & 4 \\ 3 & 6 \end{bmatrix}.$$

Compare with the exact expression for A^+ obtained using (10.28).

Unfortunately all the methods described in this section possess disadvantages. In particular, the procedures which rely on the full-rank factorization (10.26), or the singular value decomposition $A = U\Sigma V$ can suffer from ill-conditioning—for example, if some of the singular values of A are small, then in (10.7) $A^*A = V^*\Sigma^2 V$ may be nearly singular. For a discussion of these and other difficulties, consult Chapter 12 of Campbell and Meyer (1979).

10.2   Other inverses

The Moore–Penrose inverse is by no means the only generalized inverse of interest for applications, and some others are now discussed.

10.2.1 (i, j, k) inverses

Return to the four defining equations (10.11) to (10.14) for the MP-inverse of the $m \times n$ matrix A. These can be rewritten as

$$AXA = A \quad (1) \tag{10.42}$$
$$XAX = X \quad (2) \tag{10.43}$$
$$(AX)^* = AX \quad (3) \tag{10.44}$$
$$(XA)^* = XA \quad (4). \tag{10.45}$$

The solution of (10.42) to (10.45) is $X = A^+$, and its uniqueness can be readily established, as the reader is invited to attempt:

Exercise 10.23 Let X and Y be two matrices satisfying (10.42) to (10.45). Show that $X = Y$.

In this section generalized inverses will be defined which satisfy only *some* of the four conditions (10.42) to (10.45). Specifically, we denote by $A\{i, j, k\}$ the set of $n \times m$ matrices X satisfying the ith, jth and kth equations listed in (10.42) to (10.45), and any member of $A\{i, j, k\}$ is called an (i, j, k)-inverse of A. For example, X is a $(1, 2)$-inverse of A if it satisfies $AXA = A$ and $XAX = X$. Such a matrix X may or may not satisfy either of the other two conditions (3) and (4). Notice that none of the sets $A\{i, j, k\}$ is empty, because by definition A^+ automatically belongs to all of these sets. There is no general agreement over names for the various types of inverse—Rao and Mitra (1971) give a fairly comprehensive list. Some of those used are *g-inverse* or *pseudoinverse* for a (1)-inverse; *reflexive g*-inverse for a $(1, 2)$ inverse; *least-squares g*-inverse for a $(1, 3)$ inverse; *minimum-norm g*-inverse for a $(1, 4)$ inverse; *least-squares reflexive* or *normalized g*-inverse for a $(1, 2, 3)$-inverse; *minimum-norm reflexive* or *weak g*-inverse for a $(1, 2, 4)$ inverse. As the terminology indicates, inverses of this type of interest in applications are invariably members of $A(1)$. The term 'reflexive inverse' is used because by the defining equations (10.42) and (10.43), if X is a $(1, 2)$-inverse of A then A is a $(1, 2)$-inverse of X.

A fairly widely used notation is to write A^- for a (1)-inverse, which thus satisfies

$$AA^-A = A \tag{10.46}$$

Example 10.8   When

$$A = \begin{bmatrix} 4 & 1 & 1 \\ 3 & 1 & 2 \end{bmatrix} \tag{10.47}$$

it is easy to verify that each of

$$\begin{bmatrix} 1 & -1 \\ -3 & 4 \\ 0 & 0 \end{bmatrix}, \qquad \begin{bmatrix} 2 & 1 \\ -8 & -6 \\ 1 & 2 \end{bmatrix} \tag{10.48}$$

satisfies (10.46), and so is a (1)-inverse of (10.47).

The preceding example illustrates that a (1)-inverse is not unique, and indeed in general if A^- is a particular matrix satisfying (10.46), then all members of $A\{1\}$ are given by

$$A^- + Y - A^-AYAA^-, \tag{10.49}$$

where Y varies over all possible $n \times m$ matrices.

Exercise 10.24   (a) Show that (10.49) is indeed a (1)-inverse by verifying that it satisfies (10.46). (b) Show that if B is any other (1)-inverse of A then it can be expressed in the form (10.49) with $Y = B - A^-$.

Exercise 10.25   Obtain the general expression for the members of $A\{1\}$ when A is the matrix in (10.47). Hence obtain the second matrix in (10.48).

Exercise 10.26   Show that (a) A^-AA^- is a $(1, 2)$-inverse of A; (b) $(A^-)^{\mathsf{T}} = (A^{\mathsf{T}})^-$; (c) If A is hermitian then $(A^-)^*$ is a (1)-inverse of A.

Exercise 10.27   Show that if X is a $(1, 2)$-inverse of A, then (a) X^* is a $(1, 2)$-inverse of A^*; (b) $(1/k)X$ is a $(1, 2)$-inverse of kA, for any non-zero scalar k.

Exercise 10.28   Verify that $\mathrm{diag}[A^-, B^-]$ is a (1)-inverse for $\mathrm{diag}[A, B]$, where A and B are square matrices.

Another representation of $A\{1\}$ is obtained by starting with the representation $A = PNQ$ given in (5.12), which expresses A in terms of its normal form N. For convenience, suppose that the dimensions m, n of A and its rank r are such that

$$N = \begin{bmatrix} I_r & 0 \\ 0 & 0 \end{bmatrix}.$$

Then

$$Q^{-1}\begin{bmatrix} I_r & U \\ V & W \end{bmatrix} P^{-1} \qquad (10.50)$$

is a (1)-inverse of A, where U, V and W have appropriate dimensions but are otherwise arbitrary.

Exercise 10.29 Show that (10.50) is a (1)-inverse of A by verifying that (10.46) is satisfied.

Exercise 10.30 Show that (10.50) is a $(1, 2)$-inverse of A if and only if $W = VU$.

Exercise 10.31 Show that if P and Q in (10.50) are unitary, then (10.50) is a $(1, 2, 3)$ inverse of A if and only if $U = 0$, $W = 0$; and (10.50) is the MP-inverse of A if and only if $U = 0$, $V = 0$, $W = 0$.

There are a number of other expressions for (i, j, k)-inverses of A using partitioned forms. Instead of reducing A to normal form, suppose that we apply row and column operations (if necessary) so as to put A into the form

$$\begin{bmatrix} A_1 & A_2 \\ A_3 & A_4 \end{bmatrix}, \qquad (10.51)$$

where A_1 is non-singular and has order $r = R(A)$.

Exercise 10.32 Verify that the $n \times m$ matrix

$$\begin{bmatrix} A_1^{-1} & 0 \\ 0 & 0 \end{bmatrix} \qquad (10.52)$$

is a (1)-inverse of A in (10.51) provided $A_4 = A_3 A_1^{-1} A_2$.

Exercise 10.33 Use (10.52) to obtain the first expression in (10.48).

When the condition on A_4 in Exercise 10.32 holds, then we can write (10.51) as

$$A = \begin{bmatrix} I_r \\ A_3 A_1^{-1} \end{bmatrix} [A_1 \quad A_2], \qquad (10.53)$$

which is a factorization of A as a product of two matrices each having full rank. One generalization of (10.52) is as follows:

Exercise 10.34 Verify that

$$\begin{bmatrix} A_1^{-1} & -A_1^{-1}A_2W \\ 0 & W \end{bmatrix} \tag{10.54}$$

is a (1)-inverse of A in (10.53), where W has appropriate dimensions but is otherwise arbitrary.

Equation (10.54) has an interesting implication: if A is square then so is W, so by choosing W to be non-singular the matrix in (10.54) is also non-singular since its determinant is $\det A_1^{-1} \det W$ (see Problem 4.5(b)). Hence, unlike the MP-inverse in Section 10.1.2, it is possible to have non-singular (1)-inverses of A even if A is itself singular. Similarly, when A is hermitian then a (1)- or (1, 2)-inverse need not necessarily be hermitian, again differing from the Moore–Penrose case. A simple illustration of these points is now presented.

Example 10.9 Consider the matrix

$$A = \begin{bmatrix} 1 & 0 \\ 0 & 0 \end{bmatrix}. \tag{10.55}$$

It is easy to check that A is a (1)-inverse of itself, so using (10.49) with $A^- = A$ produces the general expression

$$\begin{bmatrix} 1 & y_1 \\ y_2 & y_3 \end{bmatrix} \tag{10.56}$$

for (1)-inverses of A, where y_1, y_2 and y_3 are arbitrary. Although A in (10.55) is singular, the matrix in (10.56) is non-singular provided $y_3 \neq y_1 y_2$. Similarly, although (10.55) is symmetric, (10.56) will be symmetric only when $y_1 = y_2$.

Suppose that A has rank r, and that its rows and columns have been permuted (if necessary) to make the leading $r \times r$ submatrix non-singular. Then to compute a (1)-inverse of A, apply row operations to the augmented matrix $A_a = [A, I_m]$ (assuming $m \geqslant n$) to reduce it to the form $[B, C]$ where

$$B = \begin{bmatrix} I_r & B_1 \\ 0 & 0 \end{bmatrix}$$

is $m \times n$. Then the first n rows of C form a (1)-inverse of A. If $m < n$ the procedure can be applied to A^T, to give A^- using the result in Exercise 10.26(b). When A is square and non-singular, the method becomes that of gaussian elimination for computing A^{-1}, described in Section 4.4.

Example 10.8 (continued) Augmenting the transpose of A in (10.47) produces

$$A_a = \begin{bmatrix} 4 & 3 & \vdots & 1 & 0 & 0 \\ 1 & 1 & \vdots & 0 & 1 & 0 \\ 1 & 2 & \vdots & 0 & 0 & 1 \end{bmatrix}.$$

Since $R(A) = 2$, we obtain

$$A_a \rightarrow \begin{bmatrix} 1 & 1 & \vdots & 0 & 1 & 0 \\ 4 & 3 & \vdots & 1 & 0 & 0 \\ 1 & 2 & \vdots & 0 & 0 & 1 \end{bmatrix} \quad (R1) \leftrightarrow (R2)$$

$$\rightarrow \begin{bmatrix} 1 & 1 & \vdots & 0 & 1 & 0 \\ 0 & -1 & \vdots & 1 & -4 & 0 \\ 0 & 1 & \vdots & 0 & -1 & 1 \end{bmatrix} \quad \begin{matrix} (R2) - 4(R1) \\ (R3) - (R1) \end{matrix}$$

$$\rightarrow \begin{bmatrix} 1 & 0 & \vdots & 1 & -3 & 0 \\ 0 & 1 & \vdots & -1 & 4 & 0 \\ 0 & 0 & \vdots & 1 & -5 & 1 \end{bmatrix} \quad \begin{matrix} (R1) + (R2) \\ (R3) + (R2) \\ (-1)(R2) \end{matrix}.$$

$$\underbrace{\phantom{\begin{bmatrix} 1 & 0 \end{bmatrix}}}_{B} \qquad \underbrace{\phantom{\begin{bmatrix} 1 & -3 \end{bmatrix}}}_{C}$$

The first two rows in the right-hand partition give

$$X = \begin{bmatrix} 1 & -3 & 0 \\ -1 & 4 & 0 \end{bmatrix}$$

as a (1)-inverse of A^T, so a (1)-inverse of A in (10.47) is X^T. This is the same as the first expression in (10.48).

Exercise 10.35 Use the procedure described above to find a (1)-inverse of

$$\begin{bmatrix} 1 & -5 \\ 0 & 2 \\ 4 & 3 \end{bmatrix}.$$

In Exercise 10.12 and Problem 10.1(b) expressions are given for A^+ in terms of the MP-inverse of the positive semidefinite hermitian matrices A^*A and AA^*, and in Problem 10.6(b) similar formulae are given for other generalized inverses. It is therefore useful to record that for the positive semidefinite hermitian matrix,

$$A = \begin{bmatrix} A_1 & A_2 \\ A_2^* & A_3 \end{bmatrix} \tag{10.57}$$

(where A_1 and A_3 are hermitian), then

$$B = \begin{bmatrix} A_1^{(g)} + A_1^{(g)}A_2A_4^{(g)}A_2^*A_1^{(g)} & -A_1^{(g)}A_2A_4^{(g)} \\ -A_4^{(g)}A_2^*A_1^{(g)} & A_4^{(g)} \end{bmatrix} \qquad (10.58)$$

is a (1)-inverse of A if $A_i^{(g)} = A_i^-$, and a $(1, 2)$-inverse if $A_i^{(g)} \in A_i\{1, 2\}$, where $A_4 = A_3 - A_2^*A_1^{(g)}A_2$. Notice that it follows from Exercise 10.27(a) that when B in (10.58) is a (1)- or $(1, 2)$-inverse of A then so is B^*. In addition, if in (10.57) A_3 is non-singular and $R(A) = R(A_1) + R(A_3)$ then B is a $(1, 2, 3)$-inverse of A if $A_i^{(g)} \in A_i\{1, 2, 3\}$, and $B = A^+$ if $A_1^{(g)} = A_1^+$, $A_4^{(g)} = A_4^+$.

Exercise 10.36 When A and A_1 in (10.57) are non-singular, show that (10.58) agrees with the standard formula for A^{-1} given in Section 4.3.2, assuming A_4 is non-singular.

To round off this discussion on partitioned forms, it is worth recording a generalization of the expression for the determinant of a partitioned matrix, given in Problem 4.7. Consider

$$U = \begin{bmatrix} W & X \\ Y & Z \end{bmatrix} \begin{matrix} n \\ m \end{matrix}.$$
$$\begin{matrix} n & m \end{matrix}$$

It can be shown that provided either

$$\text{rank}(W, X) = \text{rank}(W) \qquad (10.59)$$

or

$$\text{rank}\begin{bmatrix} W \\ Y \end{bmatrix} = \text{rank}(W),$$

then

$$\det U = \det W \det(Z - YW^-X) \qquad (10.60)$$

for every $W^- \in W\{1\}$ (see Problem 10.11). The matrix $Z - YW^-X$ is called the *generalized Schur complement* of W in U, and when W is non-singular reduces to the standard case defined in Problem 4.7.

In general the eigenvalues and eigenvectors of a square matrix A and an (i, j, k)-inverse are not related in any simple way (as we have already seen for the MP-inverse). However, when A is hermitian then any matrix $B \in A\{1, 3\}$ has the property that $\lambda(\neq 0)$ is an eigenvalue of A if and only if $(1/\lambda)$ is an eigenvalue of B (see Problem 10.8).

10.2.2 Drazin inverse

The inverses considered so far in this chapter are primarily of interest because of their application to the solution of sets of linear equations, and this will be discussed in Section 10.3. The *Drazin* (D) *inverse*, which has applications to the theory of finite Markov chains and elsewhere, is rather different in that it does not in general satisfy the condition $AXA = A$ in (10.42) and so $\notin A\{1\}$. However, as will be seen, the D-inverse does have some properties like those of an ordinary inverse. For an $n \times n$ matrix A the D-inverse is denoted by A^D and is the *unique* solution of the equations

$$A^{k+1}A^D = A^k \tag{10.61}$$

$$A^D A A^D = A^D \tag{10.62}$$

$$AA^D = A^D A, \tag{10.63}$$

where in (10.61) k is the *index* of A, being the smallest non-negative integer such that $R(A^k) = R(A^{k+1})$. Notice that when A is non-singular then $k = 0$, using the convention that $A^0 = I$, and $A^D = A^{-1}$. Furthermore, by definition A^D commutes with A just as does the ordinary inverse, whereas the MP and other inverses do not in general have this property.

Exercise 10.37 Show that $A^D A^{k+1} = A^k$.

Exercise 10.38 Prove that $(A^D)^{k+1}A^k = A^D = A^k(A^D)^{k+1}$, and ·hence deduce that $(A^D)^{p+1}A^p = A^D$ for any integer $p \geqslant k$.

It can be shown that any square matrix A is similar to diag$[C, N]$, where C is non-singular and N is nilpotent of *index k* (defined in Problem 4.15), and then

$$A = P\begin{bmatrix} C & 0 \\ 0 & N \end{bmatrix}P^{-1}, \qquad A^D = P\begin{bmatrix} C^{-1} & 0 \\ 0 & 0 \end{bmatrix}P^{-1}. \tag{10.64}$$

Exercise 10.39 Verify that A^D in (10.64) satisfies the equations (10.61), (10.62) and (10.63).

Exercise 10.40 Use (10.64) to deduce that A^D is a (1)-inverse of A if and only if $k \leqslant 1$.

In the special cases when $k = 0$ or 1, Exercise 10.40 shows that $A^D \in A\{1\}$, in which case the term *group inverse* is used, with notation $A^\#$ (the name is used because in this case A belongs to a multiplicative group with multiplicative inverse $A^\#$). In other words, $A^\#$ is the *unique* (1, 2) inverse which commutes with A, and is characterized by (10.64) in which $N = 0$. There are applications in statistics (Basilevsky, 1983).

Exercise 10.41 If A is the $n \times n$ singular matrix all of whose elements are 1, show that $A^\# = (1/n^2)A$.

Exercise 10.42 Show that for an $n \times n$ matrix A, the D-inverse is equal to the MP-inverse if and only if A is an EP-matrix.

It was seen in Section 10.1.2 that in general the eigenvalues of A^+ are not the reciprocals of those of A, as is the case for the usual inverse A^{-1}. However, we now show that this property does hold for A^D. Suppose that A has an eigenvalue λ ($\neq 0$) and corresponding eigenvector u; it is convenient to let A be in the form (10.64) and to write

$$u = P\begin{bmatrix} u_1 \\ u_2 \end{bmatrix}. \tag{10.65}$$

Substituting (10.64) and (10.65) into $Au = \lambda u$ gives

$$P\begin{bmatrix} Cu_1 \\ Nu_2 \end{bmatrix} = \lambda P\begin{bmatrix} u_1 \\ u_2 \end{bmatrix}$$

and since P is non-singular this reduces to $Cu_1 = \lambda u_1$, $Nu_2 = \lambda u_2$. Because all the eigenvalues of N are zero (see Problem 6.19(b)) it follows that $u_2 = 0$; and also $C^{-1}u_1 = \lambda^{-1}C^{-1}u_1$ since C is non-singular. Combining these facts together gives

$$\begin{bmatrix} C^{-1} & 0 \\ 0 & 0 \end{bmatrix}\begin{bmatrix} u_1 \\ 0 \end{bmatrix} = \frac{1}{\lambda}\begin{bmatrix} u_1 \\ 0 \end{bmatrix}$$

and premultiplying both sides by P gives $A^Du = \lambda^{-1}u$, where A^D is expressed in (10.64). Thus A^D has λ^{-1} and u as corresponding eigenvalue and eigenvector. Indeed, the property also holds for generalized eigenvectors of A (defined in Section 8.1).

Exercise 10.43 If u in (10.65) is an eigenvector of A corresponding to a zero eigenvalue, show that it is also an eigenvector of A^D.

Another way in which A^D behaves like the ordinary inverse is that it can be expressed as a polynomial in A. Once again, this property does not in general carry over to the MP or (i, j, k)-inverses. When A is square and non-singular the Cayley–Hamilton theorem produces (6.46). When A is $n \times n$ and singular let its characteristic polynomial be $\lambda^p c(\lambda)$, where

$$c(\lambda) = \lambda^{n-p} + k_1 \lambda^{n-p-1} + \cdots + k_{n-p-1}\lambda + k_{n-p}$$

and p is the multiplicity of the zero eigenvalue of A, so that $k_{n-p} \neq 0$. It follows that $p \geq k$ (the index of A), and we exclude the trivial case $p = n$ when A is nilpotent and $A^D = 0$. From the Cayley–Hamilton theorem $A^p c(A) = 0$, and multiplying both sides by $(A^D)^{p+1}$ and using Exercise 10.38 gives $A^D c(A) = 0$. Now define the polynomial

$$d(\lambda) = \frac{k_{n-p} - c(\lambda)}{k_{n-p}\lambda}. \tag{10.66}$$

Exercise 10.44 By considering $Ad(A)$ show that

$$A^D = A^D A d(A). \tag{10.67}$$

Next, raise both sides of (10.67) to the $(k + 1)$th power, and multiply both sides on the left by A^k to produce

$$A^k(A^D)^{k+1} = A^k(A^D)^{k+1}A^{k+1}[d(A)]^{k+1}. \tag{10.68}$$

Notice that the commutativity property (10.63) is used in obtaining the right side in (10.68). A further application of Exercise 10.38 to (10.68) gives finally

$$A^D = A^D A^{k+1}[d(A)]^{k+1}$$
$$= A^k[d(A)]^{k+1}, \tag{10.69}$$

the last step following from Exercise 10.37. Equation (10.69) represents an expression for A^D as a polynomial in A. In fact k in (10.69) can be replaced by p, which is known from the characteristic polynomial of A. The expression (10.69) can be combined with Leverrier's algorithm for determining the characteristic polynomial of A (see Problem 6.29) to produce a method for computing A^D (Campbell and Meyer 1979).

Exercise 10.45 Consider the 3×3 singular matrix A in (6.24). Use (10.69) to show that $A^D = A^3/81 = A/9$.

An interesting application of (10.69) is that if A and B are two $n \times n$ matrices which commute, then $(AB)^D = B^D A^D$. This is because A^D, B^D can be expressed as polynomials in A, B respectively, and these polynomials will also commute. It will be recalled from Section 10.1.2 that conditions for $(AB)^+$ to equal $B^+ A^+$ are more stringent.

Exercise 10.46 If $AB = BA$, deduce that $A^D B = B A^D$, $AB^D = B^D A$.

10.3 Solution of linear equations

We have remarked previously that a major motivation for studying generalized inverses is their application to the linear equations (10.1), or the more general equation (10.9). Indeed, the most general situation is the matrix equation

$$AXC = B \tag{10.70}$$

where A, B, C are given rectangular matrices. A particular case was expressed (with a slightly different notation) in terms of Kronecker products in Problem 5.11. A necessary and sufficient condition for (10.70) to have a solution (i.e. to be consistent) is that

$$AA^- BC^- C = B. \tag{10.71}$$

If (10.71) holds for some pair of (1)-inverses A^-, C^- of A, C respectively, then the general solution of (10.70) is

$$X = A^- BC^- + Y - A^- AYCC^-, \tag{10.72}$$

where Y is an arbitrary matrix having the same dimensions as X.

Exercise 10.47 Confirm that (10.72) satisfies (10.70) provided the equation is consistent. Verify the condition (10.71) by using (10.46).

The particular case when (10.70) reduces to the standard set of equations in (10.1) is most important. We obtain:

The equations $Ax = b$ in (10.1) are consistent if and only if $AA^- b = b$ for some $A^- \in A\{1\}$, and in this case the general solution is

$$x = A^- b + (I_n - A^- A)y, \tag{10.73}$$

where y is an arbitrary column n-vector. This provides an alternative to the consistency theorem in Section 5.4.3.

Example 10.10 Consider the equations $Ax = b$ where A is the matrix in (10.47) and $b = [8, 5]^T$. Using the first expression for A^- in (10.48) gives $AA^- = I_2$, so the consistency condition is obviously satisfied. The general solution (10.73) then becomes

$$x = \begin{bmatrix} 1 & -1 \\ -3 & 4 \\ 0 & 0 \end{bmatrix} \begin{bmatrix} 8 \\ 5 \end{bmatrix} + \left\{ I - \begin{bmatrix} 1 & 0 & -1 \\ 0 & 1 & 5 \\ 0 & 0 & 0 \end{bmatrix} \right\} \begin{bmatrix} y_1 \\ y_2 \\ y_3 \end{bmatrix}$$

$$= \begin{bmatrix} 3 + y_3 \\ -4 - 5y_3 \\ y_3 \end{bmatrix}, \tag{10.74}$$

where y_3 is arbitrary.

Exercise 10.48 Use the second expression for A^- in (10.48) to obtain a solution of the equations in Example 10.10, and show that this is equivalent to (10.74).

An alternative way of regarding (10.73) is that it is equivalent to A^-b as A^- runs over the complete set of (1)-inverses of A generated by (10.49).

Exercise 10.49 Use the general expression for A^- obtained in Exercise 10.25 to derive the solution (10.74).

When the equations $Ax = b$ are consistent, then a solution in (10.73) has minimum norm whenever $A^- \in A\{1, 4\}$, which explains the name 'minimum-norm g-inverse' for a $(1, 4)$-inverse. Note that the minimum-norm solution is unique even though the minimum norm inverse may not be.

Exercise 10.50 (a) If $X \in A\{1, 4\}$ show that $XAA^* = A^*$. (b) If $X_1, X_2 \in A\{1, 4\}$ use the result in Problem 10.7(a) to show that $(X_1 - X_2)A = 0$. (c) Hence deduce that the minimum-norm solution of $Ax = b$ is unique.

Recall from the discussion at the beginning of this chapter that when the equations $Ax = b$ are inconsistent we seek a least-squares solution. This is given by $x = Xb$ where $X \in A\{1, 3\}$, which accounts for the name 'least-squares g-inverse' in Section 10.2.1. The least-squares solution may not be unique, but the minimum value of $\|Ax - b\|^2$ is unique. The least-squares solution having minimum norm is given by (10.10) as $x = A^+b$, and an illustration of this case was given in Example 5.13.

Example 10.10 *(continued)* The solution in (10.74) has euclidean norm given by

$$\|x\|^2 = (3 + y_3)^2 + (-4 - 5y_3)^2 + y_3^2$$
$$= 27y_3^2 + 46y_3 + 25$$
$$= 27\left(y_3 + \frac{23}{27}\right)^2 + \frac{146}{27}$$

and so takes its minimum value when $y_3 = -23/27$, showing that (10.74) is then unique.

In general the solution (10.74) corresponds to $x = A^- b$, where by using Exercise 10.49

$$A^- = \begin{bmatrix} 1 + \alpha & -1 + \beta \\ -3 - 5\alpha & 4 - 5\beta \\ \alpha & \beta \end{bmatrix} \tag{10.75}$$

with α an arbitrary parameter and $\beta = (y_3 - 8\alpha)/5$. In the minimum norm case $\beta = -8\alpha/5 - 23/135$ and (10.75) becomes

$$A^- = \begin{bmatrix} 1 + \alpha & -\dfrac{158}{135} - \dfrac{8\alpha}{5} \\ -3 - 5\alpha & \dfrac{131}{27} + 8\alpha \\ \alpha & -\dfrac{23}{135} - \dfrac{8\alpha}{5} \end{bmatrix} \tag{10.76}$$

and with A in (10.47) this gives

$$A^- A = \begin{bmatrix} \left(\dfrac{22}{45} - \dfrac{4\alpha}{5}\right) & \left(-\dfrac{23}{135} - \dfrac{3\alpha}{5}\right) & \left(-\dfrac{181}{135} - \dfrac{11\alpha}{5}\right) \\ \left(\dfrac{23}{9} + 4\alpha\right) & \left(\dfrac{50}{27} + 3\alpha\right) & \left(\dfrac{181}{27} + 11\alpha\right) \\ \left(-\dfrac{23}{45} - \dfrac{4\alpha}{5}\right) & \left(-\dfrac{23}{135} - \dfrac{3\alpha}{5}\right) & \left(-\dfrac{46}{135} - \dfrac{11\alpha}{5}\right) \end{bmatrix}.$$

For A^- to be a $(1, 4)$ inverse it must satisfy (10.45) so that $A^- A$ must be symmetric, and this is the case when $\alpha = -16/27$, as the reader can verify.

Exercise 10.51 For the matrix A in Exercise 10.15 determine a general expression for A^- using (10.52) and (10.49), and find the values of the parameters for which A^- becomes A^+. Determine the minimum-norm least-squares solution of $Ax = [4, 10, 5]^T$.

Finally, return to the general matrix equation (10.70). When this is consistent, a minimum-norm solution is $X_1 B X_2$, where $X_1 \in A\{1, 4\}$ and $X_2 \in C\{1, 3\}$.

10.4 Some applications

We briefly indicate in this section some additional applications of generalized inverses. Many more can be found in the references listed at the end of the chapter.

10.4.1 Linear feedback control

Consider a linear control system like that in (4.36) but now having a vector $u = [u_1, u_2, \ldots, u_m]^\mathsf{T}$ of control variables, so that the describing linear differential equations take the form

$$\frac{\mathrm{d}x}{\mathrm{d}t} = Ax + Bu, \tag{10.77}$$

where B is a given constant $n \times m$ matrix. As was seen in Problem 5.1, the system is controllable if and only if the matrix in (5.76) has rank n. If each control variable is expressed as a linear combination of the state variables, then in matrix form we can write

$$u = K_1 x, \tag{10.78}$$

where K_1 is a constant $m \times n$ matrix. The expression (10.78) is called *linear state feedback*, and when substituted into (10.77) produces

$$\frac{\mathrm{d}x}{\mathrm{d}t} = (A + BK_1)x, \tag{10.79}$$

which is known as the *closed-loop system*. It can be shown that provided the system is controllable, it is always possible to choose K_1 so that the closed loop matrix $A + BK_1$ in (10.79) has an arbitrary characteristic polynomial. This in turn allows a considerable degree of freedom in choosing the eigenvalues of $A + BK_1$; these then determine the time-behaviour of the solution $x(t)$ of (10.79) as has been seen in Sections 6.5 and 8.1. Methods for computing K_1 are outside the scope of this book. What is of interest here is the case when it is not possible to

measure the state variables x_i, but only linear combinations
called the *output variables* $y_1, \ldots, y_r$. In matrix terms

$$y = Cx, \tag{10.80}$$

where C is a constant $r \times n$ matrix and $y = [y_1, \ldots, y_r]^T$, and the
feedback takes the form $u = K_2 y = K_2 Cx$ where K_2 is a constant
$m \times r$ matrix. One way of achieving the same effect with output
feedback as with the state feedback (10.78) is to require that
$BK_1 = BK_2 C$ so that the closed-loop matrix $A + BK_1$ will be
unaltered. This condition will be satisfied if

$$K_2 C = K_1. \tag{10.81}$$

Equation (10.81) is a special case of (10.70), and by (10.71) a
necessary and sufficient condition for a solution K_2 to exist is that

$$K_1 C^- C = K_1. \tag{10.82}$$

In this case the general solution of (10.81) is

$$K_2 = K_1 C^- + Y(I_r - CC^-), \tag{10.83}$$

where Y is an arbitrary $m \times r$ matrix.

10.4.2   Singular systems

It sometimes happens in setting up models of continuous-time
linear systems that the differential equations take the form

$$E \frac{\mathrm{d}x}{\mathrm{d}t} = Ax, \tag{10.84}$$

where E is square but singular. When E is non-singular (10.84)
can be written as

$$\frac{\mathrm{d}x}{\mathrm{d}t} = (E^{-1}A)x$$

and the problem is then the same as that solved previously
in Sections 6.5, 8.1 and 9.3. Otherwise, for the equations
to be consistent, the initial conditions $x(0) = x_0$ must satisfy
$(I - \hat{E}\hat{E}^D)x_0 = 0$, and then the unique solution of (10.84) is

$$x(t) = \exp(\hat{E}^D \hat{A} t)x_0, \tag{10.85}$$

where $\hat{E} = (\lambda E - A)^{-1}E$, $\hat{A} = (\lambda E - A)^{-1}A$ and λ is such that $\lambda E - A$ is non-singular. When $E = I$ then (10.85) reduces to the standard formula (9.46). When A is non-singular we can take $\lambda = 0$ and $\hat{E} = A^{-1}E$, $\hat{A} = I$.

A similar development holds for linear difference equations, studied previously in Sections 6.5 and 9.3. Campbell and Meyer (1979) give details of these and other applications of the Drazin inverse.

10.4.3 Linear programming

One version of the linear programming problem (see Example 1.5) is to maximise the linear function

$$z = c^\mathsf{T}x = \sum_{i=1}^{n} c_i x_i$$

subject to the linear constraints

$$a \leqslant Ax \leqslant b, \tag{10.86}$$

where A is an $m \times n$ matrix and a, b, c and x are column n-vectors. Assume that the problem is *feasible,* i.e. there is at least one vector x satisfying (10.86), that $R(A) = m$ and that the optimal value of z is finite. Optimal solutions of the problem are then given by

$$x = \sum_{i \in J} g_i a_i + \sum_{i \in K} g_i b_i + \sum_{i \in L} g_i [\theta_i b_i + (1 - \theta_i)a_i] + (I_n - A^- A)y,$$

where A^- has columns $g_1, g_2, \ldots, g_m$, $0 \leqslant \theta_i \leqslant 1$, y is an arbitrary column n-vector and

$$J = \{i \mid c^\mathsf{T}g_i < 0\}, \qquad K = \{i \mid c^\mathsf{T}g_i > 0\}, \qquad L = \{i \mid c^\mathsf{T}g_i = 0\}.$$

Most of the references listed contain applications to linear programming problems.

10.4.4 Estimation of parameters

An important class of linear models in statistics is described by

$$Y = X\beta + \varepsilon, \tag{10.87}$$

where X is $n \times m$, Y and ε are column n-vectors and β is a column m-vector of parameters to be estimated from the observations forming the elements of X and Y. In the so-called ordinary least-squares model it is assumed that $R(X) = m < n$, and that the error vector satisfies

$$E(\varepsilon) = 0, \qquad E(\varepsilon\varepsilon^\mathsf{T}) = \sigma^2 I_n,$$

where $E(\cdot)$ denotes expected value. In this case it can be proved (Basilevsky, 1983) that $\hat{\beta} = (X^\mathsf{T}X)^{-1}X^\mathsf{T}Y$ is an unbiased estimator of β and has minimum variance. However, when $R(X) < m$ the matrix $X^\mathsf{T}X$ is singular, and then the set of all solutions of (10.87) is given by

$$\tilde{\beta} = X^- Y + (I_m - X^- X)z,$$

where $X^- = (X^\mathsf{T}X)^- X^\mathsf{T}$ and z is an arbitrary column m-vector. Furthermore, an unbiased estimator $c^\mathsf{T}\tilde{\beta}$ of the scalar product $c^\mathsf{T}\beta$ has minimum variance if and only if $c^\mathsf{T}\tilde{\beta} = c^\mathsf{T}X^{1,3}$ where $X^{1,3}$ is a least-squares g-inverse of X.

Rao and Mitra (1971) is a major source for further applications to statistics.

Problems

10.1 Prove that for any A

(a) $A^*AA^+ = A^* = A^+AA^*$;

(b) $A^+ = A^*(A^+)^*A^+ = A^*(AA^*)^+$.

10.2 Prove that if A is $m \times n$ then both $I_n - A^+A$ and $I_m - AA^+$ are hermitian and idempotent (defined in Problem 4.14).

10.3 Determine the reduced row echelon form of

$$A = \begin{bmatrix} 0 & 0 & 1 & 3 & -2 \\ 0 & 1 & 2 & 6 & 0 \\ 0 & 2 & 3 & 9 & 2 \\ 0 & 1 & 1 & 3 & 2 \end{bmatrix}$$

and hence use (10.26) to compute A^+.

10.4 Deduce that A^*A is an EP-matrix. Hence, using Exercise 10.12 and Problem 10.1(b), show that if A is normal (i.e. $A^*A = AA^*$) then $A^+A = AA^+$, so A is an EP-matrix.

10.5 Prove that for any $m \times n$ matrix A, the matrices AA^-, A^-A, $I_m - AA^-$ and $I_n - A^-A$ are all idempotent (defined in Problem 4.14). Use (5.11) to prove that $R(AA^-) = R(A) = R(A^-A)$. Hence use Problem 5.15 to show that

$$R(I - AA^-) = n - R(A), \qquad R(I - A^-A) = m - R(A).$$

10.6 If X_1, X_2 are any two (1)-inverses of A, show that $X_1 A X_2 \in A\{1, 2\}$.

10.7 (a) By considering

$$\|BA - CA\|^2 = \text{tr}[(BA - CA)(BA - CA)^*]$$

show that if $BAA^* = CAA^*$ then $BA = CA$.

(b) Hence show that (i) $A = AA^*(AA^*)^- A$; (ii) $A^*(AA^*)^- \in A\{1, 2, 4\}$.

It can be shown similarly that $(A^*A)^- A^* \in A\{1, 2, 3\}$.

10.8 By considering $(A^*AA^*)^-$, use Problem 10.1(a) and Problem 10.7(a) to show that $A^+ = A^*(A^*AA^*)^- A^*$.

10.9 Let A be hermitian and $B \in A\{1, 3\}$.

(a) Show that $AB^*A = A$.

(b) Hence show that if μ is a non-zero eigenvalue of B then $1/\mu$ is an eigenvalue of A.

(c) Show similarly that if λ is a non-zero eigenvalue of A then $1/\lambda$ is an eigenvalue of B.

10.10 For any $n \times n$ matrix A show that $(A^2)^D = (A^D)^2$.

10.11 If A and B are two $n \times n$ matrices, use Problem 10.10 to show that

$$(AB)^D = A[(BA)^2]^D B$$

even if $AB \neq BA$.

10.12 It can be shown that (10.59) implies $WW^-X = X$. Use this to verify (10.60) by considering the determinant of the product

$$\begin{bmatrix} W & 0 \\ Y & I \end{bmatrix} \begin{bmatrix} I & W^-X \\ 0 & Z - YW^-X \end{bmatrix}$$

10.13 Use (2.76) to show that $A^- \otimes B^- \in (A \otimes B)\{1\}$.

10.14 Use (5.11) to show that $R(A) \leq R(A^-)$, and that if $X \in A\{2\}$ then $R(A) \geq R(X)$. Hence deduce that if $A^- \in A\{1, 2\}$ then $R(A) = R(A^-)$. The converse of this result also holds.

10.15 Show that a necessary and sufficient condition for the consistent equations $AX = B$, $XC = D$ to have a common solution X, where A, B, C, D are rectangular matrices having appropriate dimensions, is $AD = BC$. Verify that in this case a general form of the common solution is

$$X = A^-B + DC^- - A^-ADC^- + (I - A^-A)Y(I - CC^-),$$

where Y is an arbitrary matrix having the same dimensions as X.

10.16 Consider a system described by (10.77) and (10.80),

$$\frac{dx}{dt} = \begin{bmatrix} 0 & 1 & 0 \\ -2 & 3 & 0 \\ 5 & 1 & 3 \end{bmatrix} x + \begin{bmatrix} 0 & 0 \\ 1 & 3 \\ 0 & 1 \end{bmatrix} u$$

$$y = \begin{bmatrix} 0 & 0 & 7 \\ 7 & 9 & 0 \end{bmatrix} x.$$

Use (5.76) to show that it is controllable. Verify that linear state feedback $u = K_1 x$ with

$$K_1 = \begin{bmatrix} -7 & -9 & 21 \\ 0 & 0 & -7 \end{bmatrix}$$

makes the closed loop eigenvalues equal to $-3, -3, -4$.

Use (10.52) to determine a (1)-inverse of C which satisfies (10.82). Hence use (10.83) to obtain a linear output feedback $u = K_2 y$ which produces the same closed-loop matrix. Show that in this problem the method produces a unique output feedback control.

10.17 Consider the differential equations (10.84) with

$$E = \begin{bmatrix} 1 & 0 & -2 \\ -1 & 0 & 2 \\ 2 & 3 & 2 \end{bmatrix}, \quad A = \begin{bmatrix} 0 & -1 & -2 \\ 27 & 22 & 17 \\ -18 & -14 & -10 \end{bmatrix}.$$

Show that E and A are both singular, but $E - A$ is non-singular. Verify that

$$\hat{E}^D = \frac{1}{27} \begin{bmatrix} -27 & -41 & -28 \\ 54 & 77 & 46 \\ -27 & -34 & -14 \end{bmatrix}$$

is the Drazin inverse of $\hat{E} = (E - A)^{-1}E$ by confirming that (10.61)–(10.63) hold. Hence show that for consistency the initial conditions must satisfy

$$9x_1(0) + 7x_2(0) + 5x_3(0) = 0.$$

10.18 The relationship between the vector of currents $i = [i_1, i_2, \ldots, i_n]^\mathsf{T}$ and that of voltages $v = [v_1, v_2, \ldots, v_n]^\mathsf{T}$ for an n-terminal network is given by $i = Yv$, where Y is the real indefinite admittance matrix. The matrix Y is called *doubly centred*, since it satisfies $Ye = 0$, $e^\mathsf{T}Y = 0$, where $e = [1, 1, \ldots, 1]^\mathsf{T}$. Deduce that Y is singular.

Suppose that the solution is given by $v = Xi$.

(a) Show that if $Z = Y + ee^\mathsf{T}/n$ is assumed non-singular then $Z^{-1}Y = I - ee^\mathsf{T}/n$. Hence deduce that $Z^{-1} \in Y\{1\}$.

(b) If $X = Z^{-1} - ee^\mathsf{T}/n$, show that $XY = YX = I - ee^\mathsf{T}/n$. Hence deduce that $X = Y^+$, and that X is also doubly centred.

References

Basilevsky (1983), Ben-Israel and Greville (1980), Boullion and Odell (1971), Campbell (1982), Campbell and Meyer (1979), Gregory and Krishnamurthy (1984), Pringle and Rayner (1971), Nashed (1976), Rao and Mitra (1971), Searle (1982).

11 Polynomials, stability, and matrix equations

In Problem 4.8 it was seen how when the Laplace transform is applied to a set of constant-coefficient linear differential equations then the solution involves the characteristic polynomial of the system matrix. Readers familiar with the Laplace transform will be aware that any nth order linear differential equation can be transformed into a problem involving an nth degree polynomial; the same applies to linear difference equations when the z-transform is used (Ogata, 1987). In this chapter we show how properties of polynomials can be investigated using matrix techniques. In particular, it will be seen that the gaussian elimination procedure once again proves useful.

11.1 Companion matrices

In Section 6.3.4 we associated with the nth degree polynomial

$$a(\lambda) = \lambda^n + a_1\lambda^{n-1} + a_2\lambda^{n-1} + \cdots + a_n \qquad (11.1)$$

an $n \times n$ *companion matrix*

$$C = \begin{bmatrix} 0 & & I_{n-1} \\ -a_n & \cdots & -a_2 & -a_1 \end{bmatrix}, \qquad (11.2)$$

for which

$$\det(\lambda I_n - C) = a(\lambda).$$

The last row of C consists of the coefficients of $a(\lambda)$ in reverse order and with opposite signs. In fact, it was seen in Problem 6.12 that there are three other forms of companion matrix, all of which have the same characteristic polynomial (11.1). These are C^T, KCK and KC^TK, where K is the reverse unit matrix defined in Problem 2.4, having 1's along the secondary diagonal.

Example 11.1 When $n = 3$ the four forms of companion matrix whose characteristic polynomial is

$$\lambda^3 + a_1\lambda^2 + a_2\lambda + a_3$$

are

$$C = \begin{bmatrix} 0 & 1 & 0 \\ 0 & 0 & 1 \\ -a_3 & -a_2 & -a_1 \end{bmatrix}, \quad C^T = \begin{bmatrix} 0 & 0 & -a_3 \\ 1 & 0 & -a_2 \\ 0 & 1 & -a_1 \end{bmatrix}$$

$$KCK = \begin{bmatrix} -a_1 & -a_2 & -a_3 \\ 1 & 0 & 0 \\ 0 & 1 & 0 \end{bmatrix}, \quad KC^TK = \begin{bmatrix} -a_1 & 1 & 0 \\ -a_2 & 0 & 1 \\ -a_3 & 0 & 0 \end{bmatrix}.$$

Thus the coefficients of $a(\lambda)$ form respectively the last row, last column, first row or first column of the appropriate companion matrix.

The polynomial in (11.1) is called *monic*, since its leading coefficient is 1. If the leading coefficient is $a_0 \neq 0$, then the monic form is $a_0^{-1}a(\lambda)$, and the coefficients in the various companion forms become $-a_i/a_0$. We saw in Section 6.4.4 that provided all the roots $\lambda_1, \ldots, \lambda_n$ of $a(\lambda)$ are distinct then C can always be transformed according to

$$V_n^{-1}CV_n = \mathrm{diag}[\lambda_1, \lambda_2, \ldots, \lambda_n], \qquad (11.3)$$

where the matrix V_n in the similarity transformation (11.3) is the Vandermonde matrix defined in (4.80). A companion matrix is a basic tool which enables polynomial problems to be handled using matrices, and in this chapter we shall mainly use the form (11.2).

Exercise 11.1 Apply Gershgorin's Theorem (stated in Problem 6.4) to C defined in (11.2) and its transpose. Hence show that the roots of $a(\lambda)$ in (11.1) lie in the union of the following discs in the complex z-plane:

$$|z| \leqslant 1, \qquad |z| \leqslant a_n, \qquad |z| \leqslant 1 + |a_i|, \qquad i = 2, 3, \ldots, n-1$$

$$|z + a_1| \leqslant 1, \qquad |z + a_1| \leqslant \sum_{j=2}^{n} |a_j|.$$

Exercise 11.2 In the case when $n = 3$, verify that C defined in (11.2) satisfies the identity $C^T T = TC$, where

$$T = \begin{bmatrix} a_2 & a_1 & 1 \\ a_1 & 1 & 0 \\ 1 & 0 & 0 \end{bmatrix}. \qquad (11.4)$$

In fact, it is easy to verify that the result in Exercise 11.2 holds in general, the matrix T having the triangular form

$$T = \begin{bmatrix} a_{n-1} & a_{n-2} & \cdots & a_1 & 1 \\ a_{n-2} & a_{n-3} & \cdots & 1 & 0 \\ a_{n-3} & \cdot & \cdots & & \cdot \\ \cdot & \cdot & \cdots & & \\ a_1 & 1 & & & 0 \\ 1 & 0 & & & \end{bmatrix}. \tag{11.5}$$

Notice that (11.4) and (11.5) display the '*striped*' form of T: the elements along sloping lines parallel to the secondary diagonal are all equal. Such matrices are said to have *Hankel* form, and will be studied in detail in Section 13.3. Because of the 1's along the diagonal in (11.5) it follows that T is non-singular, so we can write the relationship between C and C^T in the similarity form

$$C^T = TCT^{-1} \tag{11.6}$$

where T is defined in (11.5).

Exercise 11.3 Compute the inverse of (11.4). Show that in general the inverse of T^{-1} is lower triangular relative to the secondary diagonal. Show also that T^{-1} has Hankel form, and that if the elements along the sloping lines parallel to and below the principal diagonal are denoted by $x_1, x_2, \ldots, x_{n-1}$, respectively, then $x_1 = -a_1$,

$$x_i = -\sum_{j=1}^{i-1} a_{i-j} x_j - a_i, \qquad i \geq 2.$$

11.2 Resultant matrices

The notion of a resultant associated with the polynomial $a(\lambda)$ in (11.1) and a second polynomial

$$b(\lambda) = b_0 \lambda^m + b_1 \lambda^{m-1} + \cdots + b_m \tag{11.7}$$

was introduced in Section 6.3.4. In general, a *resultant* is a scalar function of all the coefficients a_i and b_i which is non-zero if and only if the polynomials are relatively prime (i.e. have no common factor of degree greater than zero). A *resultant matrix* is a matrix whose determinant is a resultant.

11.2.1 Companion matrix form

We have already seen in Section 6.3.4 that the matrix

$$b(C) = b_0 C^m + b_1 C^{m-1} + \cdots + b_{m-1} C + b_m I \qquad (11.8)$$

is a resultant matrix. It is convenient to assume here that $m \leq n$, for if $m > n$ then the roles of $a(\lambda)$ and $b(\lambda)$ can be interchanged, so we would use a companion matrix for $b(\lambda)$. If $m = n$ then $b(\lambda)$ can be replaced by $b(\lambda) - b_0 a(\lambda)$, which has degree $n - 1$. We recall from (6.62) and (6.63) that the rows $r_1, r_2, \ldots, r_n$ of $b(C)$ are given by

$$r_1 = \begin{cases} [b_m, b_{m-1}, \ldots, b_1, b_0, 0, \ldots, 0], & m < n \\ [b_n - b_0 a_n, \ldots, b_2 - b_0 a_2, b_1 - b_0 a_1], & m = n \end{cases} \qquad (11.9)$$

$$r_{i+1} = r_i C, \qquad i = 1, 2, \ldots, n - 1. \qquad (11.10)$$

In other words, $b(C)$ has rows $r_1, r_1 C, r_1 C^2, \ldots, r_1 C^{n-1}$.

11.2.2 Sylvester matrix

A disadvantage of using $b(C)$ is the need to compute its rows using (11.10). A simpler form of resultant matrix is due to Sylvester, and is defined by

$$S = \begin{bmatrix} 1 & a_1 & a_2 & \cdot & \cdot & \cdot & a_n & 0 & \cdot & \cdot & 0 & 0 \\ 0 & 1 & a_2 & \cdot & \cdot & \cdot & a_{n-1} & a_n & \cdot & \cdot & 0 & 0 \\ \cdot & \cdot & \cdot & \cdot & \cdot & \cdot & \cdot & \cdot & \cdot & \cdot & \cdot & \cdot \\ 0 & 0 & 0 & \cdot & 1 & a_1 & \cdot & \cdot & \cdot & \cdot & a_{n-1} & a_n \\ b_0 & b_1 & b_2 & \cdot & \cdot & b_m & 0 & \cdot & \cdot & \cdot & 0 & 0 \\ 0 & b_0 & b_1 & \cdot & \cdot & \cdot & b_m & 0 & \cdot & \cdot & 0 & 0 \\ \cdot & \cdot & \cdot & \cdot & \cdot & \cdot & \cdot & \cdot & \cdot & \cdot & \cdot & \cdot \\ 0 & 0 & \cdot & \cdot & b_0 & b_1 & \cdot & \cdot & \cdot & \cdot & b_{m-1} & b_m \end{bmatrix} \begin{array}{l} \Big\} \; m \text{ rows} \\ \\ \Big\} \; n \text{ rows} \end{array} \qquad (11.11)$$

In (11.11) there are m rows of a's, each row being shifted one position to the right relative to the row above, and similarly for the n rows of b's.

Example 11.2 The expression for S when $n = m = 2$ was given in Problem 4.1. Similarly when $n = 3$, $m = 2$ we have

$$S = \begin{bmatrix} 1 & a_1 & a_2 & a_3 & 0 \\ 0 & 1 & a_1 & a_2 & a_3 \\ \hdashline b_1 & b_1 & b_2 & 0 & 0 \\ 0 & b_0 & b_1 & b_2 & 0 \\ 0 & 0 & b_0 & b_1 & b_2 \end{bmatrix}.$$

It is convenient to partition S in (11.11) into the form

$$\begin{matrix} m & \quad n \end{matrix}$$
$$S = \begin{bmatrix} S_1 & S_2 \\ S_3 & S_4 \end{bmatrix} \begin{matrix} m \\ n \end{matrix} \tag{11.12}$$

(as illustrated in Example 11.2) because it can then be shown that

$$\begin{bmatrix} I_m & 0 \\ -S_3 S_1^{-1} & I_n \end{bmatrix} S = \begin{bmatrix} S_1 & S_2 \\ 0 & Kb(C)K \end{bmatrix} \tag{11.13}$$

where K is the $n \times n$ reverse unit matrix. Notice that S_1 is upper triangular with 1's along the principal diagonal, so $\det S_1 = 1$. By the result in Problem 4.5, the determinant of the right side of (11.13) is

$$\det S_1 \det Kb(C)K = (\det K)^2 \det b(C), \quad \text{by (4.33)},$$
$$= \det K^2 \det b(C)$$
$$= \det b(C),$$

since $K^2 = I$ (see Problem 2.4(a)). Again by the product formula (4.33), the determinant of the left side of (11.13) is just $\det S$. Hence $\det S = \det b(C)$, so S is indeed a resultant matrix.

Exercise 11.4 Repeat Exercise 6.22 where $a(\lambda) = \lambda^3 + 6\lambda^2 + 11\lambda + 6$, $b(\lambda) = \lambda^2 - \lambda - 2$ by evaluating the determinant of the appropriate Sylvester matrix. Also, verify that (11.13) holds in this case.

Exercise 11.5 Show using (11.13) that $b(C) = K(S/S_1)K$, where (S/S_1) is the Schur complement (defined in Problem 4.7) of S_1 in the matrix S in (11.12). Hence deduce that (S/S_1) is a resultant matrix for $a(\lambda)$ and $b(\lambda)$.

11.2.3 Bezoutian matrix

Another important resultant matrix, associated with Bézout, is the $n \times n$ symmetric matrix Z whose elements z_{ij} are defined by

$$\sum_{i=1}^{n} \sum_{j=1}^{n} z_{ij} \lambda^{i-1} \mu^{j-1} = \frac{a(\lambda)b(\mu) - a(\mu)b(\lambda)}{\lambda - \mu}. \tag{11.14}$$

Notice that because of the way the notation has been set up, if the degree m of $b(\lambda)$ is less than n, then we write

$$b(\lambda) = b_{n-m}\lambda^m + b_{n-m+1}\lambda^{m-1} + \cdots + b_{n-1}\lambda + b_n. \quad (11.15)$$

Example 11.3 When $n = 3$, expanding (11.14) and comparing terms shows that

$$Z = \begin{bmatrix} |a_2b_2| & |a_1b_3| & |a_0b_3| \\ |a_1b_3| & (|a_0b_3| + |a_1b_2|) & |a_0b_2| \\ |a_0b_3| & |a_0b_2| & |a_0b_1| \end{bmatrix}, \quad (11.16)$$

where $a_0 = 1$, and it is convenient to use the notation

$$|a_ib_j| = \begin{vmatrix} a_i & b_i \\ a_j & b_j \end{vmatrix} = a_ib_j - a_jb_i. \quad (11.17)$$

If, for example, $m = 2$ then in (11.15) we have $b(\lambda) = b_1\lambda^2 + b_2\lambda + b_3$ and hence $b_0 = 0$, so $|a_0b_3| = a_0b_3 - a_3b_0 = b_3$, and similarly $|a_0b_2| = b_2$, $|a_0b_1| = b_1$.

Exercise 11.6 Construct the bezoutian matrix (11.16) for the polynomials in Exercise 11.4.

It is possible to obtain an explicit expression for the elements z_{ij} in general, but the easiest way to show that Z is a resultant matrix is to establish that it can be factorized as

$$Z = Tb(C) \quad (11.18)$$

where T is the triangular matrix encountered in (11.5).

Example 11.3 (continued) For simplicity let $b(\lambda)$ have degree 2. From (11.9) and (11.10) the matrix $b(C)$, where C is given in Example 11.1, has rows

$r_1 = [b_3, b_2, b_1]$,

$r_2 = r_1 C = [-b_1a_3, b_3 - b_1a_2, b_2 - b_1a_1]$,

$r_3 = r_2 C = [-a_3(b_2 - b_1a_1), -b_1a_3 - a_2(b_2 - b_1a_1),$

$$b_3 - b_1a_2 - a_1(b_2 - b_1a_1)].$$

It is left as an exercise for the reader to verify that premultiplying $b(C)$ by T in (11.4) gives Z in (11.16).

Exercise 11.7 Verify that (11.18) holds for the bezoutian matrix in Exercise 11.6.

Exercise 11.8 Prove that the matrix T in (11.5) has determinant equal to $(-1)^{n(n-1)/2}$. (Hint: expand det T by its last column, using (4.24) and (4.17).)

In view of the Exercise 11.8, equating the determinants on either side of (11.18) shows that $\det Z = \pm \det b(C)$, showing that Z is a resultant matrix. A further interesting property of Z can also be deduced immediately from the factorization (11.18). Since from (11.6) we have $C^T T = TC$, it follows that

$$C^T Z = C^T T b(C)$$
$$= TCb(C). \qquad (11.19)$$

Moreover $b(C)$ is a polynomial in C, and hence commutes with C (since all powers of a matrix commute with each other—see Section 2.2.3). Equation (11.19) therefore becomes

$$C^T Z = Tb(C)C = ZC.$$

If Z is non-singular (i.e. $a(\lambda)$ and $b(\lambda)$ are relatively prime) then pre- and postmultiplying this last equation by Z^{-1} gives

$$Z^{-1}C^T = CZ^{-1}$$

and it follows from Problem 11.1 that Z^{-1} has Hankel form.

Exercise 11.9 For the polynomials

$$a(\lambda) = \lambda^3 + 6\lambda^2 + 11\lambda + 6, \qquad b(\lambda) = \lambda^2 + \lambda + 2$$

compute Z and Z^{-1}, and hence confirm that the latter has Hankel form.

Combining together the identities (11.13) and (11.18) produces, after some manipulations, another useful expression for Z:

$$Z = S_1(S_4 K) - S_3(S_2 K), \qquad (11.20)$$

where S_1, S_2, S_3, S_4 are the $n \times n$ block partitions of the Sylvester matrix in (11.12) with $m = n$ and $b(\lambda)$ as in (11.15). Recall (from Problem 2.4) that postmultiplication of a matrix by K simply reverses the order of its columns. Thus (11.20) expresses Z in terms of four triangular matrices, as is now illustrated.

Example 11.4 For the case $n = 3$, the matrix Z in (11.16) is expressed by (11.20) as

$$Z = \begin{bmatrix} 1 & a_1 & a_2 \\ 0 & 1 & a_1 \\ 0 & 0 & 1 \end{bmatrix} \begin{bmatrix} 0 & 0 & b_3 \\ 0 & b_3 & b_2 \\ b_3 & b_2 & b_1 \end{bmatrix} - \begin{bmatrix} b_0 & b_1 & b_2 \\ 0 & b_0 & b_1 \\ 0 & 0 & b_0 \end{bmatrix} \begin{bmatrix} 0 & 0 & a_3 \\ 0 & a_3 & a_2 \\ a_3 & a_2 & a_1 \end{bmatrix}.$$

$$\quad\quad S_1 \qquad\qquad S_4 K \qquad\qquad S_3 \qquad\qquad S_2 K \qquad (11.21)$$

Exercise 11.10 Express Z in Exercise 11.9 in the form (11.21).

11.3 Greatest common divisor

For the two polynomials $a(\lambda)$ in (11.1) and $b(\lambda)$ in (11.7), any other polynomial $p(\lambda)$ which divides both of them without remainder is called a *common divisor*. Amongst the set of such divisors, those having highest degree k are called *greatest common divisors* (*g.c.d.*'s), and can be expressed as

$$d(\lambda) = d_0\lambda^k + d_1\lambda^{k-1} + \cdots + d_k. \tag{11.22}$$

A g.c.d. is unique up to multiplication by a scalar, so when $d(\lambda)$ is monic (i.e. $d_0 = 1$) we can refer to it as *the* g.c.d. If $k = 0$ (i.e. $d(\lambda) = $ constant) then $a(\lambda)$ and $b(\lambda)$ are called *relatively prime* (or *coprime*). We now indicate how resultant matrices can be used to determine $d(\lambda)$.

11.3.1 Computation via row operations

First return to the $n \times n$ resultant matrix $b(C) = B$, say, defined in (11.8) in terms of the companion matrix C of $a(\lambda)$. It can be shown that the degree k of the g.c.d. in (11.22) is equal to $n - R(B)$. As we saw in Section 5.3.2, the rank of B can be determined by reducing it to triangular form using gaussian elimination. However, it is possible to modify this procedure so as to simultaneously find both the degree k *and* the coefficients of $d(\lambda)$ in (11.22). First note that if $t = n - m \geqslant 1$ then it is easy to use (11.10) to show that the first t rows of B take the form

$$\begin{aligned}
r_1 &= [b_m, b_{m-1}, \ldots, b_1, b_0, 0, \ldots, 0] \\
r_2 &= [0, b_m, b_{m-1}, \ldots, b_1, b_0, \ldots, 0] \\
&\ \vdots \\
r_t &= [0, 0, \ldots, b_m, b_{m-1}, \ldots, b_1, b_0].
\end{aligned} \tag{11.23}$$

It is convenient to suppose that $b_0 = 1$ (i.e. to replace $b(\lambda)$ in (11.7) by $b(\lambda)/b_0$). The remaining m rows of B are then computed from (11.10). We can write B in the partitioned form

$$B = \begin{bmatrix} B_1 & B_2 \\ B_3 & B_4 \end{bmatrix} \begin{matrix} t \\ m \end{matrix},$$

with column labels m and t above B_1 and B_2 respectively.

where in view of (11.23), B_2 is lower triangular with 1's along the principal diagonal, and so is non-singular. Then apply elementary row operations, described in Section 5.3.1, so as to reduce B_4 to zero and B_3 to upper triangular form X, say, relative to the *secondary* diagonal, ending up with

$$\begin{array}{cc} m & t \\ \begin{bmatrix} B_1 & B_2 \\ X & 0 \end{bmatrix} & \begin{array}{c} t \\ m \end{array} \end{array}. \tag{11.24}$$

Recall that these operations do not alter the rank of B, which in (11.24) is seen to equal the sum of the ranks of B_2 and of X (compare with Problem 5.9). Since $R(B_2) = t$, it follows that $k = n - t - R(X) = m - R(X)$. In other words, the degree k of the g.c.d. is equal to the number of zero pivots on the secondary diagonal of X. In particular, if X is non-singular then the polynomials are relatively prime. Furthermore, a g.c.d. itself can be read off from the last non-vanishing row of X, which has the form

$$[d_k, d_{k-1}, \ldots, d_1, d_0, 0, \ldots, 0]. \tag{11.25}$$

Notice that to obtain (11.22), the coefficients in (11.25) are read from right to left. When $m = n - 1$ then $B_2 = [1]$, and the reduced matrix (11.24) is simply upper triangular relative to the secondary diagonal.

Example 11.5 Let

$$a(\lambda) = \lambda^3 - 2\lambda^2 - 3\lambda + 10 \quad [\equiv (\lambda + 2)(\lambda^2 - 4\lambda + 5)] \tag{11.26}$$

$$b(\lambda) = \lambda^2 + 3\lambda + 2 \quad\quad [\equiv (\lambda + 1)(\lambda + 2)], \tag{11.27}$$

so that $n = 3$, $m = 2$. The companion matrix (11.2) is

$$C = \begin{bmatrix} 0 & 1 & 0 \\ 0 & 0 & 1 \\ -10 & 3 & 2 \end{bmatrix}$$

and from (11.9) and (11.10) the rows of $b(C)$ are

$$r_1 = [2, 3, 1], \quad\quad r_2 = r_1 C = [-10, 5, 5],$$

$$r_3 = r_2 C = [-50, 5, 15].$$

The gaussian elimination progresses as follows:

$$B = \begin{bmatrix} 2 & 3 & \vdots & 1 \\ \text{-----} & \text{---} & \vdots & \text{---} \\ -10 & 5 & \vdots & 5 \\ -50 & 5 & \vdots & 15 \end{bmatrix} \begin{matrix} r_1 \\ r_2 \\ r_3 \end{matrix} = \begin{matrix} 2 & 1 \\ \begin{bmatrix} B_1 & B_2 \\ B_3 & B_4 \end{bmatrix} & \begin{matrix} 1 \\ 2 \end{matrix} \end{matrix}$$

$$\rightarrow \begin{bmatrix} 2 & 3 & 1 \\ -20 & -10 & 0 \\ -80 & -40 & 0 \end{bmatrix} \quad \begin{matrix} (R2) - 5(R1) \\ (R3) - 15(R1) \end{matrix}$$

$$\rightarrow \begin{bmatrix} 2 & 3 & \vdots & 1 \\ \text{------} & \text{---} & \vdots & \text{-} \\ -20 & -10 & \vdots & 0 \\ 0 & 0 & \vdots & 0 \end{bmatrix} \quad (R3) - 4(R2). \qquad (11.28)$$
$$X$$

Equation (11.28) now has the required form (11.24), which here is triangular since $m = n - 1$. We therefore deduce immediately that because X has one zero pivot, the degree of a g.c.d. of (11.25) and (11.26) is one; and furthermore the last non-zero row of X gives by (11.25)

$$d(\lambda) = -10\lambda - 20 = -10(\lambda + 2).$$

Hence the monic g.c.d. is $\lambda + 2$, which agrees with the factorized forms of $a(\lambda)$ and $b(\lambda)$ displayed in (11.26) and (11.27).

Next suppose that $a(\lambda)$ in (11.26) is replaced by

$$a(\lambda) = \lambda^4 + \lambda^3 - 9\lambda^2 + \lambda + 30$$

$$= (\lambda + 2)(\lambda + 3)(\lambda^2 - 4\lambda + 5), \qquad (11.29)$$

so that $t = 4 - 2 = 2$. The first two rows of B can be written down using (11.23), and the other two rows are given by (11.9) and (11.10) as

$r_3 = r_2C$, $r_4 = r_3C$, where C is now the companion matrix for (11.29). We obtain

$$B = \begin{bmatrix} 2 & 3 & 1 & 0 \\ 0 & 2 & 3 & 1 \\ \hdashline -30 & -1 & 11 & 2 \\ -60 & -32 & 17 & 9 \end{bmatrix} \begin{matrix} r_1 \\ r_2 \\ r_3 \\ r_4 \end{matrix} \begin{matrix} 2 & 2 \\ \end{matrix} = \begin{bmatrix} B_1 & B_2 \\ B_3 & B_4 \end{bmatrix} \begin{matrix} 2 \\ 2 \end{matrix}$$

$$\rightarrow \begin{bmatrix} 2 & 3 & 1 & 0 \\ 0 & 2 & 3 & 1 \\ -30 & -5 & 5 & 0 \\ -60 & -50 & -10 & 0 \end{bmatrix} \quad \begin{matrix} \text{(R3)} - 2\text{(R2)} \\ \text{(R4)} - 9\text{(R2)} \end{matrix}$$

$$\rightarrow \begin{bmatrix} 2 & 3 & 1 & 0 \\ 0 & 2 & 3 & 1 \\ -40 & -20 & 0 & 0 \\ -40 & -20 & 0 & 0 \end{bmatrix} \quad \begin{matrix} \text{(R3)} - 5\text{(R1)} \\ \text{(R4)} + 10\text{(R1)} \end{matrix}$$

$$\rightarrow \begin{bmatrix} 2 & 3 & 1 & 0 \\ 0 & 2 & 3 & 1 \\ \hdashline -40 & -20 & 0 & 0 \\ 0 & 0 & 0 & 0 \end{bmatrix} \quad \text{(R4)} - \text{(R3)},$$

$$X$$

which shows that a g.c.d. of (11.27) and (11.29) is $d(\lambda) = -20\lambda - 40 = -20(\lambda + 2)$. Notice how the rows $r_1, r_2, \ldots, r_t$ are first used to obtain the zero block in the lower right corner of B, before obtaining the triangular block X.

Exercise 11.11 Use the companion matrix method to obtain the g.c.d. of the polynomials in Exercise 6.22.

Exercise 11.12 Determine the g.c.d. of the polynomials

$$a(\lambda) = \lambda^5 + 10\lambda^4 + 22\lambda^3 + 4\lambda^2 - 23\lambda - 14$$
$$b(\lambda) = \lambda^3 + 8\lambda^2 + \lambda - 42$$

using the companion matrix method.

A second approach to determining a g.c.d. uses the Sylvester matrix S defined in (11.11). It follows from (11.13) that $R(S) = R(S_1) + R(B) = m + n - k$, so the degree of a g.c.d. is $k = n + m - R(S)$; moreover if S is reduced by elementary row operations to row echeolon form, then the last non-vanishing row has the form

$$[0, 0, \ldots, 0, d_0, d_1, \ldots, d_k]$$

giving the coefficients in (11.22). The row echelon form is as described in Section 10.1.3, except that D in (10.30) need only satisfy the conditions (i).

Example 11.6 We repeat Example 11.5 using the Sylvester matrix in (11.11). For the polynomials in (11.26) and (11.27) we have

$$S = \begin{bmatrix} 1 & -2 & -3 & 10 & 0 \\ 0 & 1 & -2 & -3 & 10 \\ 1 & 3 & 2 & 0 & 0 \\ 0 & 1 & 3 & 2 & 0 \\ 0 & 0 & 1 & 3 & 2 \end{bmatrix}$$

$$\rightarrow \begin{bmatrix} 1 & -2 & -3 & 10 & 0 \\ 0 & 1 & -2 & -3 & 10 \\ 0 & 0 & 15 & 5 & -50 \\ 0 & 0 & 5 & 5 & -10 \\ 0 & 0 & 1 & 3 & 2 \end{bmatrix} \quad \begin{matrix} \\ \\ (R3) - (R1) \\ (R4) - (R2) \\ (R3) - 5(R2) \end{matrix}$$

$$\rightarrow \begin{bmatrix} 1 & -2 & -3 & 10 & 0 \\ 0 & 1 & -2 & -3 & 10 \\ 0 & 0 & 1 & 3 & 2 \\ 0 & 0 & 0 & 1 & 2 \\ 0 & 0 & 0 & 0 & 0 \end{bmatrix} \quad \begin{matrix} (R3) \leftrightarrow (R5) \\ (R4) - 5(R3) \\ (R5) - 15(R3) \\ (R5) - 4(R4) \\ (R4) \times \left(-\dfrac{1}{10} \right). \end{matrix} \quad (11.30)$$

The last non-zero row in the row echelon form (11.30) shows that $d(\lambda) = \lambda + 2$.

The reader is invited to apply the same procedure to the Sylvester matrix for the polynomials in (11.27) and (11.29).

Exercise 11.13 Repeat Exercise 11.11 using the Sylvester matrix.

11.3.2 Euclid's algorithm

Both the above methods using resultant matrices are cumbersome for high-degree polynomials, when neither is recommended as an efficient computational procedure. In practice, versions of the following algorithm associated with *Euclid* are used (Knuth,

1969). For a pair of polynomials $a(\lambda)$, $b(\lambda)$ in (11.1) and (11.7), perform the sequence of divisions

$$a(\lambda) = b(\lambda)q_1(\lambda) + f_2(\lambda)$$
$$b(\lambda) = f_2(\lambda)q_2(\lambda) + f_3(\lambda)$$
$$f_2(\lambda) = f_3(\lambda)q_3(\lambda) + f_4(\lambda)$$
$$\vdots$$

(11.31)

The polynomials $f_2(\lambda)$, $f_3(\lambda)$, $f_4(\lambda)$, . . . are called the *euclidean remainders*, and have successively decreasing degrees. If $a(\lambda)$ and $b(\lambda)$ are relatively prime then the last remainder is a constant; if a zero remainder $f_N(\lambda)$, say, is encountered then $f_{N-1}(\lambda)$ is a g.c.d.

Example 11.7 For the polynomials

$$a(\lambda) = \lambda^3 + 4\lambda^2 - 5\lambda + 2, \qquad b(\lambda) = 2\lambda^2 - 3\lambda + 1$$

direct polynomial division gives

$$a(\lambda) = b(\lambda)\left(\frac{1}{2}\lambda + \frac{11}{4}\right) + \frac{11}{4}\lambda - \frac{3}{4}$$
$$b(\lambda) = \left(\frac{11}{4}\lambda - \frac{3}{4}\right)\left(\frac{8}{11}\lambda - \frac{108}{121}\right) + \frac{40}{121},$$

so $f_2(\lambda) = \frac{11}{4}\lambda - \frac{3}{4}$, $f_3(\lambda) = \frac{40}{121}$, and the polynomials are relatively prime.

It is interesting that if the resultant matrix $B = b(C)$ is reduced in the same way as in Section 11.3.1, then the rows of X in (11.24) give the complete set of euclidean remainders, not just the g.c.d. For simplicity consider the case when the degree of $b(\lambda)$ is $n - 1$, so that the reduced matrix in (11.24) is upper triangular relative to the secondary diagonal. Denote the ith row of the complete $n \times n$ reduced matrix (11.24) by

$$x_{i1}, x_{i2}, \ldots, x_{i,n-i+1}, 0, \ldots, 0, \qquad i = 1, 2, 3, \ldots.$$

Then the euclidean remainders can be read off immediately in the form

$$f_i(\lambda) = \frac{(-1)^{i-1}}{f_{i-1,1}}(x_{i1} + x_{i2}\lambda + x_{i3}\lambda^2 + \cdots + x_{i,n-i+1}\lambda^{n-i}), \qquad i \geqslant 2,$$

(11.32)

where $f_{i-1,1}$ is the leading coefficient of $f_{i-1}(\lambda)$ and $f_{11} = b_0$.

Example 11.7 (continued) The companion matrix (11.2) for $a(\lambda)$ is

$$C = \begin{bmatrix} 0 & 1 & 0 \\ 0 & 0 & 1 \\ -2 & 5 & -4 \end{bmatrix}$$

and by applying (11.9) and (11.10) we obtain

$$B = b(C) = \begin{bmatrix} 1 & -3 & 2 \\ -4 & 11 & -11 \\ 22 & -59 & 55 \end{bmatrix} \begin{matrix} r_1 \\ r_2 = r_1 C \\ r_3 = r_2 C \end{matrix}.$$

The reduction of B to the required triangular form is

$$B \rightarrow \begin{bmatrix} 1 & -3 & 2 \\ \dfrac{3}{2} & -\dfrac{11}{2} & 0 \\ -\dfrac{11}{2} & \dfrac{47}{2} & 0 \end{bmatrix} \begin{matrix} \\ (\text{R2}) + \dfrac{11}{2}(\text{R1}) \\ (\text{R3}) - \dfrac{55}{2}(\text{R1}) \end{matrix}$$

$$\rightarrow \begin{bmatrix} 1 & -3 & 2 \\ \dfrac{3}{2} & -\dfrac{11}{2} & 0 \\ \dfrac{10}{11} & 0 & 0 \end{bmatrix} \begin{matrix} \\ \\ (\text{R3}) + \dfrac{47}{11}(\text{R2}). \end{matrix}$$

The second row of this reduced form gives, on referring to (11.32):

$$f_2(\lambda) = -\frac{1}{2}\left(\frac{3}{2} - \frac{11}{2}\lambda\right) = \frac{11}{4}\lambda - \frac{3}{4}$$

since $f_{11} = b_0 = 2$. Similarly, from the last row, since $f_{21} = 11/4$, we have

$$f_3(\lambda) = \frac{(-1)^2}{\left(\dfrac{11}{4}\right)} \cdot \frac{10}{11} = \frac{40}{121},$$

agreeing with the expressions given above.

Exercise 11.14 Use the companion matrix method to find the euclidean remainders when

$$a(\lambda) = \lambda^4 - 3\lambda^3 + 5\lambda^2 + 7\lambda + 2, \qquad b(\lambda) = 2\lambda^3 + 5\lambda^2 + \lambda + 3.$$

Check your results by performing the polynomial divisions as set out in (11.31).

11.3.3 Diophantine equations

Of interest in applications, including feedback control problems (Kučera, 1979) is the so-called *diophantine equation*

$$a(\lambda)x(\lambda) + b(\lambda)y(\lambda) = c(\lambda), \qquad (11.33)$$

where $a(\lambda)$, $b(\lambda)$ are defined in (11.1) and (11.7) respectively, $c(\lambda)$ is a given polynomial and the polynomials $x(\lambda)$, $y(\lambda)$ are to be determined. To begin with, consider the case when $c(\lambda) = 1$. For convenience, we can set $b_0 = 1$ in (11.7), and take

$$x(\lambda) = x_0\lambda^{m-1} + x_1\lambda^{m-2} + \cdots + x_{m-1} \qquad (11.34)$$

$$y(\lambda) = -x_0\lambda^{n-1} + y_1\lambda^{n-2} + \cdots + y_{n-1}. \qquad (11.35)$$

The leading coefficient $-x_0$ in (11.35) ensures that the coefficient of λ^{m+n-1} in (11.33) is zero. In fact, the equation

$$a(\lambda)x(\lambda) + b(\lambda)y(\lambda) = 1 \qquad (11.36)$$

has a solution if and only if $a(\lambda)$ and $b(\lambda)$ are relatively prime. A nice way of demonstrating this is to establish a relationship between (11.36) and the matrix equation

$$C^{\mathsf{T}}F - FD = E, \qquad (11.37)$$

where C is the $n \times n$ companion matrix for $a(\lambda)$ in (11.2), D is the $m \times m$ companion matrix in the same form associated with $b(\lambda)$, $E = [e_{ij}]$ is an $n \times m$ matrix with all elements zero except $e_{11} = 1$, and $F = [f_{ij}]$ with last row $[-x_{m-1}, -x_{m-2}, \ldots, -x_0]$ and last column $[y_{n-1}, y_{n-2}, \ldots, y_1, -x_0]^{\mathsf{T}}$. This is best understood from a specific case.

Example 11.8 Let $m = 2$, $n = 3$, substitute the expressions for $a(\lambda)$, $b(\lambda)$, $x(\lambda)$, $y(\lambda)$ into (11.36), and compare coefficients of powers of λ:

$$\begin{aligned}
\lambda^3\colon\ & x_1 + a_1x_0 + y_1 - b_1x_0 = 0 \\
\lambda^2\colon\ & a_1x_1 + a_2x_0 + b_1y_1 - b_2x_0 = 0 \\
\lambda\colon\ & a_3x_0 + a_2x_1 + b_1y_2 + b_2y_1 = 0 \\
\lambda^0\colon\ & a_3x_1 + b_2y_2 = 1.
\end{aligned} \qquad (11.38)$$

Next consider (11.37), which here becomes

$$\begin{bmatrix} 0 & 0 & -a_3 \\ 1 & 0 & -a_2 \\ 0 & 1 & -a_1 \end{bmatrix} \begin{bmatrix} f_{11} & y_2 \\ f_{21} & y_1 \\ -x_1 & -x_0 \end{bmatrix} - \begin{bmatrix} f_{11} & y_2 \\ f_{21} & y_1 \\ -x_1 & -x_0 \end{bmatrix} \begin{bmatrix} 0 & 1 \\ -b_2 & -b_1 \end{bmatrix} = \begin{bmatrix} 1 & 0 \\ 0 & 0 \\ 0 & 0 \end{bmatrix},$$

and equate elements on either side to obtain:

$$\left.\begin{array}{l} (1,1)\colon a_3x_1 + b_2y_2 = 1 \\ (3,2)\colon y_1 + a_1x_0 + x_1 - b_1x_0 = 0 \end{array}\right\}. \tag{11.39}$$

The $(2,1)$ and $(1,2)$ elements give respectively

$$f_{11} + a_2x_1 + b_2y_1 = 0$$

$$a_3x_0 - f_{11} + b_1y_2 = 0$$

and when added together these produce

$$a_3x_0 + a_2x_1 + b_1y_2 + b_2y_1 = 0. \tag{11.40}$$

Similarly, adding the expressions from the $(3,1)$ and $(2,2)$ elements gives

$$a_1x_1 + a_2x_0 + b_1y_1 - b_2x_0 = 0. \tag{11.41}$$

Clearly (11.39), (11.40) and (11.41) are identical to the four equations in (11.38), showing that the solutions of the polynomial equation (11.36) and the matrix equation (11.37) are equivalent. The argument is easily extended to the general case.

From Section 6.3.5 we know that the solution of (11.37) is unique if and only if there are no eigenvalues λ_i of C^T and μ_j of $-D$ such that $\lambda_i + \mu_j = 0$. Since the eigenvalues of $-D$ are $-\mu_j$, this is equivalent to requiring that C^T and D have no eigenvalues in common. However, by definition C^T and D have characteristic polynomials $a(\lambda)$ and $b(\lambda)$ respectively, so this condition on eigenvalues is in turn equivalent to $a(\lambda)$ and $b(\lambda)$ having no common roots. In other words, we have established that the eqn (11.36) has a unique solution in the form (11.34) and (11.35) if and only if $a(\lambda)$ and $b(\lambda)$ are relatively prime.

Some important properties of diophantine equations are developed in the following three exercises.

Exercise 11.15 Prove that if $\hat{x}(\lambda), \hat{y}(\lambda)$ is a particular solution of (11.36), then $\hat{x} - t(\lambda)b(\lambda)$, $\hat{y} + t(\lambda)a(\lambda)$ is also a solution for any arbitrary polynomial $t(\lambda)$.

Exercise 11.16 Prove that if $d(\lambda)$ is a g.c.d. of $a(\lambda)$ and $b(\lambda)$ then it is always possible to find polynomials $x(\lambda)$, $y(\lambda)$ satisfying

$$a(\lambda)x(\lambda) + b(\lambda)y(\lambda) = d(\lambda), \tag{11.42}$$

where the degrees of $x(\lambda)$ and $y(\lambda)$ are $m - k - 1$ and $n - k - 1$ respectively.

Exercise 11.17 Prove that (11.33) has a solution if and only if the g.c.d. $d(\lambda)$ of $a(\lambda)$ and $b(\lambda)$ is a factor of $c(\lambda)$.

It is interesting that the method of Sections 11.3.1 and 11.3.2 for determining a greatest common divisor $d(\lambda)$ and the set of euclidean remainders from B, the companion matrix form of resultant, can also be applied to solve (11.42). Recall that B was reduced by elementary row operations to the triangular form (11.24). The idea we now use was introduced in Section 4.4 for the calculation of the inverse of a matrix: apply the row operations to the augmented matrix $[B \vdots I_n]$ so as to obtain the reduced form

$$\begin{bmatrix} B_1 & B_2 & \vdots & Y \\ X & 0 & \vdots & \end{bmatrix}, \tag{11.43}$$

where the first n columns in (11.43) are exactly as before in (11.24). Appending the unit matrix to B means that Y acts as a 'record' of the row operations which have been applied; in particular it follows that Y is lower triangular. We saw in (11.25) that the last non-vanishing row in X gives the coefficients d_k, $d_{k-1}, \ldots, d_0$ of a g.c.d.; the corresponding row in Y consists of the coefficients of $y(\lambda)$ in (11.42) in the form

$$[y_{n-k-1}, y_{n-k-2}, \ldots, y_1, 1, 0, \ldots, 0].$$

Again, as in (11.25), read these coefficients from right to left for descending powers of λ.

Example 11.9 Return to the polynomials (11.26) and (11.27) in Example 11.5. Apply the same row operations as before but now to $[B \vdots I_3]$ to end up with

$$\begin{bmatrix} 2 & 3 & 1 & \vdots & 1 & 0 & 0 \\ -20 & -10 & 0 & \vdots & -5 & 1 & 0 \\ 0 & 0 & 0 & \vdots & 5 & -4 & 0 \end{bmatrix}.$$
$$\underbrace{}_{Y}$$

From the second row, reading from right to left for each half, we see that in (11.42)

$$d(\lambda) = -10\lambda - 20, \qquad y(\lambda) = \lambda - 5.$$

It is then easy to check that

$$x(\lambda) = \frac{d(\lambda) - b(\lambda)y(\lambda)}{a(\lambda)} = -1.$$

Exercise 11.18 Solve (11.42) for $y(\lambda)$ and $x(\lambda)$ when $a(\lambda)$ and $b(\lambda)$ are the two polynomials in Exercise 11.12.

11.4 Location of roots

11.4.1 *Relative to the imaginary axis*

It has already been pointed out (Section 9.3) that when $a(\lambda)$ is the characteristic polynomial for a set of linear differential equations then the necessary and sufficient condition for asymptotic stability is that all the roots of $a(\lambda)$ have negative real parts—in other words, they must all lie to the left of the imaginary axis in the complex λ-plane. In this case $a(\lambda)$ is itself said to be *asymptotically stable.* We shall assume that the coefficients a_i in (11.1) are all real, as is usually the case in practical applications. One method of testing whether this condition on the roots holds relies on constructing the $n \times n$ *Hurwitz matrix*

$$H = \begin{bmatrix} a_1 & a_3 & a_5 & \cdots & a_{2n-1} \\ a_0 & a_2 & a_4 & \cdots & a_{2n-2} \\ 0 & a_1 & a_3 & \cdots & a_{2n-3} \\ 0 & a_0 & a_2 & \cdots & a_{2n-4} \\ \vdots & & & \cdots & \\ 0 & 0 & 0 & \cdots & a_n \end{bmatrix} \qquad (11.44)$$

where $a_r = 0$, $r > n$ and a_0 (>0) is the leading coefficient of $a(\lambda)$, which need not necessarily be 1.

Example 11.10   When $n = 5$ then (11.44) becomes

$$
H = \begin{bmatrix}
a_1 & a_3 & a_5 & 0 & 0 \\
a_0 & a_2 & a_4 & 0 & 0 \\
0 & a_1 & a_3 & a_5 & 0 \\
0 & a_0 & a_2 & a_4 & 0 \\
0 & 0 & a_1 & a_3 & a_5
\end{bmatrix}
\begin{matrix}
(1) \\ (4) \\ (2) \\ (5) \\ (3)
\end{matrix}.
\tag{11.45}
$$

If the rows in (11.45) are rearranged into the order indicated to the right, then comparison with (11.11) reveals that H is a Sylvester resultant matrix associated with the polynomials

$$
a_1\lambda^2 + a_3\lambda + a_5, \qquad a_0\lambda^3 + a_2\lambda^2 + a_4\lambda
$$

(compare with Example 11.2). It is easily confirmed that this fact holds in general.

Hurwitz theorem   The polynomial $a(\lambda)$ is asymptotically stable if and only if all the leading principal minors H_1, H_2, H_3, . . . , H_n of H (shown within dashed lines in (11.44)) are positive.

Example 11.11   When $n = 3$ then

$$
H = \begin{bmatrix}
a_1 & a_3 & 0 \\
a_0 & a_2 & 0 \\
0 & a_1 & a_3
\end{bmatrix}
\tag{11.46}
$$

and the leading principal minors have been written out in Problem 4.2. In fact, as we saw in Section 4.2, the best way to evaluate these minors is to triangularize H using gaussian elimination. For H in (11.46) it is easy to check that with $a_0 = 1$ (for convenience)

$$
\begin{bmatrix}
1 & 0 & 0 \\
(-1/a_1) & 1 & 0 \\
-a_1/(a_3 - a_1 a_2) & a_1^2/(a_3 - a_1 a_2) & 1
\end{bmatrix}
H = \begin{bmatrix}
a_1 & a_3 & 0 \\
0 & r_{21} & 0 \\
0 & 0 & a_3
\end{bmatrix},
\tag{11.47}
$$

where $r_1 = (a_1 a_2 - a_3)/a_1$. It follows by taking determinants of both sides of (11.47) that

$$
H_1 = a_1, \qquad H_2 = a_1 r_{21}, \qquad H_3 = a_1 a_3 r_{21}.
$$

Hence the stability condition is satisfied if and only if the diagonal elements on the right side in (11.47) are all positive.

It is not difficult to show that the method illustrated in Example 11.11 for $n = 3$ holds in general. The stability condition is equivalent to requiring that all the diagonal elements in the triangularized form of the Hurwitz matrix must be positive.

These elements, denoted by $r_{01}, r_{11}, r_{21}, \ldots, r_{n1}$ can be computed from the following *Routh array* $\{r_{ij}\}$: The first two rows are defined by alternate coefficients of $a(\lambda)$:

$$\{r_{01}, r_{02}, r_{03}, \ldots\} = \{a_0, a_2, a_4, \ldots\}$$
$$\{r_{11}, r_{12}, r_{13}, \ldots\} = \{a_1, a_3, a_5, \ldots\} \quad . \qquad (11.48)$$

Each subsequent row is then obtained from the preceding two rows by the formula

$$r_{ij} = \frac{-\begin{vmatrix} r_{i-2,1} & r_{i-2,j+1} \\ r_{i-1,1} & r_{i-1,j+1} \end{vmatrix}}{r_{i-1,1}}, \qquad i = 2, 3, \ldots, \qquad (11.49)$$

where r_{ij} is the element in row i and column j. This can be represented schematically as follows:

$$
\begin{array}{lllll}
r_{i-2,1} & \cdot\cdot & \cdot & r_{i-2,j+1} & \cdot\cdot & \leftarrow \text{row } i-2 \\
r_{i-1,1} & \cdot\cdot & \cdot & r_{i-1,j+1} & \cdot\cdot & \leftarrow \text{row } i-1 \\
\cdot & \cdot\cdot & r_{ij} & \cdot & \cdot\cdot & \leftarrow \text{row } i \\
\uparrow & & \uparrow & \uparrow & & \\
\text{column } 1 & & \text{column } j & \text{column } j+1 & &
\end{array}
$$

Thus, starting with the two rows in (11.48), the third row can be constructed, and then the fourth row from rows two and three, and so on. Unlike the Hurwitz condition, which requires the computation of minors of increasing orders, the Routh array involves evaluating only 2×2 determinants.

Routh theorem The necessary and sufficient condition for asymptotic stability of $a(\lambda)$ is that all the elements $r_{01}, r_{11}, r_{21}, \ldots, r_{n1}$ in the *first* column of the Routh array must be positive. If a zero or negative element is obtained in the first column then the procedure need not be continued, since $a(\lambda)$ will not be asymptotically stable.

Example 11.11 (continued) When $n = 3$ the Routh array is obtained from (11.48) and (11.49):

$$
\begin{array}{lcc}
\{r_{0j}\} & a_0 & a_2 \\
\{r_{1j}\} & a_1 & a_3 \\
\{r_{2j}\} & \dfrac{a_1 a_2 - a_0 a_3}{a_1} & 0 \\
\{r_{3j}\} & a_3 &
\end{array}
$$

so the necessary and sufficient conditions for asymptotic stability of
$a_0\lambda^3 + a_1\lambda^2 + a_2\lambda + a_3$ are

$$a_0 > 0, \qquad a_1 > 0, \qquad a_1a_2 - a_0a_3 > 0, \qquad a_3 > 0.$$

Example 11.12 Apply the Routh test to the polynomial

$$a(\lambda) = \lambda^4 + 3\lambda^3 + 2\lambda^2 + 4\lambda + 1. \tag{11.50}$$

Equations (11.48) and (11.49) give

$$
\begin{array}{llll}
\{r_{0j}\} & 1 & 2 & 1 \\[2mm]
\{r_{1j}\} & 3 & 4 & 0 \\[2mm]
\{r_{2j}\} & \dfrac{2}{3} & 1 & 0 \\[3mm]
\{r_{3j}\} & -\dfrac{1}{2} & &
\end{array}
\tag{11.51}
$$

where

$$
r_{21} = \frac{-\begin{vmatrix} 1 & 2 \\ 3 & 4 \end{vmatrix}}{3} = \frac{2}{3}, \qquad
r_{22} = \frac{-\begin{vmatrix} 1 & 1 \\ 3 & 0 \end{vmatrix}}{3} = 1,
$$

$$
r_{31} = \frac{-\begin{vmatrix} 3 & 4 \\ \dfrac{2}{3} & 1 \end{vmatrix}}{(2/3)} = -\frac{1}{2}.
$$

Since r_{31} is negative we conclude that $a(\lambda)$ in (11.50) is not asymptotically stable, so the procedure is terminated.

At this point it is worth pointing out that a *necessary* condition for asymptotic stability of a real polynomial is that all its coefficients must be positive (see Problem 11.4). However, that this is not sufficient is seen from Example 11.12, where all the coefficients in (11.50) are positive, but $a(\lambda)$ is not asymptotically stable.

Exercise 11.19 Determine whether or not the following polynomials are asymptotically stable.

(a) $\lambda^3 + 18\lambda^2 + 2\lambda + 1$;

(b) $\lambda^4 + 2\lambda^3 + 10\lambda^2 + \lambda + 4$;

(c) $\lambda^4 + \lambda^3 + 5\lambda^2 + 5\lambda + 4$;

(d) $\lambda^3 + 3\lambda^2 - 2\lambda + 1$;

(e) $4\lambda^5 + 5\lambda^4 + 26\lambda^3 + 30\lambda^2 + 7\lambda + 2$.

If none of the first column elements in the Routh array is zero then it provides more than just a test for asymptotic stability. The numbers of roots with positive and negative real parts are respectively V and $n - V$, where V is the number of variations in sign in the first column of the Routh array. A variation is counted whenever two consecutive elements r_{i1} and $r_{i+1,1}$ differ in sign. Hence we can determine the location of the roots in the complex plane relative to the imaginary axis (i.e. the numbers of roots to the right and left of the axis). A polynomial having any roots with positive real parts is called *unstable*.

Example 11.12 (continued)   Complete the array (11.51):

$$\{r_{2j}\} \quad \frac{2}{3} \quad 1$$

$$\{r_{3j}\} \quad -\frac{1}{2} \quad 0$$

$$\{r_{4j}\} \quad 1 \quad .$$

Here $V = 2$, since there is a sign change from r_{21} to r_{31}, and from r_{31} to r_{41}. Hence $a(\lambda)$ in (11.50) has 2 roots with positive real parts, and $4 - 2 = 2$ roots with negative real parts.

The cases when a zero element is encountered in the first column lie outside the scope of this book. For details of this, and other methods for determining root location, see Barnett (1983).

Exercise 11.20   Determine the location of the roots of the polynomial in Exercise 11.19(d) relative to the imaginary axis.

Exercise 11.21   Determine the location of the roots of the polynomial $\lambda^4 + 2\lambda^3 + 3\lambda^2 + 4\lambda + 5$ relative to the imaginary axis.

11.4.2   *Relative to the unit circle*

We now turn to the situation where $a(\lambda)$ is the characteristic polynomial for a set of linear *difference* equations. It was pointed out at the end of Section 9.3 that in this case the necessary and sufficient condition for asymptotic stability is that all the roots have modulus less than 1, and $a(\lambda)$ is then called a *convergent* polynomial. In geometrical terms, this means that all the roots must lie inside the *unit circle* $|\lambda| = 1$ in the complex plane, having centre the origin and radius 1. If there are any roots of $a(\lambda)$ outside the unit circle, then $a(\lambda)$ is called *divergent*.

The analogue of the Hurwitz matrix (11.44) is more complicated, although again it is related to a Sylvester-type resultant matrix. This is best seen from a simple case.

Example 11.13 When $n = 4$ the *Schur–Cohn matrix* for

$$a(\lambda) = a_0\lambda^4 + a_1\lambda^3 + a_2\lambda^2 + a_3\lambda + a_4$$

is

$$\Delta = \begin{bmatrix} a_0 & a_1 & a_2 & a_3 & 0 & 0 & 0 & a_4 \\ 0 & a_0 & a_1 & a_2 & 0 & 0 & a_4 & a_3 \\ 0 & 0 & a_0 & a_1 & 0 & a_4 & a_3 & a_2 \\ 0 & 0 & 0 & a_0 & a_4 & a_3 & a_2 & a_1 \\ 0 & 0 & 0 & a_4 & a_0 & a_1 & a_2 & a_3 \\ 0 & 0 & a_4 & a_3 & 0 & a_0 & a_1 & a_2 \\ 0 & a_4 & a_3 & a_2 & 0 & 0 & a_0 & a_1 \\ a_4 & a_3 & a_2 & a_1 & 0 & 0 & 0 & a_0 \end{bmatrix}. \tag{11.52}$$

Notice that if the last four columns in (11.52) are reversed in order, and then the last four rows are also reversed in order, the matrix takes precisely the form of a Sylvester resultant matrix in (11.11), associated with $a(\lambda)$ and the *reverse polynomial* $\lambda^4 a(1/\lambda) = a_4\lambda^4 + a_3\lambda^3 + a_2\lambda^2 + a_1\lambda + a_0$.

This relationship with a resultant matrix holds in general, when Δ can be written in the partitioned form

$$\Delta = \begin{bmatrix} \Delta_1 & \Delta_2 \\ \Delta_2 & \Delta_1 \end{bmatrix}, \tag{11.53}$$

where

$$\Delta_1 = \begin{bmatrix} a_0 & a_1 & a_2 & \cdots & & a_{n-1} \\ 0 & a_0 & a_1 & \cdots & & a_{n-2} \\ \cdot & \cdot & \cdot & \cdots & & a_{n-3} \\ & & & \cdot & \cdots & \cdot \\ & 0 & & & a_0 & a_1 \\ & & & \cdot & \cdots & 0 & a_0 \end{bmatrix}$$

$$\Delta_2 = \begin{bmatrix} & & & \cdots & 0 & a_n \\ & 0 & & & a_n & a_{n-1} \\ & & \cdot & \cdots & \cdot & \cdot \\ 0 & a_n & & \cdots & a_3 & a_2 \\ a_n & a_{n-1} & & \cdots & a_2 & a_1 \end{bmatrix}.$$

Notice that Δ_2 has Hankel form, and that $\Delta_1 K$ (which is Δ_1 with the order of its columns reversed) is identical to the Hankel matrix T in (11.5) (with $a_0 = 1$).

Schur–Cohn theorem The necessary and sufficient condition for asymptotic stability is that all the *inner matrices*, indicated within dashed lines in (11.52) for $n = 4$, must have positive determinants.

As for the Hurwitz matrix, a tabular scheme has been developed to avoid calculating determinants of increasing orders. Define its initial two rows by

$$\begin{aligned} \{c_{11}, c_{12}, \ldots, c_{1,n+1}\} &= \{a_0, a_1, a_2, \ldots, a_n\} \\ \{d_{11}, d_{12}, \ldots, d_{1,n+1}\} &= \{a_n, a_{n-1}, a_{n-2}, \ldots, a_0\} \end{aligned}. \quad (11.54)$$

The rows are then constructed in pairs, the second member of each pair consisting simply of the first in reversed order, as illustrated in (11.54). The first row in each pair has its i, j element given by

$$c_{ij} = \begin{vmatrix} c_{i-1,1} & c_{i-1,j+1} \\ d_{i-1,1} & d_{i-1,j+1} \end{vmatrix}, \quad (11.55)$$

which has the schematic representation

$$\begin{array}{cccccc} c_{i-1,1} & \cdots & \cdot & c_{i-1,j+1} & \cdots \\ d_{i-1,1} & \cdots & \cdot & d_{i-1,j+1} & \cdots \\ \cdot & \cdots & c_{ij} & \cdot & \cdots \\ \uparrow & & \uparrow & \uparrow & \end{array}$$

$\left.\begin{array}{l} \\ \\ \end{array}\right\}$ $(i-1)$th pair of rows

column 1   column j column $j + 1$

Hence each pair of rows is constructed from the preceding pair. Provided no entry d_{i1} in the first column is zero, there are respectively k and $n - k$ roots inside and outside the unit circle, where k is the number of negative products in the sequence

$$P_k = (-1)^k d_{21} d_{31} \cdots d_{k+1,1}, \quad k = 1, 2, \ldots, n. \quad (11.56)$$

In particular, the condition for $a(\lambda)$ to be convergent is

$$d_{21} > 0, \quad d_{i1} < 0, \quad i = 3, 4, \ldots, n + 1.$$

Example 11.14 For the polynomial

$$a(\lambda) = 2\lambda^3 + 4\lambda^2 - 5\lambda + 3 \quad (11.57)$$

the formulae (11.54) and (11.55) produce

$$
\begin{array}{llrrr}
\{c_{1j}\} & & 2 & 4 & -5 & 3 \\
\{d_{1j}\} & & 3 & -5 & 4 & 2 \\
\{c_{2j}\} & & -22 & 23 & -5 \\
\{d_{2j}\} & & -5 & 23 & -22 \\
\{c_{3j}\} & & -391 & 459 \\
\{d_{3j}\} & & 459 & -391 \\
c_{41} & -57800 \\
& = d_{41},
\end{array}
$$

where

$$
c_{21} = \begin{vmatrix} 2 & 4 \\ 3 & -5 \end{vmatrix}, \qquad c_{22} = \begin{vmatrix} 2 & -5 \\ 3 & 4 \end{vmatrix}, \qquad c_{23} = \begin{vmatrix} 2 & 3 \\ 3 & 2 \end{vmatrix},
$$

$$
c_{31} = \begin{vmatrix} -22 & 23 \\ -5 & 23 \end{vmatrix}, \qquad c_{32} = \begin{vmatrix} -22 & -5 \\ -5 & -22 \end{vmatrix},
$$

$$
c_{41} = \begin{vmatrix} -391 & 459 \\ 459 & -391 \end{vmatrix}.
$$

The sequence (11.56) is

$$
P_1 = -d_{21} > 0, \qquad P_2 = d_{21}d_{31} < 0, \qquad P_3 = -d_{21}d_{31}d_{41} < 0,
$$

so $k = 2$. Hence $a(\lambda)$ in (11.57) has two roots inside the unit circle, and one outside.

Again we refer the reader to Barnett (1983) for the cases when some of the P_k are zero, and for further results on the difference equations problem.

Exercise 11.22 Determine the location of the roots of $\lambda^4 + 2\lambda^3 + \lambda^2 + 3\lambda + 2$ relative to the unit circle.

11.4.3 Bilinear transformation

An interesting alternative way of looking at the two problems of root location is to relate them to each other through the *bilinear transformation*

$$
\lambda = \frac{\mu + 1}{\mu - 1}, \qquad \mu = \frac{\lambda + 1}{\lambda - 1}. \tag{11.58}
$$

This is a one-to-one mapping between the left half of the

complex λ-plane, i.e. the region $\text{Re}(\lambda) < 0$, and the unit disc $|\mu| < 1$ in the complex μ-plane. Let $a(\lambda)$ in (11.1) be the polynomial which is being investigated in the λ-plane, and apply (11.58) to it to obtain

$$a\left(\frac{\mu + 1}{\mu - 1}\right) = \frac{b(\mu)}{(\mu - 1)^n}, \qquad (11.59)$$

where we define

$$b(\mu) = b_0 \mu^n + b_1 \mu^{n-1} + \cdots + b_n. \qquad (11.60)$$

The polynomial $b(\mu)$ in (11.60) has the same distribution of roots relative to the unit circle in the μ-plane as does $a(\lambda)$ relative to the imaginary axis in the λ-plane. The coefficients b_i in (11.60) can be easily determined from

$$[b_n, b_{n-1}, \ldots, b_1, b_0] = [a_n, a_{n-1}, \ldots, a_1, a_0]\Gamma. \quad (11.61)$$

In (11.61) the $(n + 1) \times (n + 1)$ matrix $\Gamma = [\gamma_{ij}]$ has all its last column elements $\gamma_{1,n+1}, \gamma_{2,n+1}, \ldots, \gamma_{n+1,n+1}$ equal to 1. The elements in the first row of Γ are simply the binomial coefficients in the expansion of $(-1 + \mu)^n$. The remaining elements of Γ are obtained from

$$\gamma_{ij} = \gamma_{i,j+1} + \gamma_{i-1,j+1} + \gamma_{i-1,j}, \qquad i = 2, 3, \ldots, n + 1$$
$$j = n, n - 1, \ldots, 1. \quad (11.62)$$

The formula (11.62) means that γ_{ij} is the sum of three adjacent elements as follows:

Example 11.15 When $n = 3$, the first row of Γ is given by the coefficients in the expansion of $(-1 + \mu)^3$, i.e. $-1, 3, -3, 1$. The last column of Γ consists of all 1's, so we have

$$\Gamma = \begin{bmatrix} -1 & 3 & -3 & 1 \\ x_3 & x_2 & x_1 & 1 \\ x_6 & x_5 & x_4 & 1 \\ x_9 & x_8 & x_7 & 1 \end{bmatrix}.$$

It is now possible to build up Γ in the order $x_1, x_2, \ldots, x_8, x_9$. For example, (11.62) gives

$$x_1 = 1 + 1 - 3 = -1, \qquad x_2 = x_1 - 3 + 3 = -1$$

and so on, to produce

$$\Gamma = \begin{bmatrix} -1 & 3 & -3 & 1 \\ 1 & -1 & -1 & 1 \\ -1 & -1 & 1 & 1 \\ 1 & 3 & 3 & 1 \end{bmatrix}. \tag{11.63}$$

For example, if $a(\lambda) = \lambda^3 + 6\lambda^2 + 11\lambda + 6$ then (11.61) gives

$$[b_3, b_2, b_1, b_0] = [6, 11, 6, 1]\Gamma$$
$$= [0, 4, -20, 24],$$

so $b(\mu) = 24\mu^3 - 20\mu^2 + 4\mu$. It is easy to check that in fact

$$a(\lambda) = (\lambda + 1)(\lambda + 2)(\lambda + 3), \qquad b(\mu) = 4\mu(3\mu - 1)(2\mu - 1),$$

so $a(\lambda)$ is asymptotically stable and $b(\mu)$ is convergent.

Exercise 11.23 Use (11.63) to obtain $b(\mu)$ when $a(\lambda)$ is the polynomial in Exercise 11.19(a).

Exercise 11.24 Construct Γ when $n = 4$, and hence obtain $b(\mu)$ when $a(\lambda)$ is the polynomial in Exercise 11.21. What is the location of the roots of $b(\mu)$ relative to the unit circle?

If the transformation (11.58) is applied the other way round, starting with a given polynomial $b(\mu)$, we obtain

$$b\left(\frac{\lambda + 1}{\lambda - 1}\right) = \frac{a(\lambda)}{(\lambda - 1)^n}, \tag{11.64}$$

where $a(\lambda) = a_0\lambda^n + a_1\lambda^{n-1} + \cdots + a_n$. Because of the symmetry between (11.59) and (11.64), the a_i can be obtained in terms of the b_i by postmultiplying both sides of (11.61) by Γ^{-1}, and it is easy to prove (see Problem 11.5) that $\Gamma^{-1} = \Gamma/2^n$.

Exercise 11.25 Obtain $a(\lambda)$ when

$$b(\mu) = 2\mu^3 + 4\mu^2 - 5\mu + 3.$$

By applying the Routh test to $a(\lambda)$, determine the location of the roots of $b(\mu)$ relative to the unit circle. Compare your results with Example 11.14.

11.5 Matrix equations and stability

11.5.1 Liapunov equations

Return again to the set of linear differential equations in the form

$$\dot{x} = Ax, \tag{11.65}$$

where A is a real constant $n \times n$ matrix. A different approach to the root location problem was outlined in Section 7.5.4 using Liapunov theory. It follows that all the roots of $a(\lambda) = \det(\lambda I - A)$ have negative real parts if and only if the solution for the real symmetric matrix P of the so-called *Liapunov matrix equation*

$$A^{\mathsf{T}}P + PA = -Q \tag{11.66}$$

is positive definite, where the real symmetric matrix Q in (11.66) is also positive definite. Since the eigenvalues of A^{T} are the same as those of A, it follows from the result at the end of Section 6.5.3 that (11.66) will have a unique solution provided there are no eigenvalues λ_i, λ_j of A such that $\lambda_i + \lambda_j = 0$.

We can adopt a more general attitude: define the *inertia* In(A) of A to be the triple (π_A, ν_A, δ_A) where π_A, ν_A, δ_A are the numbers of eigenvalues of A (i.e. roots of $a(\lambda)$) having positive, negative and zero real parts, respectively. The name arises because Sylvester's law of inertia (Section 7.3) can be written in the form

$$\text{In}(H) = \text{In}(T^*HT),$$

where H is an arbitrary hermitian matrix and T is non-singular. Thus, for example, if A is the matrix of an asymptotically stable system (11.65) then In$(A) = (0, n, 0)$, in which case it is called a *stability matrix*. Similarly, with this notation a positive definite symmetric matrix has inertia $(n, 0, 0)$. Notice that because the eigenvalues of $-A$ are the negative of those of A then In$(-A) = (\nu_A, \pi_A, \delta_A)$.

Inertia theorem   Provided $\delta_A = 0$, then the solution P of (11.66) satisfies In$(A) = $ In$(-P)$. In particular, A is a stability matrix if and only if P is positive definite.

Thus the problem of determining the location of the roots of the characteristic polynomial (i.e. the eigenvalues) of A relative to the imaginary axis is solved when the inertia of P, the solution of (11.66), is found.

Exercise 11.26 When

$$A = \begin{bmatrix} 0 & 1 & 0 \\ 3 & 0 & 1 \\ 0 & -2 & 1 \end{bmatrix}$$

and $Q = \mathrm{diag}[0, 0, 2]$, verify that the solution P of (11.66) is also diagonal. It will be seen later (Problem 11.11) that the inertia result still holds in this case, even though Q is only positive semidefinite. Hence determine the location of the eigenvalues of A.

In general, to determine the inertia of P, recall (Section 6.3.2) that since P is real and symmetric, all its eigenvalues are real. Furthermore, we saw in Section 7.4.2 that π_p and ν_P are respectively equal to the numbers of positive and negative coefficients in any reduction of the quadratic form $x^{\mathrm{T}}Px$ to a sum of squares; finally, these coefficients are easily obtained by triangularizing P using gaussian elimination (see Example 7.11).

Exercise 11.27 Let A be the companion matrix in the form (11.2) for the polynomial $a(\lambda)$ in Exercise 11.19(d). Solve (11.66) for P with $Q = 2I_3$. Reduce P to triangular form, and hence determine $\mathrm{In}(-P)$ and the root location of $a(\lambda)$ relative to the imaginary axis (compare your result with Exercise 11.20).

Exercise 11.28 By replacing A in (11.66) by $A + kI$, where k is a real number, deduce that the real parts of the eigenvalues of A are less than $-k$ if and only if the solution P of

$$A^{\mathrm{T}}P + PA + 2kP = -Q$$

is positive definite.

Now turn to the set of linear difference equations, which we have on several occasions considered in the form

$$X(k + 1) = AX(k), \qquad k = 0, 1, 2, \ldots, \qquad (11.67)$$

The discrete analogue of the Liapunov stability techniques introduced in Section 7.5.4 is to consider, as before, a quadratic form $V = X^{\mathrm{T}}PX$. We now construct not its derivative, but its difference

$$V(k+1) - V(k) = X^T(k+1)PX(k+1) - X^T(k)PX(k), \quad (11.68)$$

which for asymptotic stability of (11.67) is required to be negative definite, with V itself positive definite. Substituting (11.67) into (11.68) gives

$$V(k+1) - V(k) = X^T(k)(A^TPA - P)X(k)$$
$$= -X^T(k)QX(k),$$

where

$$A^TPA - P = -Q \quad (11.69)$$

and (11.69) is the matrix equation corresponding to (11.66), with Q as before real, symmetric and positive definite. From Problem 6.27, eqn (11.69) will have a unique solution provided there are no eigenvalues $\lambda_i \lambda_j$ of A such that $\lambda_i \lambda_j = 1$. Define the *c-inertia* of A to be $\text{In}_c(A) = [\pi_A^c, \nu_A^c, \delta_A^c]$, being respectively the numbers of eigenvalues of A outside, inside and on the unit circle $|\lambda| = 1$. The inertia theorem in this case is that provided $\delta_A^c = 0$ then the solution of (11.69) satisfies the condition $\text{In}_c(A) = \text{In}(-P)$. In particular, the difference equations (11.67) are asymptotically stable if and only if the solution P of (11.69) is positive definite. In this case $A^k \to 0$ as $k \to \infty$, A is called a *convergent* matrix, and $\text{In}_c(A) = (0, n, 0)$. Thus we have a direct analogy with the differential equation case: the problem of determining the root location of the characteristic polynomial of A relative to the unit circle can be solved by determining the inertia of the solution P of the matrix equation (11.69).

Exercise 11.29 Solve (11.69) when A is the companion matrix in the form (11.2) for the polynomial $a(\lambda) = \lambda^2 + 3\lambda - 2$ and $Q = 24I_2$. Hence determine the location of the roots of $a(\lambda)$ relative to the unit circle. Check by computing the roots of $a(\lambda)$.

Exercise 11.30 Verify that the solution of (11.69) when $Q = I$ can be written in the form of an infinite series:

$$P = I + A^TA + (A^T)^2A^2 + (A^T)^3A^3 + \cdots. \quad (11.70)$$

Deduce that this series converges if A is a convergent matrix.

Exercise 11.31 Show that an efficient way of computing the sum of the
series in (11.70) is by defining the iteration

$$P_{k+1} = P_k + (A^T)^m P_k A^m, \qquad k = 1, 2, 3, \dots,$$

where $m = 2^{k-1}$ and $P_1 = I$. How many iterations are required to produce
over 4000 terms in the series (11.70)?

To close this section two useful extensions are worth noting.
Firstly, if A has complex elements, then the only modification
which needs to be made is that A^T in (11.66) or (11.69) is
replaced by A^*; the results on inertia then apply as stated.
Secondly, if Q in (11.66) or (11.69) is only positive semidefinite,
the inertia results still hold provided Q is such that the
controllability matrix (first defined in (5.76)) for A^* and Q has
rank n, that is

$$R[Q, A^*Q, (A^*)^2Q, \dots, (A^*)^{n-1}Q] = n. \qquad (11.71)$$

This appearance of a condition relating to a property of control
systems is rather surprising.

11.5.2   Riccati equation

We return to the linear control problem introduced in Section
10.4.1 in the form

$$\dot{x} = Ax + Bu \qquad (11.72)$$

where the initial state $x(0)$ is specified. It was noted that
provided the system is controllable then it is always possible to
apply linear state feedback control $u = Kx$ so that the matrix
$A + BK$ of the closed-loop system has an arbitrary characteristic
polynomial. In practice it is usually essential to ensure that the
closed-loop system is asymptotically stable, and one way of
achieving this is to determine the feedback matrix K from the
solution of the following quadratic matrix equation, called after
Riccati:

$$PBR^{-1}B^TP - A^TP - PA - Q = 0. \qquad (11.73)$$

In (11.73) the matrices P, Q and R are real and symmetric, with
R positive definite and Q positive semidefinite. It can be shown
that there is a unique positive definite solution P of (11.73)
provided the controllability condition in (11.71) is satisfied (here

$A^* = A^T$ since A is real). In this case, setting

$$K = -R^{-1}B^TP \qquad (11.74)$$

produces a closed-loop matrix

$$A_c = A - BR^{-1}B^TP \qquad (11.75)$$

which is asymptotically stable. The derivation of these results lies outside the scope of this book, and relies on the theory of optimal control systems (Barnett and Cameron 1985). The feedback control defined by (11.74) actually minimizes the *performance index*

$$\int_0^\infty (x^TQx + u^TRu) \, dt. \qquad (11.76)$$

The first quadratic form in the integrand in (11.76) ensures that $x(t) \to 0$ as $t \to \infty$, and the second quadratic form provides a measure of the total control energy involved in transferring the system from $x(0)$ to the origin. It should be emphasized that in general A in (11.72) will not be a stability matrix, so the control (11.74) has a *stabilizing* effect.

Exercise 11.32 Obtain the positive definite solution of the Riccati equation (11.73) with

$$A = \begin{bmatrix} -1 & 0 \\ 1 & 0 \end{bmatrix}, \qquad B = \begin{bmatrix} 1 \\ 0 \end{bmatrix}, \qquad Q = \begin{bmatrix} 0 & 0 \\ 0 & 1 \end{bmatrix}, \qquad R = \frac{1}{10},$$

$$P = \begin{bmatrix} p_1 & p_2 \\ p_2 & p_3 \end{bmatrix}.$$

Hence determine the feedback control with matrix in (11.74), and verify that the resulting closed-loop system is asymptotically stable.

Exercise 11.33 Consider the system (11.72) with a feedback control $u = Kx$, where K is given by (11.74). Assuming that Q in (11.73) is positive definite, construct the Liapunov equation (11.66) with A replaced by $A + BK$, and P satisfying (11.73). Hence deduce that the closed loop system is asymptotically stable.

The Riccati equation which arises for control systems described by linear difference equations is discussed in Barnett (1983) and Ogata (1987).

11.5.3   Solution via eigenvectors

The general problem of accurate numerical solution of Liapunov and Riccati-type matrix equations lies outside the scope of this book. We restrict our discussion to two interesting, and related, approaches.

Consider first the Riccati equation (11.73), and construct the $2n \times 2n$ matrix

$$H = \begin{matrix} n \quad\quad n \\ \begin{bmatrix} A & -BR^{-1}B^{\mathsf{T}} \\ -Q & -A^{\mathsf{T}} \end{bmatrix} \begin{matrix} n \\ n \end{matrix} \end{matrix}. \tag{11.77}$$

This is an example of a *hamiltonian* matrix, which satisfies $H = LH^{\mathsf{T}}L$, where

$$L = \begin{bmatrix} 0 & -I_n \\ I_n & 0 \end{bmatrix}. \tag{11.78}$$

Exercise 11.34   Show that $L^{\mathsf{T}} = L^{-1} = -L$ and that $L^2 = -I_{2n}$.

Exercise 11.35   Verify that H in (11.77) is hamiltonian.

Exercise 11.36   Show that if θ_i is an eigenvalue of *any* hamiltonian matrix then so is $-\theta_i$. (Hint: show that $u_i^{\mathsf{T}}L$ is a left eigenvector of H.)

Suppose that

$$Z = \begin{matrix} n \quad\; n \\ \begin{bmatrix} Z_1 & Z_2 \\ Z_3 & Z_4 \end{bmatrix} \begin{matrix} n \\ n \end{matrix} \end{matrix} \tag{11.79}$$

is any matrix which transforms H in (11.77) into its Jordan form $\mathscr{J}$, i.e. $Z^{-1}HZ = \mathscr{J}$. It can be shown that since the system (11.72) is assumed to be controllable, it is always possible to choose Z in (11.79) such that Z_1 is non-singular.

Exercise 11.37   By considering the identity $HZ = Z\mathscr{J}$, where H and Z are defined in (11.77) and (11.79) respectively, obtain the equations

$$AZ_1 - BR^{-1}B^{\mathsf{T}}Z_3 = Z_1J_1 \tag{11.80}$$

$$-QZ_1 - A^{\mathsf{T}}Z_3 = Z_3J_1 \tag{11.81}$$

where

$$\mathscr{J} = \begin{bmatrix} J_1 & J_2 \\ 0 & J_3 \end{bmatrix}. \tag{11.82}$$

Exercise 11.38 Premultiply (11.80) by $Z_3 Z_1^{-1}$, and postmultiply both (11.80) and (11.81) by Z_1^{-1}. Hence show that

$$Z_3 Z_1^{-1} B R^{-1} B^{\mathsf{T}} Z_3 Z_1^{-1} - A^{\mathsf{T}} Z_3 Z_1^{-1} - Z_3 Z_1^{-1} A - Q = 0. \quad (11.83)$$

Equation (11.83) reveals that $P = Z_3 Z_1^{-1}$ satisfies the Riccati equation (11.73). This solution is expressed in terms of the first n columns of the matrix Z in (11.79) which transforms H into its Jordan form. For simplicity, now assume that all the eigenvalues $\theta_1, \ldots, \theta_{2n}$ of H are distinct, so that

$$\mathcal{J} = \text{diag}[\theta_1, \theta_2, \ldots, \theta_{2n}].$$

It can be shown that the controllability condition ensures that none of the θ_i is purely imaginary. Recall (Section 8.1) that Z will now consist of the (right) eigenvectors of H. However, there remains the problem of deciding in what order to select these eigenvectors. This can be resolved by considering the closed-loop system matrix $A_c = A - BR^{-1}B^{\mathsf{T}}P$ in (11.75). Postmultiplying (11.80) by Z_1^{-1} produces

$$A - BR^{-1}B^{\mathsf{T}}Z_3 Z_1^{-1} = Z_1 J_1 Z_1^{-1},$$

which shows that A_c with $P = Z_3 Z_1^{-1}$ is similar to J_1. Hence, in order to make the closed-loop matrix asymptotically stable we simply select as the first n eigenvalues of H those having negative real parts—this can always be done, since it follows from the result in Exercise 11.36 that the $2n$ eigenvalues of H are $\pm\theta_1, \pm\theta_2, \ldots, \pm\theta_n$. It can be shown that $P = Z_3 Z_1^{-1}$ is the same whatever the selected order of those eigenvalues having negative real parts.

Exercise 11.39 When

$$A = \begin{bmatrix} 0 & -1 \\ 0 & 0 \end{bmatrix}, \quad B = \begin{bmatrix} 1 & 0 \\ 0 & 1 \end{bmatrix}, \quad Q = \begin{bmatrix} 4 & 2 \\ 2 & 1 \end{bmatrix}, \quad R = \begin{bmatrix} 1 & 0 \\ 0 & 1 \end{bmatrix},$$

show that H in (11.77) has eigenvalues $\pm 1, \pm 2$. By computing appropriate eigenvectors of H determine the solution of the Riccati eqn (11.73) which makes the closed-loop matrix A_c in (11.75) have eigenvalues $-1, -2$.

When $B = 0$ the Riccati eqn (11.73) reduces to the Liapunov eqn (11.66). Since H in (11.77) is then block triangular, its eigenvalues are just those of A (i.e. $\lambda_1, \ldots, \lambda_n$) and $-A^T$ (i.e. $-\lambda_1, \ldots, -\lambda_n$), and the method is equivalent to transforming A to diagonal form (see Problems 11.19 and 11.20).

Exercise 11.40 Suppose that, as in (6.74), we write

$$T^{-1}AT = \text{diag}[\lambda_1, \lambda_2, \ldots, \lambda_n] = \Lambda$$

and let $X = [x_{ij}]$ be the symmetric solution of the equation

$$\Lambda X + X\Lambda = -Q_1,$$

where $Q_1 = T^T QT = [q_{ij}]$. Show that

$$x_{ij} = \frac{-q_{ij}}{\lambda_i + \lambda_j},$$

assuming $\lambda_i + \lambda_j \neq 0$ for all i and j (recall that this is the condition for the solution to be unique). Hence express the solution P of (11.66) in terms of X and left eigenvectors v_i of A.

11.5.4 Solution via matrix sign function

We begin with the Liapunov equation obtained when $B = 0$ in (11.73), namely

$$A^T P + PA = -Q. \tag{11.84}$$

The hamiltonian matrix H in (11.77) becomes

$$H = \begin{bmatrix} A & 0 \\ -Q & -A^T \end{bmatrix}. \tag{11.85}$$

If P is the solution of (11.84) then it follows that we can write

$$H = \underbrace{\begin{bmatrix} I & 0 \\ P & I \end{bmatrix}}_{X} \begin{bmatrix} A & 0 \\ 0 & -A^T \end{bmatrix} \underbrace{\begin{bmatrix} I & 0 \\ -P & I \end{bmatrix}}_{X^{-1}}. \tag{11.86}$$

This is easily verified by multiplying out the terms in (11.86). From the definition of the matrix sign function in (9.54), we deduce from (11.86) that

$$\text{sgn}(H) = X \, \text{sgn}\begin{bmatrix} A & 0 \\ 0 & -A^T \end{bmatrix} X^{-1}$$

$$= X \begin{bmatrix} \text{sgn}(A) & 0 \\ 0 & -\text{sgn}(A^T) \end{bmatrix} X^{-1}. \tag{11.87}$$

Assume now that A is a stability matrix, so by Exercise 9.27 we have $\text{sgn}(A) = -I = \text{sgn}(A^{\mathrm{T}})$. This reduces (11.87) to

$$\text{sgn}(H) = \begin{bmatrix} I & 0 \\ P & I \end{bmatrix} \begin{bmatrix} -I & 0 \\ 0 & I \end{bmatrix} \begin{bmatrix} I & 0 \\ -P & I \end{bmatrix}$$

$$= \begin{bmatrix} -I & 0 \\ -2P & I \end{bmatrix}. \tag{11.88}$$

Hence, if $\text{sgn}(H)$ is known then the required solution P of (11.85) can be obtained from (11.88). To compute $\text{sgn}(H)$ we can use the iterative scheme (9.61), which here becomes

$$H_{r+1} = \frac{1}{2}(H_r + H_r^{-1}), \qquad r = 1, 2, 3, \ldots \tag{11.89}$$

with $H_1 = H$, and $H_r \to \text{sgn}(H)$ as $r \to \infty$. It follows (see Problem 4.6) that H^{-1} will have the same lower block triangular form as H, so $H_2, H_2^{-1}, H_3, H_3^{-1}, \ldots$ all have this same form, and we can therefore write

$$H_r = \begin{bmatrix} A_r & 0 \\ -Q_r & -A_r^{\mathrm{T}} \end{bmatrix}, \qquad r \geqslant 1, \tag{11.90}$$

where $A_1 = A$, $Q_1 = Q$.

Exercise 11.41 Show that the inverse of the matrix in (11.90) is

$$H_r^{-1} = \begin{bmatrix} A_r^{-1} & 0 \\ -(A_r^{-1})^{\mathrm{T}}Q_rA_r^{-1} & -(A_r^{-1})^{\mathrm{T}} \end{bmatrix}. \tag{11.91}$$

Substituting (11.90) and (11.91) into (11.89) produces the following expression involving $n \times n$ matrices:

$$A_{r+1} = \frac{1}{2}(A_r + A_r^{-1}) \tag{11.92}$$

$$Q_{r+1} = \frac{1}{2}[Q_r + (A_r^{-1})^{\mathrm{T}}Q_rA_r^{-1}]. \tag{11.93}$$

Comparing (11.88) and (11.90) reveals that the solution of (11.84) is

$$P = \frac{1}{2}\lim_{r \to \infty} Q_r. \tag{11.94}$$

Exercise 11.42 Use (11.92), (11.93) and (11.94) to solve (11.84) when

$$A = \begin{bmatrix} -2 & 1 \\ 3 & -4 \end{bmatrix}, \qquad Q = \begin{bmatrix} 1 & 0 \\ 0 & 1 \end{bmatrix}.$$

Now return to the Riccati equation (11.73), which has the associated hamiltonian matrix H in (11.77) whose Jordan form we denoted by $\mathscr{J}$. Since $H = Z\mathscr{J}Z^{-1}$, where Z is defined in (11.79), it follows as before that

$$\text{sgn}(H) = Z \,\text{sgn}(\mathscr{J})Z^{-1}.$$

Moreover, since $\mathscr{J}$ has eigenvalues $\pm\theta_i$ (Exercise 11.36) with none purely imaginary, we have $\text{sgn}(\mathscr{J}) = \text{diag}[-I, I]$. Define

$$H_s = \text{sgn}(H) + \text{diag}[I, -I] \tag{11.95}$$

$$= Z\begin{bmatrix} -I & 0 \\ 0 & I \end{bmatrix} Z^{-1} + ZZ^{-1}\begin{bmatrix} I & 0 \\ 0 & -I \end{bmatrix}. \tag{11.96}$$

Substitute into (11.96) the partitioned form (11.79) for Z together with

$$Z^{-1} = \begin{bmatrix} W_1 & W_2 \\ W_3 & W_4 \end{bmatrix}$$

to obtain

$$H_s = Z\left(\begin{bmatrix} -W_1 & -W_2 \\ W_3 & W_4 \end{bmatrix} + \begin{bmatrix} W_1 & -W_2 \\ W_3 & -W_4 \end{bmatrix}\right)$$

$$= \begin{bmatrix} Z_1 & Z_2 \\ Z_3 & Z_4 \end{bmatrix}\begin{bmatrix} 0 & -2W_2 \\ 2W_3 & 0 \end{bmatrix}. \tag{11.97}$$

Hence if we write

$$H_s = \begin{bmatrix} H_{s1} & H_{s2} \\ H_{s3} & H_{s4} \end{bmatrix}\begin{matrix} n \\ n \end{matrix}, \tag{11.98}$$

then comparing terms in (11.97) and (11.98) gives

$$H_{s2} = -2Z_1 W_2, \qquad H_{s4} = -2Z_3 W_2,$$

so that

$$H_{s4}H_{s2}^{-1} = (2Z_3 W_2)\left(\frac{1}{2} W_2^{-1} Z_1^{-1}\right)$$

$$= Z_3 Z_1^{-1}.$$

This shows that the positive definite solution $Z_3 Z_1^{-1}$ of the Riccati equation can be computed as $H_{s4}H_{s2}^{-1}$, where H_s is defined in (11.95). The advantage of this approach is that $\text{sgn}(H)$ can be determined from the iterative algorithm (9.61), so there is no need to compute eigenvectors of H.

Exercise 11.43 Use the above procedure to compute the positive definite solution of

$$P \begin{bmatrix} 0 & 0 \\ 0 & 1 \end{bmatrix} P - \begin{bmatrix} 0 & 0 \\ 1 & -2 \end{bmatrix} P - P \begin{bmatrix} 0 & 1 \\ 0 & -2 \end{bmatrix} - \begin{bmatrix} 4 & 0 \\ 0 & 1 \end{bmatrix} = 0.$$

Exercise 11.44 Replace the Jordan form (11.82) of H in (11.77) by

$$\mathscr{J} = \begin{bmatrix} J_1 & 0 \\ J_2 & J_3 \end{bmatrix}.$$

Following a similar procedure to that in Exercises 11.37 and 11.38, show that $P = Z_4 Z_2^{-1}$ is also a solution of the Riccati equation (11.73), assuming Z_2 is non-singular. Hence deduce from (11.97) and (11.98) that $H_{s3} H_{s1}^{-1}$ is a solution of (11.73), and compute this for the equation in Exercise 11.43.

Problems

11.1 Show that if X is an $n \times n$ matrix satisfying the equation $XC^T = CX$, where C is the companion matrix defined in (11.2), then it must have Hankel form.

11.2 Let S be the Sylvester matrix defined in (11.11).

(a) By using the result $\det S = \det b(C)$ show that

$$|\det S| = b_0^n \prod_{i=1}^{m} \prod_{j=1}^{n} (\mu_i - \lambda_j),$$

where λ_i, μ_j are the roots of $a(\lambda)$, $b(\lambda)$ respectively.

(b) Denote by S_p the corresponding matrix for $a(\lambda^p)$, $b(\lambda^p)$, where p is any positive integer. Prove that $S_p = S \otimes I_p$, and hence deduce using Problem 6.28(b) that $\det S_p = (\det S)^p$.

11.3 Consider the polynomials $a(\lambda)$ in (11.1) and $b(\lambda)$ in (11.7) when $m = n - 1$. Use the result in Exercise 11.5, together with (11.18), to show that the bezoutian matrix Z and Schur complement (S/S_1) are related according to $Z = S_1(S/S_1)K$, where S is the Sylvester matrix in (11.12).

11.4 By considering the pairs of complex conjugate roots of a real polynomial $a(\lambda)$, show that if it is asymptotically stable then all its coefficients must be positive.

11.5 Prove that the matrix $\Gamma = [\gamma_{ij}]$ defined by (11.61) and (11.62) satisfies the condition $\Gamma^2 = 2^n I$. Hence determine expressions for $\det \Gamma$ and Γ^{-1}.

11.6 Let $a'(\lambda)$ be the derivative with respect to λ of the polynomial in (11.1). Consider the matrix $D = a'(C)$, where C is the companion matrix (11.2). It has been seen in Problem 6.14 that $\det D$ is a *discriminant* for $a(\lambda)$.

Let $a(\lambda)$ have h_i distinct factors with multiplicity i. By considering the degrees of $a(\lambda)$, and of the g.c.d. of $a(\lambda)$ and $a'(\lambda)$, show that the number of distinct factors of $a(\lambda)$ is equal to $R(D)$.

11.7 Write the Liapunov eqn (11.66) in the form

$$\left(PA + \frac{1}{2}Q\right) + \left(PA + \frac{1}{2}Q\right)^{\mathsf{T}} = 0.$$

Hence deduce that provided A is non-singular, the solution is given by $P = (S - \frac{1}{2}Q)^{-1}$, where S is the real skew symmetric matrix satisfying

$$A^{\mathsf{T}}S + SA = \frac{1}{2}(A^{\mathsf{T}}Q - QA). \qquad (11.99)$$

What is the reduction in the number of unknowns and equations in solving (11.99) rather than (11.66)?

11.8 (a) Let A_1 be a matrix, none of whose eigenvalues μ_i is equal to 1, and define

$$A = (A_1 + I)(A_1 - I)^{-1}. \qquad (11.100)$$

Show that the eigenvalues of A are given by

$$\lambda_i = \frac{\mu_i + 1}{\mu_i - 1}$$

(b) Prove that $\mathrm{Re}(\lambda_i) < 0$ if and only if $|\mu_i| < 1$. This shows that A is a stability matrix if and only if A_1 is a convergent matrix.

11.9 Apply the transformation (11.100) to the Liapunov eqn (11.66), and hence obtain the form (11.69) for the location of the eigenvalues of A_1 relative to the unit circle.

11.10 Consider the *Schwartz matrix*

$$A = \begin{bmatrix} 0 & 1 & 0 & 0 & & & \\ -b_n & 0 & 1 & 0 & & \mathbf{0} & \\ 0 & -b_{n-1} & 0 & 1 & & & \\ & & & & \ddots & & \\ & & & & & 0 & 1 \\ \mathbf{0} & & & & & 0 & -b_2 & -b_1 \end{bmatrix}, \qquad (11.101)$$

where the b_i are all real and non-zero. In eqn (11.66) take

$$Q = 2b_1^2 \operatorname{diag}[0, 0, \ldots, 0, 1].$$

Verify that the controllability condition (11.71) is satisfied, and that the solution of (11.66) is

$$P = \operatorname{diag}[b_1 b_2 \cdots b_n, \, b_1 b_2 \cdots b_{n-1}, \ldots, b_1 b_2, \, b_1].$$

A simple case when $n = 3$ was considered in Exercise 11.26. Deduce that A in (11.101) has inertia $(k, n - k, 0)$ where k is the number of negative terms in the sequence b_1, $b_1 b_2$, $b_1 b_2 b_3, \ldots, b_1 b_2 b_3 \cdots b_n$.

11.11 Let H be an $n \times n$ hermitian matrix partitioned in the form

$$H = \begin{bmatrix} H_1 & H_2 \\ H_2^* & H_3 \end{bmatrix},$$

where H_1 is $k \times k$ and non-singular. By considering the matrix T^*HT, where

$$T = \begin{bmatrix} I_k & -H_1^{-1}H_2 \\ 0 & I_{n-k} \end{bmatrix}$$

show that $\operatorname{In}(H) = \operatorname{In}(H_1) + \operatorname{In}(H/H_1)$, where (H/H_1) is the Schur complement of H_1 in H (defined in Problem 4.7).

11.12 Let $A = S - kI$, where S is real skew-symmetric and k is a real positive scalar. Use (11.66) to show that AD is a stability matrix when D is a diagonal matrix with positive elements.

11.13 Consider the quadratic matrix equation

$$XEX + DX + XF + G = 0,$$

where all the matrices are $n \times n$, and define

$$H = \begin{bmatrix} -F & -E \\ G & D \end{bmatrix}.$$

Let

$$T = \begin{bmatrix} T_1 & T_2 \\ T_3 & T_4 \end{bmatrix}$$

be any matrix such that $T^{-1}HT$ is in Jordan form. Show that, provided T_1 is non-singular, a solution of the equation is $X = T_3 T_1^{-1}$.

11.14 Find in each case for what values of the parameter k:

   (a) the polynomial $\lambda^4 + 4\lambda^3 + 2\lambda^2 + \lambda + k$ is asymptotically stable;
   (b) the polynomial $3\lambda^3 + 4\lambda^2 + k + 1$ is convergent.

11.15 The equations of motion describing a speedometer for a motor vehicle are

$$J_1 \frac{d^2\theta_1}{dt^2} = -k_1(\theta_1 - u) - \mu\left(\frac{d\theta_1}{dt} - \frac{d\theta_2}{dt}\right)$$

$$J_2 \frac{d^2\theta_2}{dt^2} = \mu\left(\frac{d\theta_1}{dt} - \frac{d\theta_2}{dt}\right) - k_2\theta_2,$$

(11.102)

where

   J_1, J_2 = moments of inertia of the damper elements;
   k_1, k_2 = torsional spring constants;
   θ_1, θ_2 = angular displacements for the input and reactor dampers;
   μ = proportionality constant for the reaction damper;
   u = angular position of the wheel.

By defining $x_1 = \theta_1$, $x_2 = \dot{\theta}_1$, $x_3 = \theta_2$, $x_4 = \dot{\theta}_2$, write the equations in the form

$$\dot{x} = Ax + bu,$$

where $x = [x_1, x_2, x_3, x_4]^T$, A is 4×4 and b is 4×1.

   Determine the characteristic polynomial of A, and by applying the Routh test show that when $u = 0$ the system is asymptotically stable provided $J_1 k_2 \neq J_2 k_1$.

11.16 Consider the control system

$$\dot{x}(t) = \begin{bmatrix} 1 & -1 \\ 2 & 3 \end{bmatrix} x(t) + \begin{bmatrix} 1 \\ 0 \end{bmatrix} u(t).$$

   (a) Show that the system is controllable (see (4.37)).
   (b) Use Problem 6.25 to transform it into

$$\dot{z} = Cz + \begin{bmatrix} 0 \\ 1 \end{bmatrix} u$$

(11.103)

   where C is a 2×2 companion matrix in the form (11.2).
   (c) Show that when $u = 0$ the system is unstable.
   (d) If linear feedback

$$u = \alpha_1 z_1 + \alpha_2 z_2$$

   is applied to (11.103) show that the closed-loop system is

asymptotically stable if and only if $\alpha_1 < 5$, $\alpha_2 < -4$. Hence show that linear feedback

$$u = \beta_1 x_1 + \beta_2 x_2$$

stabilizes the original system if and only if $\beta_1 < -4$, $2\beta_2 - 3\beta_1 < 5$.

11.17 A linear control system is described by the equations

$$\dot{x} = x_2$$
$$\dot{x}_2 = -4x_1 - 4x_2 + u.$$

Use (11.73) and (11.74) to determine the linear feedback control $u(t)$ which minimizes the performance index

$$\int_0^\infty (20x_1^2 + 5x_2^2 + u^2)\, dt.$$

11.18 Deduce that the control system described by

$$\dot{x} = C^T x + r_1^T u$$

where C is defined in (11.2) and r_1 in (11.9), is controllable (see (4.37)) if and only if the polynomials $a(\lambda)$ and $b(\lambda)$ in (11.2) and (11.7) are relatively prime.

11.19 Assume that all the eigenvalues of an $n \times n$ matrix A are distinct, and let u_i be a right eigenvector corresponding to an eigenvalue λ_i. By setting $B = 0$ in (11.77) and (11.83), show that the solution of the Liapunov equation $A^T P + PA = -Q$ when A is a stability matrix can be expressed as

$$P = [c_1, c_2, \ldots, c_n] U^{-1},$$

where $U = [u_1, u_2, \ldots, u_n]$ and $c_i = -(A^T + \lambda_i I)^{-1} Q u_i$.

11.20 Show that the solution in the preceding problem can be written as $P = -(U^T)^{-1} P_1 U^{-1}$ where $P_1 = [p_{ij}]$ and

$$p_{ij} = \frac{u_i^T Q u_j}{\lambda_i + \lambda_j} = p_{ji}.$$

Hence show that this form of solution is the same as that obtained by diagonalizing A. (Compare with Exercise 11.40.)

References

Barnett (1983, 1984), Barnett and Cameron (1985), Barnett and Storey (1970), Fiedler (1986), Jury (1982), Knuth (1969), Kučera (1979), Ogata (1987), Willems (1970).

12 Polynomial and rational matrices

Example 12.1 Consider the pair of equations (11.102) in Problem 11.15 which describe a certain simple dynamical model. Suppose that the initial conditions for θ_1, $\dot\theta_1$, θ_2, $\dot\theta_2$ are all zero. Applying the Laplace transform to these equations then produces

$$J_1 s^2 \bar\theta_1 = -k_1(\bar\theta_1 - \bar u) - \mu(s\bar\theta_1 - s\bar\theta_2)$$
$$J_2 s^2 \bar\theta_2 = \mu(s\bar\theta_1 - s\bar\theta_2) - k_2\bar\theta_2, \tag{12.1}$$

where $\bar\theta_1(s) = \mathcal{L}\{\theta_1(t)\}$, and so on. After rearrangement (12.1) can be written as

$$\begin{bmatrix} J_1 s^2 + \mu s + k_1 & -\mu s \\ -\mu s & J_2 s^2 + \mu s + k_2 \end{bmatrix} \begin{bmatrix} \bar\theta_1 \\ \bar\theta_2 \end{bmatrix} = \begin{bmatrix} k_1 \\ 0 \end{bmatrix} \bar u. \tag{12.2}$$

The 2×2 matrix $A(s)$ on the left in (12.2) is an example of a *polynomial matrix*, whose elements are themselves polynomials. As for a scalar polynomial, $A(s)$ can be written in terms of powers of s:

$$A(s) = \begin{bmatrix} J_1 & 0 \\ 0 & J_2 \end{bmatrix} s^2 + \begin{bmatrix} \mu & -\mu \\ -\mu & \mu \end{bmatrix} s + \begin{bmatrix} k_1 & 0 \\ 0 & k_2 \end{bmatrix}$$
$$= A_0 s^2 + A_1 s + A_2.$$

Assuming that $d(s) = \det A(s) \neq 0$, the solution of (12.2) can be written as

$$\begin{bmatrix} \bar\theta_1 \\ \bar\theta_2 \end{bmatrix} = A^{-1}(s) \begin{bmatrix} k_1 \\ 0 \end{bmatrix} \bar u,$$

where by the simple formula (4.7)

$$A^{-1}(s) = \begin{bmatrix} \dfrac{J_2 s^2 + \mu s + k_2}{d} & \dfrac{\mu s}{d} \\[2ex] \dfrac{\mu s}{d} & \dfrac{J_1 s^2 + \mu s + k_1}{d} \end{bmatrix} \tag{12.3}$$

and

$$d = d(s) = (J_1 s^2 + \mu s + k_1)(J_1 s^2 + \mu s + k_2) - \mu^2 s^2.$$

Since each element of $A^{-1}(s)$ in (12.3) is a rational function (a ratio of two polynomials), it is an example of a *rational matrix*. Both polynomial and rational matrices arise in a number of areas including control and network problems. In this chapter we give only an introduction to the subject, further details of the theory and applications being available in the references listed.

12.1 Basic properties of polynomial matrices

To conform with our notation for polynomials we shall use λ for the indeterminate, so a general $m \times n$ polynomial matrix $A(\lambda)$ can be written as

$$A(\lambda) = A_0\lambda^N + A_1\lambda^{N-1} + \cdots + A_{N-1}\lambda + A_N, \qquad (12.4)$$

where each A_i is a constant $m \times n$ matrix. The *degree* of $A(\lambda)$ is N, assuming that the *leading coefficient matrix* A_0 is non-zero. The basic operations of addition, subtraction and multiplication of two or more polynomial matrices are defined in exactly the same way as was done for scalar matrices in Section 2.2.

Exercise 12.1 Determine $A(\lambda) + B(\lambda)$ and $A(\lambda)B(\lambda)$ when

$$
\left.
\begin{aligned}
A(\lambda) &= \begin{bmatrix} 2\lambda^2 + 1 & \lambda^2 + 3\lambda \\ -2\lambda^2 + \lambda - 3 & -\lambda^2 + 4\lambda + 2 \end{bmatrix} \\
B(\lambda) &= \begin{bmatrix} 4\lambda^2 + \lambda & -\lambda^2 + 2\lambda - 3 \\ -8\lambda^2 + 4 & 2\lambda^2 - \lambda + 5 \end{bmatrix}
\end{aligned}
\right\}. \qquad (12.5)
$$

The reader who completes Exercise 12.1 will find that the degree of the product is only three, whereas each of $A(\lambda)$ and $B(\lambda)$ has degree two. This is because in (12.5)

$$A_0 = \begin{bmatrix} 2 & 1 \\ -2 & -1 \end{bmatrix}, \qquad B_0 = \begin{bmatrix} 4 & -1 \\ -8 & 2 \end{bmatrix}$$

and $A_0B_0 = 0$.

Exercise 12.2 Prove that the product of any two non-zero $n \times n$ constant matrices is zero only if both are singular. Hence deduce that if $A(\lambda)$, $B(\lambda)$ are each $n \times n$ and have degrees N_1, N_2 respectively then the degree of AB is $N_1 + N_2$ provided at least one of A_0, B_0 is non-singular.

As we saw in Section 4.1, the determinant of an $n \times n$ matrix consists of a sum of $n!$ terms, each of which is the product of n

elements of the matrix. By definition, each element of $A(\lambda)$ in (12.4) has degree at most N, so when $m = n$ the degree of $\det A(\lambda)$ is at most nN.

Example 12.2 For the matrix $A(\lambda)$ in (12.5)

$$\det A(\lambda) = (2\lambda^2 + 1)(-\lambda^2 + 4\lambda + 2) - (\lambda^2 + 3\lambda)(-2\lambda^2 + \lambda - 3)$$
$$= 13\lambda^3 + 3\lambda^2 + 13\lambda + 2.$$

Here $nN = 2 \cdot 2 = 4$, but the degree of $\det A(\lambda)$ is only three.

In general, when $A(\lambda)$ in (12.4) is $n \times n$ it follows from the definition of determinant that the coefficient of the highest degree term λ^{nN} in the expansion of $\det A(\lambda)$ is $\det A_0$. Example 12.2 illustrates that when A_0 is singular the degree of $\det A(\lambda)$ is less than nN. When A_0 is non-singular then $A(\lambda)$ is called *regular,* and in particular $A(\lambda)$ is *monic* when $A_0 = I_n$. The inverse of $A(\lambda)$ will not in general be a polynomial matrix, as we saw in Example 12.1. However, recall from (4.46) that

$$A^{-1}(\lambda) = \frac{\operatorname{adj} A(\lambda)}{\det A(\lambda)}. \tag{12.6}$$

The elements of $\operatorname{adj} A(\lambda)$ are minors of $A(\lambda)$, and are therefore polynomials. Hence, if $\det A(\lambda)$ is a non-zero scalar it follows from (12.6) that $A^{-1}(\lambda)$ *will* be a polynomial matrix, in which case $A(\lambda)$ is called *unimodular* (or *invertible*). Since $\det A(\lambda)$ is then independent of λ its value will be unaltered by setting $\lambda = 0$ in (12.4), so for a unimodular matrix $\det A(\lambda) = \det A_N$. Clearly, a necessary but not sufficient condition for $A(\lambda)$ to be unimodular is that $\det A_0 = 0$, which ensures that there is no term in λ^{nN} in $\det A(\lambda)$.

Exercise 12.3 Verify that

$$A(\lambda) = \begin{bmatrix} \lambda^3 + 3\lambda^2 + 5\lambda + 4 & \lambda^2 + 2\lambda + 3 \\ 5\lambda^3 + 15\lambda^2 + 17\lambda + 12 & 5\lambda^2 + 10\lambda + 7 \end{bmatrix}$$

is unimodular, and hence write down $A^{-1}(\lambda)$.

We now suppose that $A(\lambda)$ is square, and turn to the question of division of an $n \times n$ matrix in (12.4) by a second $n \times n$ polynomial matrix

$$B(\lambda) = B_0 \lambda^M + B_1 \lambda^{M-1} + \cdots + B_M, \tag{12.7}$$

where we can assume that $M \leq N$. Provided $B(\lambda)$ is regular then there exist *unique* matrices $Q_1(\lambda)$ and $R_1(\lambda)$ such that

$$A(\lambda) = Q_1(\lambda)B(\lambda) + R_1(\lambda), \tag{12.8}$$

where the degree of $R_1(\lambda)$ is less than M. The matrix $Q_1(\lambda)$ is called a *right* quotient, and $R_1(\lambda)$ a *right* remainder, because there is another expression when $A(\lambda)$ is divided by $B(\lambda)$ on the *left*, namely

$$A(\lambda) = B(\lambda)Q_2(\lambda) + R_2(\lambda). \tag{12.9}$$

In (12.9) $Q_2(\lambda)$ and $R_2(\lambda)$ are the left quotient and remainder respectively, and the degree of $R_2(\lambda)$ is less than M. When $B(\lambda)$ is not regular, division is still possible, but the quotient and remainder are not unique.

Example 12.3 Let $A(\lambda)$ be the matrix in (12.5), and suppose

$$B(\lambda) = \begin{bmatrix} \lambda + 3 & \lambda - 1 \\ -\lambda + 2 & 4 \end{bmatrix} \tag{12.10}$$

so that $M = 1$. Since

$$\det B_0 = \det \begin{bmatrix} 1 & 1 \\ -1 & 0 \end{bmatrix} = 1,$$

$B(\lambda)$ is regular. We can determine the right quotient and remainder $Q_1(\lambda)$ and $R_1(\lambda)$ in (12.8) by multiplying (12.8) on the right by $B^{-1}(\lambda)$ to give

$$A(\lambda)B^{-1}(\lambda) = Q_1(\lambda) + R_1(\lambda)B^{-1}(\lambda).$$

This shows that $Q_1(\lambda)$ is the polynomial part of the product $A(\lambda)B^{-1}(\lambda)$. The inverse of (12.10) is

$$B^{-1}(\lambda) = \frac{1}{d(\lambda)} \begin{bmatrix} 4 & -\lambda + 1 \\ \lambda - 2 & \lambda + 3 \end{bmatrix},$$

where $d(\lambda) = \det B(\lambda) = \lambda^2 + \lambda + 14$, and hence by direct multiplication

$$A(\lambda)B^{-1}(\lambda) = \frac{1}{d(\lambda)} \begin{bmatrix} \lambda^3 + 9\lambda^2 - 6\lambda + 4 & -\lambda^3 + 8\lambda^2 + 8\lambda + 1 \\ -\lambda^3 - 2\lambda^2 - 2\lambda - 16 & \lambda^3 - 2\lambda^2 + 18\lambda + 3 \end{bmatrix}$$

$$= \begin{bmatrix} \lambda + 8 & -\lambda + 9 \\ -\lambda - 1 & \lambda - 3 \end{bmatrix} + \frac{1}{d(\lambda)} \begin{bmatrix} -28\lambda - 108 & 13\lambda - 125 \\ 13\lambda - 2 & 7\lambda + 45 \end{bmatrix}.$$

$$\underbrace{\phantom{\begin{bmatrix} \lambda + 8 & -\lambda + 9 \\ -\lambda - 1 & \lambda - 3 \end{bmatrix}}}_{Q_1(\lambda)} \tag{12.11}$$

The right quotient $Q_1(\lambda)$ is the polynomial matrix part in (12.11).

Finally, from (12.8) we have

$$R_1(\lambda) = A(\lambda) - Q_1(\lambda)B(\lambda)$$
$$= \begin{bmatrix} -41 & -28 \\ 6 & 13 \end{bmatrix},$$

which has degree zero ($<M$).

Exercise 12.4 Find the left quotient and left remainder when $A(\lambda)$ in (12.5) is divided by $B(\lambda)$ in (12.10).

We can see that the preceding discussion still applies even when both $A(\lambda)$ and $B(\lambda)$ are rectangular. For division of $A(\lambda)$ by $B(\lambda)$ on the right (left) the two matrices must have the same numbers of columns (rows). Thus, for example, (12.8) still holds with $A(\lambda)$, $Q_1(\lambda)$ and $R_1(\lambda)$ all being $m \times n$. In particular, if $R_1(\lambda)$ in (12.8) is identically zero, then $A(\lambda)$ is *divisible on the right* by $B(\lambda)$, and $B(\lambda)$ is a *right divisor* of $A(\lambda)$. Suppose that $A_1(\lambda)$, $A_2(\lambda)$ are two $m \times n$ polynomial matrices, and that $D(\lambda)$ is a right divisor of each of them. Then $D(\lambda)$ is called a *common right divisor* of $A_1(\lambda)$ and $A_2(\lambda)$, and $A_1 = Q_1D$, $A_2 = Q_3D$ for some polynomial matrices $Q_1(\lambda)$, $Q_3(\lambda)$. If $\bar{D}(\lambda)$, $D_1(\lambda)$, $D_2(\lambda)$, . . . are common right divisors of $A_1(\lambda)$ and $A_2(\lambda)$ such that $\bar{D} = X_iD_i$ for polynomial matrices X_i, $i = 1, 2, \ldots$, then $\bar{D}(\lambda)$ is a *greatest* common right divisor (g.c.r.d.) of $A_1(\lambda)$ and $A_2(\lambda)$. These concepts are natural generalizations of those for scalar polynomials, and apply equally to division on the left, with obvious modifications. Notice however an important difference: a g.c.r.d. of two polynomial matrices is not unique even if it is monic. For suppose that $\bar{D}(\lambda)$ and $\hat{D}(\lambda)$ are two g.c.r.d.'s of $A_1(\lambda)$ and $A_2(\lambda)$. Then by the definition of g.c.r.d. we can write $\bar{D} = X\hat{D}$, $\hat{D} = Y\bar{D}$ for some polynomial matrices X and Y. Combining these two expressions together gives $\bar{D} = XY\bar{D}$, so that $XY = I$. This in turn implies $X^{-1} = Y$, showing that the inverse of X is a polynomial matrix, i.e. $X(\lambda)$ is unimodular. In other words, a g.c.r.d. is unique only up to premultiplication by an arbitrary unimodular matrix.

12.2   Elementary operations and Smith normal form

In this section we list extensions of some results developed for matrices with scalar entries in Chapter 5.

(i)   The *rank* of a polynomial matrix is equal to the order of the largest square submatrix whose determinant is not identically zero.

(ii) Using 'line' to stand for either a row or a column of a polynomial matrix, the *elementary operations* defined in Section 5.3.1 become:

(a) interchange any two lines: $(Li) \leftrightarrow (Lj)$;

(b) multiply any line by a nonzero scalar: $(Li) \times k$;

(c) add to any line any other line multiplied by an arbitrary polynomial: $(Li) + p(\lambda)(Lj)$.

On referring back, it will be seen that only (c) differs from its scalar counterpart.

(iii) Elementary operations as defined in (ii) do not alter the rank of a polynomial rank.

(iv) An *elementary* matrix is a polynomial matrix obtained by applying a single elementary operation to I_n, and is unimodular. Specifically, if E is the matrix obtained by applying a row operation to I_n, then EA is the matrix obtained by applying the *same* operation to A. For column operations, A is postmultiplied by the appropriate elementary matrix (compare with Problem 5.7). Applying a sequence of row (column) operations is thus equivalent to premultiplying (postmultiplying) by a succession of unimodular matrices $\ldots E_3 E_2 E_1$, and this product is also unimodular.

We can now indicate one way of obtaining a g.c.r.d. of the two matrices $A(\lambda)$, $B(\lambda)$ defined in (12.4) and (12.7) respectively. Begin with the matrix

$$C(\lambda) = \begin{bmatrix} B(\lambda) \\ A(\lambda) \end{bmatrix} \begin{matrix} n \\ m \end{matrix} \qquad (12.12)$$

and reduce it, by elementary *row* operations *only*, to the form

$$\begin{bmatrix} D(\lambda) \\ 0 \end{bmatrix} \begin{matrix} n \\ m \end{matrix}.$$

Then $D(\lambda)$ is a g.c.r.d. To see why this is so, it follows from (iv) above that this reduction of (12.12) can be achieved by premultiplication by an appropriate unimodular matrix, that is

$$\begin{bmatrix} T_1 & T_2 \\ T_3 & T_4 \end{bmatrix} \begin{bmatrix} B \\ A \end{bmatrix} = \begin{bmatrix} D \\ 0 \end{bmatrix}$$

for certain polynomial matrices T_1, T_2, T_3, T_4. Multiplying out produces

$$T_1 B + T_2 A = D. \tag{12.13}$$

However, if $\hat{D}$ is any other common right divisor of A and B, then by definition $B = \hat{B}\hat{D}$, $A = \hat{A}\hat{D}$, and substituting into (12.13) gives

$$(T_1 \hat{B} + T_2 \hat{A})\hat{D} = D,$$

showing that D is indeed a greatest common right divisor. Similarly, to obtain a great common left divisor (g.c.l.d.) $D_1(\lambda)$ of $B(\lambda)$ and an $n \times m$ matrix $A(\lambda)$, reduce

$$\begin{matrix} n & \quad m \\ n[B(\lambda) & A_1(\lambda)] \end{matrix} \tag{12.14}$$

by elementary *column* operations to the form $[D_1(\lambda) \quad 0]$, where D_1 is $n \times n$.

Example 12.4   The matrix

$$D_1(\lambda) = \begin{bmatrix} 1 & 0 \\ 2 & \lambda - 1 \end{bmatrix} \tag{12.15}$$

is a g.c.l.d. of

$$A_1(\lambda) = \begin{bmatrix} \lambda + 2 \\ 6 \end{bmatrix}, \qquad B(\lambda) = \begin{bmatrix} 1 & \lambda + 1 \\ \lambda + 1 & \lambda + 3 \end{bmatrix}, \tag{12.16}$$

and, as the reader can easily check,

$$A_1(\lambda) = D_1(\lambda)\begin{bmatrix} \lambda + 2 \\ -2 \end{bmatrix}, \qquad B(\lambda) = D_1(\lambda)\begin{bmatrix} 1 & \lambda + 1 \\ 1 & -1 \end{bmatrix}.$$

Exercise 12.5   Verify directly that

$$\bar{D}_1(\lambda) = D_1(\lambda)\begin{bmatrix} \lambda + 3 & \lambda + 2 \\ \lambda + 4 & \lambda + 3 \end{bmatrix},$$

where $D_1(\lambda)$ is given in (12.15), is also a common left divisor of the matrices in (12.16).

Exercise 12.6 Determine a g.c.l.d. of the matrices in (12.16) by applying elementary column operations to (12.14). Compare your result with (12.15).

(v) Two polynomial matrices $A(\lambda)$, $B(\lambda)$ are *equivalent* if it is possible to pass from one to the other by a sequence of elementary operations. The equivalence transformation can be represented by

$$A(\lambda) = P(\lambda)B(\lambda)Q(\lambda), \qquad (12.17)$$

where $P(\lambda)$ and $Q(\lambda)$ are unimodular matrices.

For the case of regular matrices with degree one, so that $A(\lambda) = A_0\lambda + A_1$, $B(\lambda) = B_0\lambda + B_1$ with A_0 and B_0 non-singular, then there exist *constant* non-singular matrices P, Q satisfying (12.17). Indeed, it follows from this result that two square matrices A_1, B_1 are similar if and only if their characteristic matrices $\lambda I - A_1$, $\lambda I - B_1$ are equivalent. This fact was quoted in 8.2 to justify the term 'similarity invariants' for the invariant factors of a characteristic matrix.

Exercise 12.7 Prove that if A_1 and B_1 are similar then $\lambda I - A_1$, $\lambda I - B_1$ are equivalent.

(vi) The normal form for matrices with scalar entries was described in Section 5.3.3. Correspondingly, any polynomial matrix $A(\lambda)$ of rank r can be reduced by elementary row and column operations to the *Smith normal* (or *canonical*) *form* (Gantmacher 1959)

$$S(\lambda) = \left[\begin{array}{c:c} \mathrm{diag}\{i_1(\lambda), i_2(\lambda), \ldots, i_r(\lambda)\} & 0 \\ \hdashline 0 & 0 \end{array} \right], \qquad (12.18)$$

where $i_1(\lambda), \ldots, i_r(\lambda)$ are the unique *invariant factors* (or *polynomials*) of $A(\lambda)$. These are defined by

$$i_k(\lambda) = \frac{d_k(\lambda)}{d_{k-1}(\lambda)}, \qquad k = 1, 2, \ldots, r, \qquad (12.19)$$

where d_k is the kth determinantal divisor of $A(\lambda)$, being the monic greatest common divisor of all minors of $A(\lambda)$ of order k (with $d_0(\lambda) = 1$). These definitions have already been given in Section 8.2 for the special case $A(\lambda) = \lambda I - A$, with A a constant square matrix.

(vii) It follows from (12.19) that each invariant polynomial $i_k(\lambda)$ is a factor of the next one $i_{k+1}(\lambda)$, for $k = 1, 2, \ldots, r - 1$. Furthermore, two polynomial matrices having the same dimensions are equivalent if and only if they have the same invariant factors. In other words, the invariant factors of a matrix are unaltered by the application of elementary operations—hence the name 'invariant'.

(viii) Over the complex field, each invariant polynomial can be written as

$$i_k(\lambda) = (\lambda - \alpha_1)^{t_1}(\lambda - \alpha_2)^{t_2} \cdots (\lambda - \alpha_s)^{t_s}, \quad (12.20)$$

where $(\lambda - \alpha_1), \ldots, (\lambda - \alpha_s)$ are the distinct linear factors of $i_r(\lambda)$, and each t_i is a non-negative integer. The complete set of *all* the non-trivial factors displayed in (12.20), for $k = 1, 2, \ldots, r$, including repetitions, constitutes the set of *elementary divisors* of $A(\lambda)$, and depends upon the field of numbers being used. Again, these were defined for $\lambda I - A$ in Section 8.2.

Exercise 12.8 Show that when $A(\lambda)$ is $n \times n$ and regular then $\det A(\lambda) = i_1(\lambda)i_2(\lambda) \cdots i_n(\lambda)$. Hence deduce that $\alpha_1, \alpha_2, \ldots, \alpha_s$ in (12.20) are the roots of $\det A(\lambda) = 0$ (called the *latent roots* or *eigenvalues* of $A(\lambda)$).

Example 12.5 We wish to find the Smith form of the polynomial matrix

$$A(\lambda) = \begin{bmatrix} \lambda + 3 & \lambda + 2 & \lambda + 4 \\ 2\lambda^3 + 6\lambda^2 + \lambda & 2\lambda^3 + 4\lambda^2 + \lambda & 3\lambda^3 + 8\lambda^2 + \lambda \\ \lambda^2 + 4\lambda + 3 & \lambda^2 + 3\lambda + 2 & 3\lambda^2 + 7\lambda + 4 \end{bmatrix}. \quad (12.21)$$

By inspection of the elements in (12.21) it is clear that their g.c.d. is $d_1(\lambda) = 1$. Since $i_1(\lambda) = 1$, we apply the operation (C1) − (C2) to (12.21) so as to make the 1, 1 element equal to 1, producing

$$\begin{bmatrix} 1 & \lambda + 2 & \lambda + 4 \\ 2\lambda^2 & 2\lambda^3 + 4\lambda^2 + \lambda & 3\lambda^3 + 8\lambda^2 + \lambda \\ \lambda + 1 & \lambda^2 + 3\lambda + 2 & 3\lambda^2 + 7\lambda + 4 \end{bmatrix}$$

$$\rightarrow \begin{bmatrix} 1 & \lambda + 2 & \lambda + 4 \\ 0 & \lambda & \lambda^3 + \lambda \\ 0 & 0 & 2\lambda^2 + 2\lambda \end{bmatrix} \quad \begin{array}{l} (R2) - 2\lambda^2(R1) \\ (R3) - (\lambda + 1)(R1) \end{array}$$

$$\rightarrow \begin{bmatrix} 1 & 0 & 0 \\ 0 & \lambda & \lambda^3 + \lambda \\ 0 & 0 & 2\lambda^2 + 2\lambda \end{bmatrix} \quad \begin{array}{l} (C2) - (\lambda + 2)C1 \\ (C3) - (\lambda + 4)C1 \end{array}$$

$$\rightarrow \begin{bmatrix} 1 & 0 & 0 \\ 0 & \lambda & 0 \\ 0 & 0 & \lambda^2 + \lambda \end{bmatrix} \quad \begin{array}{l} (C3) - (\lambda^2 + 1)C2 \\ \frac{1}{2}(R3) \end{array}$$

This last matrix is the Smith form of $A(\lambda)$ in (12.21). The invariant factors of $A(\lambda)$ are therefore $i_1(\lambda) = 1$, $i_2(\lambda) = \lambda$, $i_3(\lambda) = \lambda(\lambda + 1)$ and the elementary divisors are λ, λ, $\lambda + 1$.

Exercise 12.19 Reduce the matrix

$$\begin{bmatrix} 2\lambda & 2\lambda - 1 & \lambda + 3 \\ -2\lambda^2 - \lambda & -2\lambda^2 & -\lambda^2 - 3\lambda \\ 2\lambda^2 + 4\lambda & 2\lambda^2 + 3\lambda - 2 & \lambda^2 + 6\lambda + 7 \end{bmatrix}$$

to its Smith normal form by elementary operations.

Exercise 12.10 For each of the following sets of invariant polynomials of a matrix, find the elementary divisors over the field of real numbers.

(a) $i_1(\lambda) = \lambda^2 + 2\lambda$, $i_2(\lambda) = \lambda^3 + 2\lambda^2$, $i_3(\lambda) = \lambda^6 + 4\lambda^5 + 4\lambda^4$;
(b) $i_1(\lambda) = \lambda + 2$, $i_2(\lambda) = \lambda^2 - 4$, $i_3(\lambda) = (\lambda^2 - 4)^2(\lambda + 3)$,
$i_4(\lambda) = (\lambda^2 - 4)^3(\lambda + 3)^2$.

Exercise 12.11 Each of the following represents the elementary divisors for a polynomial matrix of rank seven. Determine the invariant factors in each case.

(a) λ, λ, $\lambda - 2$, $\lambda - 1$, $\lambda + 1$, $\lambda + 2$;
(b) $\lambda + 2$, $(\lambda + 2)^2$, $(\lambda + 2)^2$, $(\lambda + 2)^3$, $(\lambda + 2)^4$, $(\lambda - 1)^2$

(note: $i_7 = $ least common multiple of the divisors; $i_6 = $ least common multiple of those not in i_7, and so on).

12.3 Prime matrices

12.3.1 Relative primeness

If a g.c.r.d. of $A(\lambda)$ in (12.4) and $B(\lambda)$ in (12.7) is unimodular then these two matrices are said to be *relatively right prime* (r.r.p.). This implies that it is possible to reduce the matrix $C(\lambda)$

in (12.12) by elementary row operations to the form

$$\begin{bmatrix} I_n \\ 0 \end{bmatrix}. \tag{12.22}$$

In other words, polynomal matrices $X(\lambda)$, $Y(\lambda)$, $Z_1(\lambda)$, $Z_2(\lambda)$ can be found such that

$$\begin{bmatrix} X & Y \\ Z_1 & Z_2 \end{bmatrix} \begin{bmatrix} B \\ A \end{bmatrix} = \begin{bmatrix} I \\ 0 \end{bmatrix}, \tag{12.23}$$

where the first partitioned matrix in (12.23) is unimodular. Expanding (12.23) gives

$$XB + YA = I_n. \tag{12.24}$$

The converse also holds, so a necessary and sufficient condition for $A(\lambda)$ and $B(\lambda)$ to be r.r.p. is either that there exist matrices $X(\lambda)$, $Y(\lambda)$ satisfying the polynomial matrix eqn (12.24); or, that the Smith form of $C(\lambda)$ in (12.12) is (12.22). It is interesting to note that when $m = n = 1$, so that all the terms in (12.24) are scalars, then (12.24) is precisely the diophantine eqn (11.36) for scalar polynomials which we studied in Section 11.3.3. However, in the matrix case we cannot say anything about the degrees of $X(\lambda)$, $Y(\lambda)$ in (12.24).

Example 12.6   When

$$A(\lambda) = \begin{bmatrix} 1 & 0 \\ 0 & \lambda \end{bmatrix}, \qquad B(\lambda) = \begin{bmatrix} 1 & 0 \\ 0 & \lambda \end{bmatrix}, \tag{12.25}$$

then it is trivial to observe that a g.c.r.d. is also

$$\begin{bmatrix} 1 & 0 \\ 0 & \lambda \end{bmatrix}.$$

However, when

$$A(\lambda) = \begin{bmatrix} 1 & 0 \\ 0 & \lambda \end{bmatrix}, \qquad B(\lambda) = \begin{bmatrix} \lambda & 0 \\ 0 & 1 \end{bmatrix}, \tag{12.26}$$

then the reader should check that the Smith form of

$$C = \begin{bmatrix} B \\ A \end{bmatrix} \quad \text{is} \quad \begin{bmatrix} I_2 \\ 0 \end{bmatrix},$$

showing that the two matrices in (12.26) are r.r.p. However, all four matrices in (12.25) and (12.26) have the same invariant polynomials 1, λ. This illustrates that a knowledge of the invariant polynomials of a pair of polynomial matrices is not sufficient to determine whether they are relatively prime.

In a completely analogous fashion, $B(\lambda)$ and an $n \times m$ matrix $A_1(\lambda)$ are relatively left prime if and only if $[B(\lambda) \quad A_1(\lambda)]$ has Smith form $[I_n \quad 0]$; or, if the equation corresponding to (12.24), namely

$$BX_1 + A_1 Y_1 = I_n$$

has a polynomial matrix solution $X_1(\lambda)$, $Y_1(\lambda)$. Alternatively, a convenient way of handling this case is to notice that $B(\lambda)$ and $A_1(\lambda)$ are relatively left prime if and only if $B^{\mathsf{T}}(\lambda)$ and $A_1^{\mathsf{T}}(\lambda)$ are relatively right prime.

Exercise 12.12 Deduce that if $A(\lambda)$ and $B(\lambda)$ are r.r.p. then $C(\lambda)$ in (12.12) has rank n for all values of λ.

Exercise 12.13 Use the result in Problem 12.3 to test whether the following pairs of matrices are r.r.p.:

(a) $A(\lambda) = \begin{bmatrix} 0 & (\lambda + 2)^2 \\ (\lambda + 3)^2 & (\lambda + 4) \end{bmatrix}$, $\quad B(\lambda) = \begin{bmatrix} \lambda & 0 \\ \lambda^2 & \lambda \end{bmatrix}$;

(b) $A(\lambda) = \begin{bmatrix} \lambda & \lambda + 1 \\ -\lambda^2 & \lambda \\ \lambda^2 + 2\lambda & \lambda + 2 \end{bmatrix}$, $\quad B(\lambda) = \begin{bmatrix} \lambda^2 + 5\lambda + 6 & 0 \\ 0 & \lambda + 2 \end{bmatrix}$.

12.3.2 Skew primeness

A second, and quite different, type of primeness associated with two polynomial matrices involves their determinants. Suppose that the $n \times n$ matrix $B(\lambda)$ in (12.4) having degree M is *regular*, so that we can set $B_0 = I_n$. By analogy with the companion matrix associated with a scalar polynomial (see Section 11.1) we define the $nM \times nM$ *block companion* matrix

$$C_B = \begin{bmatrix} 0 & I_n & & & \\ 0 & 0 & I_n & & 0 \\ & & & \ddots & \\ & & & & I_n \\ -B_M & -B_{M-1} & \cdots & -B_2 & -B_1 \end{bmatrix}, \quad (12.27)$$

which has the fundamental property

$$\det(\lambda I_{nM} - C_B) = \det B(\lambda).$$

When $n = 1$, (12.27) reduces to the companion form displayed in (11.2).

Return to $A(\lambda)$ in (12.4) and suppose that it is also $n \times n$. The two matrices $A(\lambda)$, $B(\lambda)$ are called *skew prime* if their determinants (which are of course scalar polynomials) are relatively prime in the usual sense. We describe two ways of expressing the condition for skew primeness, and both are interesting generalizations of scalar results which we encountered in Chapter 11. The first extends the companion matrix form of resultant in Section 11.2.1: $A(\lambda)$ and $B(\lambda)$ are skew prime if and only if the $n^2M \times n^2M$ matrix

$$A_0 \otimes C_B^N + A_1 \otimes C_B^{N-1} + \cdots + A_{N-1} \otimes C_B + A_N \otimes I_{nM} \quad (12.28)$$

is non-singular. Notice how (12.28) is formed from (12.4) by replacing powers of λ by powers of C_B, and taking the Kronecker product for each term. When $n = 1$ the A_i are scalars and (12.28) reduces to a polynomial in a companion matrix, as in (11.8).

The second approach requires $A(\lambda)$ to be regular, so we can set $A_0 = I$ in (12.4). Then a necessary and sufficient condition for skew primeness is that the equation

$$XB + AY = I_n \quad (12.29)$$

has a unique solution for the polynomial matrices $X(\lambda)$, $Y(\lambda)$ with the degree of $X(\lambda)$ less than N, and the degree of $Y(\lambda)$ less than M. It is intriguing that the only difference between (12.29) and (12.24) (when $n = m$) is in the order of the terms in the product of $A(\lambda)$ and $Y(\lambda)$, yet both equations reduce to the same scalar case (11.36) when $n = m = 1$.

Example 12.6 (continued) The two matrices in (12.26) are r.r.p. but not skew prime since each has determinant λ. This illustrates that relative primeness does not in general imply skew primeness. The converse result does however hold (see Problem 12.5).

The Sylvester and bezoutian forms of resultant described in Chapter 11 can also be extended in various ways to the polynomial matrix case, and details can be found in Barnett (1983).

Exercise 12.14 Verify directly by evaluating their determinants that

$$A(\lambda) = \begin{bmatrix} \lambda + 1 & 1 \\ 0 & \lambda + 1 \end{bmatrix}, \qquad B(\lambda) = \begin{bmatrix} \lambda & 2 \\ 2 & \lambda \end{bmatrix}$$

are skew prime. Determine the solution of (12.29) in this case.

12.4 Rational matrices

12.4.1 Smith–McMillan form

An example of a rational matrix was displayed in (12.3). Since the denominator in each element in (12.3) is the same, then multiplying both sides by $d(s)$ shows that $d(s)A^{-1}(s)$ is a polynomial matrix, to which we can apply the techniques developed so far in this chapter.

In general, let $G(\lambda)$ be an arbitrary $m \times n$ rational matrix whose elements are rational functions. If each of these rational functions is proper (i.e. the degree of the numerator is less than the degree of the denominator) then $G(\lambda) \to 0$ as $\lambda \to \infty$ and $G(\lambda)$ is called *strictly proper*. Let $g(\lambda)$ denote the least common multiple of the denominators of all the elements of $G(\lambda)$. Then $g(\lambda)G(\lambda)$ will be a polynomial matrix whose Smith form can be expressed as in (12.18), where now r is the rank of gG. Thus, as in (12.17), gG can be written as

$$gG = P(\lambda)S(\lambda)Q(\lambda), \tag{12.30}$$

where $P(\lambda)$ and $Q(\lambda)$ are unimodular polynomial matrices. Divide both sides of (12.30) by $g(\lambda)$ and cancel out any common factors between $g(\lambda)$ and the invariant polynomials in the Smith form. We then end up with

$$G(\lambda) = P(\lambda)S_M(\lambda)Q(\lambda), \tag{12.31}$$

where the *Smith–McMillan* form of $G(\lambda)$ is

$$S_M(\lambda) = \left[\begin{array}{c|c} \mathrm{diag}\{\varphi_1(\lambda),\ \varphi_2(\lambda),\ \ldots,\ \varphi_r(\lambda)\} & 0 \\ \hline 0 & 0 \end{array} \right]. \tag{12.32}$$

In (12.32):

(i) $\varphi_i(\lambda) = \varepsilon_i(\lambda)/\psi_i(\lambda)$, $i = 1, 2, \ldots, r$, and each pair of monic polynomials $\varepsilon_i(\lambda)$, $\psi_i(\lambda)$ is relatively prime;

(ii) $\varepsilon_i(\lambda)$ is a factor of $\varepsilon_{i+1}(\lambda)$, $i = 1, 2, \ldots, r - 1$;

(iii) $\psi_{i+1}(\lambda)$ is a factor of $\psi_i(\lambda)$, $i = 1, 2, \ldots, r - 1$ and $\psi_1(\lambda) = g(\lambda)$. The sum of the degrees of the $\psi_i(\lambda)$ is called the *McMillan degree* of $G(\lambda)$.

Example 12.7   The matrix

$$G(\lambda) = \begin{bmatrix} \dfrac{1}{(\lambda+3)^2} & \dfrac{1}{(\lambda+3)(\lambda+4)} \\[3mm] \dfrac{-6}{(\lambda+3)(\lambda+4)^2} & \dfrac{\lambda-1}{(\lambda+4)^2} \end{bmatrix} \tag{12.33}$$

is strictly proper since clearly $G(\lambda) \to 0$ as $\lambda \to \infty$. The least common denominator of the elements in (12.33) is

$$g(\lambda) = (\lambda+3)^2(\lambda+4)^2.$$

Multiplying both sides of (12.33) by $g(\lambda)$ gives

$$gG = \begin{bmatrix} (\lambda+4)^2 & (\lambda+3)(\lambda+4) \\ -6(\lambda+3) & (\lambda-1)(\lambda+3)^2 \end{bmatrix} \tag{12.34}$$

and $\det(gG) \neq 0$, so $r = 2$. The elements in (12.34) have no common factor, so the Smith form $S(\lambda)$ of $g(\lambda)G(\lambda)$ has diagonal elements $i_1(\lambda) = 1$ and

$$i_2(\lambda) = \det(gG) = (\lambda+1)(\lambda+2)(\lambda+3)^2(\lambda+4).$$

Hence the Smith–McMillan form of $G(\lambda)$ in (12.33) is

$$\frac{S(\lambda)}{g(\lambda)} = \begin{bmatrix} \dfrac{1}{g(\lambda)} & 0 \\[3mm] 0 & \dfrac{i_2(\lambda)}{g(\lambda)} \end{bmatrix}$$

$$= \begin{bmatrix} \dfrac{1}{(\lambda+3)^2(\lambda+4)^2} & 0 \\[3mm] 0 & \dfrac{(\lambda+1)(\lambda+2)}{\lambda+4} \end{bmatrix} \tag{12.35}$$

and we have

$$\varepsilon_1(\lambda) = 1, \qquad \psi_1(\lambda) = (\lambda+3)^2(\lambda+4)^2,$$
$$\varepsilon_2(\lambda) = (\lambda+1)(\lambda+2), \qquad \psi_2(\lambda) = \lambda+4.$$

The roots of the $\varepsilon_i(\lambda)$ are called the *zeros* of $G(\lambda)$, and are -1, -2; the roots of the $\psi_i(\lambda)$ are called the *poles* of $G(\lambda)$ and are -3 (twice), -4 (three times). The McMillan degree of $G(\lambda)$ is 5, which is different from the degree of $g(\lambda)$. Notice that although $G(\lambda)$ in (12.33) is strictly proper, its Smith–McMillan form (12.35) is not, since the 2, 2 element is not proper.

Exercise 12.15 Determine the Smith–McMillan form of each of the following rational matrices

(a) $\begin{bmatrix} \dfrac{1}{\lambda + 3} & \dfrac{\lambda + 4}{(\lambda + 2)^2} \end{bmatrix}$;

(b) $\begin{bmatrix} \dfrac{\lambda + 2}{(\lambda + 3)^2} & 0 \\ 0 & \dfrac{\lambda + 4}{(\lambda + 5)^2} \end{bmatrix}$;

(c) $\begin{bmatrix} \dfrac{\lambda + 4}{(\lambda + 3)^2} & \dfrac{1}{(\lambda + 2)(\lambda + 3)} \\ \dfrac{1}{(\lambda + 2)(\lambda + 3)} & \dfrac{1}{(\lambda + 2)^2} \end{bmatrix}$.

12.4.2 Transfer function matrices

We now return to the linear control system described in Section 10.4.1 by the equations

$$\frac{dx}{dt} = Ax + Bu, \qquad y = Cx, \qquad (12.36)$$

where $x(t)$ is the $n \times 1$ state vector, $u(t)$ the $m \times 1$ control vector and $y(t)$ the $r \times 1$ output vector. Apply Laplace transformation to (12.36), assuming that $x(0)$, $u(0)$ and $y(0)$ are all zero, to obtain

$$s\bar{x} = A\bar{x} + B\bar{u}, \qquad \bar{y} = C\bar{x}, \qquad (12.37)$$

where as before $\bar{x}(s) = \mathcal{L}\{x(t)\}$, and so on. We can write the first equation in (12.37) as

$$(sI - A)\bar{x} = B\bar{u}$$

and substituting into the second equation gives

$$\bar{y} = C(sI - A)^{-1}B\bar{u}$$
$$= G(s)\bar{u}. \qquad (12.38)$$

The $r \times m$ rational matrix $G(s)$ in (12.38) is the *transfer function matrix* of the control system, since it relates the Laplace transform of the input to that of the output. In particular, when $r = m = 1$ then $G(s)$ is a scalar transfer function, defined in

Problem 4.8. If control is applied only through the jth input, so that $\bar{u}_k = 0$, $k \neq j$, then from (12.38)

$$\bar{y}_i = g_{ij}(s)\bar{u}_j,$$

so that the i, j element of $G(s)$ is the transfer function between the ith output and jth input, when all other inputs are zero.

One problem associated with (12.38) is, when given a strictly proper rational matrix $G(s)$, to determine a *realization*, which is a set of matrices A, B, C such that

$$C(sI - A)^{-1}B = G(s). \tag{12.39}$$

Such realizations are not unique, but those in which A has least dimensions (corresponding to the smallest possible number of state variables) are called *minimal*.

Exercise 12.16 Show that if A, B, C is a realization of $G(s)$ then so is TAT^{-1}, TB, CT^{-1} for any non-singular matrix T.

It is interesting that it can be shown that a realization is minimal if and only if it is *controllable*, i.e. the controllability matrix

$$\mathscr{C} = [B, AB, A^2B, \ldots, A^{n-1}B] \tag{12.40}$$

in (5.76) has rank n, and also *observable*, i.e. the matrix having block rows C, CA, CA^2, ..., CA^{n-1} has rank n. The second property can be interpreted as meaning that it is possible to determine the state of the system from a knowledge of the output $y(t)$. Books on control theory (e.g. Barnett and Cameron 1985) should be consulted for further details.

Exercise 12.17 If

$$A = \begin{bmatrix} 1 & 0 & 0 \\ 0 & 2 & 0 \\ 0 & 0 & 2 \end{bmatrix}, \qquad B = \begin{bmatrix} 1 & 0 \\ 2 & 1 \\ -8 & 3 \end{bmatrix}, \qquad C = \begin{bmatrix} 1 & 1 & 0 \\ 1 & 0 & 1 \end{bmatrix},$$

compute $G(s)$ in (12.38). Compare $g(s)$ (defined in Section 12.4.1) and the characteristic polynomial $\det(sI - A)$.

An alternative way of looking at rational matrices, found useful in control theory, is to generalize the representation of a rational function as a ratio of the two polynomials.

Example 12.8 Return to the matrix in (12.33), and replace λ by s in accordance with the convention adopted in this section. From (12.34) we can write

$$G(s) = \frac{1}{(s+3)^2(s+4)^2} \begin{bmatrix} (s+4)^2 & (s+3)(s+4) \\ -6(s+3) & (s-1)(s+3)^2 \end{bmatrix}$$

$$= \begin{bmatrix} (s+3)^2(s+4)^2 & 0 \\ 0 & (s+3)^2(s+4)^2 \end{bmatrix}^{-1} \begin{bmatrix} (s+4)^2 & (s+3)(s+4) \\ -6(s+3) & (s-1)(s+3)^2 \end{bmatrix}$$

$$= M^{-1}(s)N(s), \tag{12.41}$$

where in (12.41) both $M(s)$ and $N(s)$ are polynomial matrices. The expression (12.41) is called a *left matrix fraction description* (m.f.d.) of $G(s)$, and $N(s)$ is the numerator, $M(s)$ is the denominator, Clearly in this example since $M(s)$ is diagonal we could write $G(s) = N(s)M^{-1}(s)$, which is a *right* m.f.d.

In general, if $G(s)$ is $m \times n$, let $g_i(s)$ denote the least common denominator of the elements in the ith column of $G(s)$ and $\tilde{N}(s)$ the $m \times n$ polynomial matrix of numerators of $G(s)$ relative to these denominators. Then a right m.f.d. is

$$\tilde{N}(s)\{\operatorname{diag}[g_1(s), g_2(s), \ldots, g_n(s)]\}^{-1}. \tag{12.42}$$

Example 12.8 (continued) For $G(s)$ in (12.33) we have

$$g_1(s) = (s+3)^2(s+4)^2, \qquad g_2(s) = (s+3)(s+4)^2$$

so a right m.f.d. is

$$G(s) = \begin{bmatrix} (s+4)^2 & (s+4) \\ -6(s+3) & (s-1)(s+3) \end{bmatrix} \begin{bmatrix} (s+3)^2(s+4)^2 & 0 \\ 0 & (s+3)(s+4)^2 \end{bmatrix}^{-1}.$$

Exercise 12.18 Determine the expression like (12.42) which expresses $G(s)$ as a *left* m.f.d. Apply this to $G(s)$ in (12.33).

Exercise 12.19 Let C be the $n \times n$ companion matrix defined in (11.2). By considering the product $(sI - C)v(s)$, where $v(s) = [1, s, s^2, \ldots, s^{n-1}]^T$, deduce that the last column of $(sI - C)^{-1}$ is $v(s)/a(s)$, where $a(s)$ is the characteristic polynomial in (11.1). Hence show that a realization of

$$(b_0 s^{n-1} + b_1 s^{n-2} + \cdots + b_{n-1})/a(s) \tag{12.43}$$

is

$$C, \quad [0, 0, \ldots, 0, 1]^T, \quad [b_{n-1}, b_{n-2}, \ldots, b_1, b_0]. \tag{12.44}$$

Verify that this realization is controllable.

The realization (12.44) of the scalar transfer function in (12.43) will be minimal if and only if the observability matrix having rows b, bC, bC^2, ..., bC^{n-1} is non-singular. However, this matrix is precisely the resultant matrix $b(C)$ in (11.8) for the two polynomials in (12.43). In other words, the realization (12.44) is minimal if and only if the polynomials in the numerator and denominator in (12.43) are relatively prime. This can be extended to the matrix case: a right m.f.d. $N(s)M^{-1}(s)$ of a given rational matrix $G(s)$ is called *irreducible* if $N(s)$ and $M(s)$ are relatively right prime.

Exercise 12.20 If $D(s)$ is a common right divisor of $M(s)$ and $N(s)$ in a right m.f.d. of $G(s)$, obtain an expression for $G(s)$ when the common factor is 'cancelled out'.

Problems

12.1 Show that if $A(\lambda)$ is the matrix in (12.4) with $m = n$, then the right and left remainders on division of $A(\lambda)$ by $\lambda I_n - K$, where K is a constant $n \times n$ matrix, are respectively

$$A_0 K^N + A_1 K^{N-1} + \cdots + A_{N-1} K + A_N$$
$$K^N A_0 + K^{N-1} A_1 + \cdots + K A_{N-1} + A_N.$$

If either of these expressions is zero then K is called a *right* or *left solvent*, respectively, of $A(\lambda)$.

12.2 If K is a right solvent of $A(\lambda)$, show that every eigenvalue of K is a latent root of $A(\lambda)$ (defined in Exercise 12.8).

If a vector x satisfies $A(\lambda_i)x = 0$, where λ_i is a latent root, it is called a right *latent vector*. Show that the right eigenvectors of K are right latent vectors of $A(\lambda)$.

12.3 Let $A(\lambda)$ and $B(\lambda)$ be as defined in (12.4) and (12.7) respectively. Let the latent roots (defined in Exercise 12.8) of $B(\lambda)$ be μ_1, μ_2, μ_3, ..., μ_x. Prove that if the matrix $C(\lambda)$ defined in (12.12) has rank n for each value of $\lambda = \mu_1, \mu_2, \ldots$ then the Smith form of $C(\lambda)$ must be (12.22). Hence deduce using Exercise 12.12 that $A(\lambda)$, $B(\lambda)$ are r.r.p. if and only if $C(\lambda)$ has rank n for $\lambda = \mu_1, \mu_2, \ldots, \mu_x$.

12.4 Return once again to the linear control system in (12.36). It can be shown (see Barnett, 1983) that the controllability matrix $\mathscr{C}$ in (12.40) has rank n if and only if $R(\lambda_i I - A, B) = n$ for $i = 1, 2, \ldots, n$, where the λ_i are the eigenvalues of A. Deduce from

Problem 12.3 that the matrices $\lambda I - A$, B are relatively left prime if and only if $R(\mathscr{C}) = n$.

12.5 Let $A(\lambda)$ and $B(\lambda)$ be two $n \times n$ skew prime matrices. By considering the Smith form of $C(\lambda)$ in (12.12), deduce that $A(\lambda)$ and $B(\lambda)$ are relatively left *and* relatively right prime.

12.6 Consider the control system in Problem 11.16. Suppose that the output variables are defined by $y_1 = x_1 + x_2$, $y_2 = x_1 - x_2$. Determine the transfer function matrix.

12.7 Determine the Smith–McMillan form and McMillan degree of

$$G(\lambda) = \frac{1}{g(\lambda)} \begin{bmatrix} \lambda + 3 & 4(\lambda + 3) \\ -1 & \lambda + 1 \end{bmatrix}$$

where $g(\lambda) = \lambda^2 + 7\lambda + 12$.

12.8 Verify that the matrix

$$A(\lambda) = \begin{bmatrix} 1 & 0 & \lambda + 1 \\ 3\lambda + 2 & 3\lambda + 5 & 5 \\ \lambda + 1 & \lambda + 2 & 2 \end{bmatrix}$$

has Smith form I_3.

Using the procedure described in Problem 5.16, find matrices $R(\lambda)$, $T(\lambda)$ such that $RAT = I_3$. Hence compute $A^{-1}(\lambda) = T(\lambda)R(\lambda)$.

12.9 Let A_1, B_1, C_1 be a realization of $G_1(s)$ and A_2, B_2, C_2 be a realization of $G_2(s)$. Use the result on partitioned form of the inverse in Section 4.3.2 to prove that

$$A = \begin{bmatrix} A_1 & B_1 C_2 \\ 0 & A_2 \end{bmatrix}, \qquad B = \begin{bmatrix} 0 \\ B_2 \end{bmatrix}, \qquad C = [C_1 \quad 0]$$

is a realization of the product $G_1(s)G_2(s)$.

References

Barnett (1983, 1984), Barnett and Cameron (1985), Chen (1984), Gohberg *et al.* (1982), Lancaster (1966), Rosenbrock (1970).

13 Patterned matrices

Matrices in which the elements exhibit a recognizable pattern are legion, and we have already encountered quite a few in this book. The simplest of all such special forms is the *diagonal* matrix, first defined in (2.36); next came the *bidiagonal* and *tridiagonal* forms, introduced in Problem 3.2; and upper and lower *triangular* matrices, used in Sections 3.3 and 8.3. We dealt extensively with *symmetric* and *hermitian* matrices in Chapter 7, these having been defined together with their *skew* counterparts in Section 2.3.2. Other special forms which we have introduced include circulants (Problem 6.24), companion matrices (Sections 6.3.4, 11.2.1), Jordan forms (Section 8.1), Hessenberg forms (Section 8.4), striped matrices (see (11.5)), the Sylvester matrix (Section 11.2.2) and the Vandermonde matrix (Problem 4.16). Our emphasis in this chapter is on patterns which can be recognized visually—thus, for example, the bezoutian matrix of Section 11.2.3 is certainly a special form of symmetric matrix, but it is not possible to determine by inspection whether or not a particular matrix is a bezoutian.

There are two reasons for devoting attention to patterned matrices. Firstly, there is almost always at least one particular property of a patterned matrix which can be determined more easily than for an arbitrary matrix. For example, the determinant of a Sylvester or Vandermonde matrix has a simple explicit expression; the characteristic polynomial of a companion matrix can be written down at once, and of a tridiagonal matrix can be computed from a simple iterative formula (Problem 6.6); the eigenvalues of a diagonal or triangular matrix are explicitly displayed; and the inverse of various forms, especially tridiagonal, triangular and striped, can be computed with much less effort than is needed in general. Our second reason for studying special forms is that they arise naturally in a wide variety of applications, including control and system theory, signal processing, statistics and numerical solution of differential equations.

Example 13.1   Consider a two-point boundary-value problem, consisting of the differential equation

$$y'' + f(x, y) = 0, \tag{13.1}$$

where $y'' = d^2y/dx^2$, which is to be solved subject to the conditions $y(0) = \alpha$, $y(1) = \beta$. This could represent, for example, a problem involving heat flow along a rod, or a vibrating stretched string (see Example 6.4). Divide the interval $[0, 1]$ into n equal parts and put $h = 1/n$. Replace the derivative by the difference approximation

$$y''(x_i) \approx \frac{y_{i+1} - 2y_i + y_{i-1}}{h^2},$$

where $y_i = y(x_i)$, $x_i = ih$, $i = 0, 1, 2, \ldots, n$ and $y_{-1} = 0$, $y_{n+1} = 0$. This transforms (13.1) into a set of linear equations in the form

$$Ay = h^2F + b, \tag{13.2}$$

where

$$y = [y_1, y_2, \ldots, y_{n-1}]^\mathsf{T}, \qquad b = [\alpha, 0, 0, \ldots, \beta]^\mathsf{T}$$
$$F = [f_1, f_2, \ldots, f_{n-1}]^\mathsf{T}, \qquad f_i = f(x_i, y_i)$$

and A is the tridiagonal matrix

$$A = \begin{bmatrix} 2 & -1 & 0 & 0 & \cdot & \cdot & \cdot & 0 \\ -1 & 2 & -1 & 0 & \cdot & \cdot & \cdot & 0 \\ 0 & -1 & 2 & -1 & \cdot & \cdot & \cdot & 0 \\ & & & & \cdot & \cdot & & \\ \cdot & \cdot & \cdot & \cdot & -1 & & 2 & -1 \\ 0 & 0 & \cdot & \cdot & \cdot & & -1 & 2 \end{bmatrix}. \tag{13.3}$$

This matrix was encountered earlier in (6.18) with $n = 5$. When $f(x, y)$ is linear, then the equations (13.2) may be solved easily, as indicated in Problem 3.2. In fact, an explicit expression for the inverse of (13.3) will be given later (Example 13.4).

13.1   Banded matrices

A *diagonal (line)* which is parallel to the principal diagonal of an $n \times n$ matrix $A = [a_{ij}]$ is one along which $|i - j|$ is constant. A *tridiagonal* matrix was defined in Problem 3.2 as a square matrix whose only non-zero elements lie on the principal diagonal, and on the diagonals immediately above and below this (i.e. the

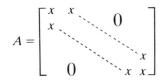

$$A = \begin{bmatrix} x & x & & & & \\ x & & & & 0 & \\ & & & & & \\ & & & & & x \\ & 0 & & & x & x \end{bmatrix}$$

Fig. 13.1 Tridiagonal matrix.

superdiagonal and subdiagonal); the 4×4 case was set out in (3.42), with another example in (13.3). Thus in terms of elements A is tridiagonal when

$$a_{ij} = 0, \qquad \text{for all } |i - j| > 1, \qquad (13.4)$$

and can be represented pictorially as in Fig. 13.1, where the non-zero elements are contained in the band shown there. More generally, a (p, q)-*banded matrix* $A = [a_{ij}]$ has all its non-zero elements contained within a *band* consisting of the principal diagonal, p diagonals above it and q diagonals below. This is shown in Fig. 13.2. In terms of elements we have

$$a_{ij} = 0 \begin{cases} j - i > p \\ i - j > q \end{cases}. \qquad (13.5)$$

Banded matrices arise frequently, especially in the numerical solution of ordinary or partial differential equations, as in Example 13.1.

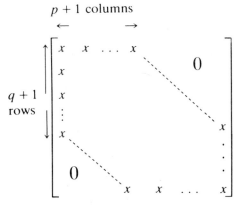

$p + 1$ columns

$q + 1$ rows

Fig. 13.2 (p, q)-banded matrix.

Example 13.2 If $n = 6$ then a $(2, 3)$-banded matrix has the form

$$
A = \begin{matrix}
\uparrow \\
4 \\
\downarrow
\end{matrix}
\overset{\leftarrow \, 3 \, \rightarrow}{
\begin{bmatrix}
x & x & x & 0 & 0 & 0 \\
x & x & x & x & 0 & 0 \\
x & x & x & x & x & 0 \\
x & x & x & x & x & x \\
0 & x & x & x & x & x \\
0 & 0 & x & x & x & x
\end{bmatrix}}, \tag{13.6}
$$

where the x's in (13.6) denote (possibly) non-zero elements. The definition (13.5) gives $a_{ij} = 0$ for diagonals with $j - i > 2$ (above the principal diagonal), and $a_{ij} = 0$ for $i - j > 3$ (diagonals below the principal diagonal).

Notice by comparing Figs. 13.1 and 13.2, or the definitions (13.4) and (13.5), that a tridiagonal matrix is a $(1, 1)$-banded matrix. Similarly, an $n \times n$ upper or lower Hessenberg matrix defined in Section 8.4 is $(n - 1, 1)$-banded or $(1, n - 1)$-banded respectively. A $(2, 2)$-banded matrix is called *pentadiagonal*. In practice we are usually interested in cases when a large proportion of the elements consist of zeros; that is, at least one of p or q is small compared with n.

We saw in Section 3.3 that any non-singular matrix can be expressed as the product of a lower and an upper triangular matrix—or, in our present terminology, as the product of a $(0, n - 1)$-banded matrix and a $(n - 1, 0)$-banded matrix. Also, Problem 3.2 showed that a $(1, 1)$-banded matrix A can be written as a $(0, 1)$-banded matrix multiplied by a $(1, 0)$-banded matrix (each of these is called *bidiagonal*). In general, the condition for this to be possible is that *all* the leading principal minors of A (including $\det A$) must be non-singular, in which case A is said to be *strongly* non-singular. The following example shows that it is not sufficient for A to be merely non-singular.

Example 13.3 The matrix

$$
A = \begin{bmatrix}
1 & 0 & 0 \\
0 & 0 & 1 \\
0 & 1 & 1
\end{bmatrix} \tag{13.7}
$$

is tridiagonal and non-singular, but its 2×2 leading principal minor is zero. The reader should verify that it is not possible to express (13.7) as the product of two bidiagonal matrices as in (3.43), namely

$$\begin{bmatrix} 1 & 0 & 0 \\ l_1 & 1 & 0 \\ 0 & l_2 & 1 \end{bmatrix} \begin{bmatrix} u_1 & v_1 & 0 \\ 0 & u_2 & v_2 \\ 0 & 0 & u_3 \end{bmatrix}.$$

Exercise 13.1 Express the tridiagonal matrix

$$\begin{bmatrix} 2 & 1 & 0 & 0 \\ 5 & -1 & 3 & 0 \\ 0 & -3 & 2 & 2 \\ 0 & 0 & 7 & 6 \end{bmatrix} \tag{13.8}$$

as the product of two bidiagonal matrices.

It is interesting that the result for tridiagonal matrices can be generalized: *any* strongly non-singular (p, q)-banded matrix can be expressed in the form LU where L is $(0, q)$-banded and U is $(p, 0)$-banded.

Exercise 13.2 Verify that the $(2, 3)$-banded matrix

$$\begin{bmatrix} 2 & 1 & 1 & 0 & 0 \\ 5 & -1 & 3 & -4 & 0 \\ 2 & -3 & 2 & 2 & 11 \\ 8 & -5 & 7 & 6 & 9 \\ 0 & -2 & 1 & 10 & 0 \end{bmatrix}$$

is strictly non-singular, and express it as the product of a $(0, 3)$-banded matrix and a $(2, 0)$-banded matrix.

For the remainder of this section we consider only tridiagonal matrices, for which various special results hold. It is convenient to use the notation

$$A = \begin{bmatrix} a_1 & b_1 & 0 & & & \\ c_1 & a_2 & b_2 & & \mathbf{0} & \\ 0 & c_2 & a_3 & \cdot & & \\ & & \cdot & \cdot & \cdot & b_{n-1} \\ & \mathbf{0} & & \cdot & c_{n-1} & a_n \end{bmatrix}. \tag{13.9}$$

Let A_k denote the $k \times k$ leading principal submatrix of (13.9), for $k = 1, 2, \ldots, n$. Clearly we have

$$\det A_1 = a_1, \qquad \det A_2 = a_1 a_2 - b_1 c_1 = a_2 \det A_1 - b_1 c_1. \tag{13.10}$$

Now consider the evaluation of

$$\det A_3 = \begin{vmatrix} a_1 & b_1 & 0 \\ c_1 & a_2 & b_2 \\ 0 & c_2 & a_3 \end{vmatrix}.$$

Expanding by the last row using the cofactors formula (4.23) gives

$$\det A_3 = -c_2 \begin{vmatrix} a_1 & 0 \\ c_1 & b_2 \end{vmatrix} + a_3 \begin{vmatrix} a_1 & b_1 \\ c_1 & a_2 \end{vmatrix}$$

$$= a_3 \det A_2 - b_2 c_2 \det A_1. \tag{13.11}$$

Proceeding in an exactly similar fashion the reader can check that

$$\det A_k = a_k \det A_{k-1} - b_{k-1} c_{k-1} \det A_{k-2}, \qquad k = 3, 4, \ldots, n. \tag{13.12}$$

The formulae (13.10), (13.11) and (13.12) provide a simple recursive procedure for computing $\det A_2$, $\det A_3$, ..., $\det A_n = \det A$. Notice that since $\lambda I - A$, where A is given in (13.9), is still tridiagonal, this scheme can also be used to calculate the characteristic polynomial of A. In fact, the reader was asked to develop such a scheme in Problem 6.6.

Exercise 13.3 Use (13.9), (13.10) and (13.12) to evaluate the determinant of the matrix in (13.8).

Exercise 13.4 Attempt Problems 6.6 (numerical part only) and 6.7.

We saw in Chapter 6 that the real symmetric matrices have all their eigenvalues real, but of course in general a real non-symmetric matrix will possess complex eigenvalues. It is interesting therefore that any tridiagonal matrix in (13.9), whether symmetric or not, has all its eigenvalues real provided all the a's, b's and c's are real and in addition satisfy

$$b_i c_i > 0, \qquad i = 1, 2, \ldots, n - 1. \tag{13.13}$$

This is proved by showing that under these conditions A is similar to a real symmetric matrix. For simplicity, we demonstrate this for the case $n = 3$. Defining

$$d_1 = 1, \qquad d_2 = (b_1/c_1)^{1/2}, \qquad d_3 = d_2(b_2/c_2)^{1/2}, \tag{13.14}$$

then (13.13) implies that each d_i is real and non-zero. If we construct $B = [b_{ij}]$ according to $B = DAD^{-1}$, where $D = $ diag$[d_1, d_2, d_3]$, then the reader can easily confirm that

$$B = \begin{bmatrix} a_1 & \dfrac{d_1}{d_2} b_1 & 0 \\[2ex] \dfrac{d_2}{d_1} c_1 & a_2 & \dfrac{d_2}{d_3} b_2 \\[2ex] 0 & \dfrac{d_3}{d_2} c_2 & a_3 \end{bmatrix}.$$

Because of (13.14) we have

$$b_{12} = (b_1 c_1)^{1/2} = b_{21}, \qquad b_{23} = (b_2 c_2)^{1/2} = b_{32},$$

showing that B is real and symmetric, as required.

Exercise 13.5 Prove that in general if (13.13) holds and $D = $ diag$[d_1, d_2, \ldots, d_n]$ where

$$d_1 = 1, \qquad d_{i+1} = d_i (b_i / c_i)^{1/2}, \qquad i = 1, 2, \ldots, n-1,$$

then DAD^{-1} is real and symmetric, where A is real and defined by (13.9).

It is worth recalling here that in Exercise 8.25 we found that any symmetric matrix is orthogonally similar to a symmetric tridiagonal matrix. Applying the QR algorithm (Section 6.6.2) becomes considerably simplified in this case, and details can be found in Golub and Van Loan (1989). In fact, the so-called Lanczos algorithm (Householder 1964) produces a similarity transformation of *any* square matrix to tridiagonal form. Other results on eigenvalues of tridiagonal matrices can be found in the book by Fiedler (1986).

We close this section with a few tridiagonal examples taken mainly from a collection of matrices useful for testing computational algorithms (Gregory and Karney 1969). In order to fully appreciate the formulae, the reader is urged to write out the matrices for, say, $n = 5$.

Example 13.4

(a) An $n \times n$ tridiagonal matrix A defined as in (13.3) has inverse $[1/(n + 1)]B$, where $B = [b_{ij}]$ and

$$b_{ij} = \begin{cases} i(n - i + 1), & i = j \\ b_{i,j-1} - i, & i < j. \\ b_{ji}, & i > j \end{cases}$$

(b) More generally, the $n \times n$ symmetric tridiagonal matrix $A = [a_{ij}]$ with

$$a_{11} = \lambda + b, \qquad a_{nn} = \lambda + a, \qquad a_{ii} = \lambda \qquad \text{for } i = 2, 3, \ldots, n-1,$$

and all other non-zero elements equal to unity has $A^{-1} = [b_{ij}]$ where

$$b_{ij} = b_{ji} = \frac{(-1)^{i+j} r_{j-1} s_{n-i}}{(\lambda + a) r_{n-1} - r_{n-2}}, \qquad i \geqslant j,$$

and

$$r_0 = 1, \qquad r_1 = \lambda + b, \qquad r_k = \lambda r_{k-1} - r_{k-2}, \qquad k = 2, 3, \ldots, n-1,$$
$$s_0 = 1, \qquad s_1 = \lambda + a, \qquad s_k = \lambda s_{k-1} - s_{k-2}, \qquad k = 2, 3, \ldots, n-1.$$

(c) The $n \times n$ symmetric tridiagonal matrix $A = [a_{ij}]$ with

$$a_{11} = 1 = a_{nn}, \qquad a_{ii} = 1 + r^2, \qquad i = 2, 3, \ldots, n-1,$$
$$a_{i,i+1} = a_{i+1,i} = -r,$$

can be expressed as a special case of (b) (see Exercise 13.7). However, a simple explicit expression for its inverse when $r^2 \neq 1$ is

$$A^{-1} = \left[\frac{r^{|i-j|}}{1 - r^2} \right].$$

(d) The tridiagonal matrix $A = [a_{ij}]$ of order $n + 1$ with all $a_{ii} = 0$,

$$a_{i,i+1} = x_i, \qquad a_{i+1,i} = y_i, \qquad i = 1, 2, \ldots, n,$$

and $x_i \neq 0$, $y_i \neq 0$ has

$$\det A = (-1)^m \prod_{i=1}^{m} (x_{2i-1} y_{2i-1}),$$

where $m = \frac{1}{2}(n+1)$ and n is odd; $\det A = 0$ if n is even. If $x_i y_i = i(n-i+1)$, $i = 1, 2, \ldots, n$ then the eigenvalues of A are

$$\pm n, \pm (n-2), \ldots, \pm 1, \qquad \text{when } n \text{ is odd}$$
$$\pm n, \pm (n-2), \ldots, \pm 2, 0, \qquad \text{when } n \text{ is even}.$$

The inverse of a general non-symmetric tridiagonal Toeplitz matrix is given at the end of Section 13.3.

Example 13.5

(a) The $n \times n$ symmetric tridiagonal matrix $A = [a_{ij}]$ with all diagonal elements $a_{ii} = a$ and all other non-zero elements equal to b (so A is Toeplitz), has eigenvalues

$$\lambda_k = a + 2b \cos \frac{k\pi}{n+1}$$

and normalized eigenvectors $[x_1^{(k)}, x_2^{(k)}, \ldots, x_n^{(k)}]^T$, where

$$x_j^{(k)} = \left(\frac{2}{n+1}\right)^{1/2} \sin \frac{kj\pi}{n+1}, \qquad j, k = 1, 2, \ldots, n.$$

(b) In the following, A is the same as in (a) except for the stated elements. Only the eigenvalues are given.

(i) $a_{11} = a - b$,

$$\lambda_k = a + 2b \cos \frac{2k\pi}{2n+1}, \qquad k = 1, 2, \ldots, n;$$

(ii) $a_{11} = a - b$, $a_{nn} = a + b$,

$$\lambda_k = a + 2b \cos \frac{(2k-1)\pi}{2n}, \qquad k = 1, 2, \ldots, n;$$

(iii) $a_{11} = a_{nn} = a + b$,

$$\lambda_k = a + 2b \cos \frac{k\pi}{n}, \qquad k = 1, 2, \ldots, n.$$

Exercise 13.6 Use the result in Example 13.4(b) to obtain the inverse of the $n \times n$ tridiagonal matrix $C = -A + \text{diag}[-1, 0, 0, \ldots, 0, 1]$, where A is defined in (13.3).

Exercise 13.7 For the matrix A in Example 13.4(c), express $(-1/r)A$ in the form of the matrix A in Example 13.4(b).

13.2 Circulant matrices

An $n \times n$ *circulant matrix* is defined by

$$C = \begin{bmatrix} c_1 & c_2 & c_3 & \cdots & c_n \\ c_n & c_1 & c_2 & \cdots & c_{n-1} \\ c_{n-1} & c_n & c_1 & \cdots & c_{n-2} \\ \cdot & \cdot & & \cdots & \cdot \\ c_2 & c_3 & c_4 & \cdots & c_1 \end{bmatrix}, \qquad (13.15)$$

where the c's can be real or complex. Each row in (13.15) consists of the elements of the preceding row shifted one position to the right, with the 'overflow' element being moved to the first position. The matrix is entirely determined by its first row, and the standard reference (Davis 1979) uses the notation $\text{circ}(c_1, c_2, \ldots, c_n)$. Two properties can be seen immediately by inspection of (13.15): firstly, the elements along each diagonal line parallel to the principal diagonal are equal, so that C is an example of a Toeplitz matrix, to be discussed in the next section; secondly, C is symmetric with respect to its secondary diagonal (the line from top right corner to bottom left corner). In fact, *any* $n \times n$ matrix $A = [a_{ij}]$ which is symmetric in this way is called *persymmetric,* and its elements satisfy the condition

$$a_{ij} = a_{n+1-j,n+1-i}. \tag{13.16}$$

Exercise 13.8 For (13.15) let $C = [c_{ij}]$ and show that $c_{ij} = c_{j-i+1}$, with $c_{-k} = c_{n-k}$ for $k \geq 0$. Hence verify that the elements of C satisfy (13.16), showing that any circulant matrix is persymmetric.

Exercise 13.9 If K_n is the reverse unit matrix defined in Problem 2.4 show that a persymmetric matrix A can be characterized by $K_n A^{\mathrm{T}} K_n = A$. Hence deduce (a) $K_n A$ is symmetric in the usual sense; (b) if A is non-singular then A^{-1} is also persymmetric.

An important special circulant matrix is

$$\Pi = \mathrm{circ}(0, 1, 0, \ldots, 0) \tag{13.17}$$

often called the *shift matrix,* because postmutiplying any matrix by Π shifts its columns one place to the right (a similar shift is applied to rows on premulticpication by Π).

Example 13.6 When $n = 4$ the definitions (13.15) and (13.17) give

$$C_1 = \begin{bmatrix} c_1 & c_2 & c_3 & c_4 \\ c_4 & c_1 & c_2 & c_3 \\ c_3 & c_4 & c_1 & c_2 \\ c_2 & c_3 & c_4 & c_1 \end{bmatrix}, \quad \Pi = \begin{bmatrix} 0 & 1 & 0 & 0 \\ 0 & 0 & 1 & 0 \\ 0 & 0 & 0 & 1 \\ 1 & 0 & 0 & 0 \end{bmatrix}. \tag{13.18}$$

Inspection of Π in (13.18) shows that it has precisely the companion form (11.2), so its characteristic polynomial is $\lambda^4 - 1$. Furthermore, the reader can check that the matrices in (13.18) commute, i.e.

$$C\Pi = \Pi C. \tag{13.19}$$

It requires little extra effort to verify that both these properties carry over to the general case: the characteristic polynomial of Π in (13.17) is $\lambda^n - 1$, so its eigenvalues are the n distinct roots of unity, namely $1, \theta, \theta^2, \ldots, \theta^{n-1}$ where $\theta = \exp(2\pi i/n)$ and $i = \sqrt{-1}$; and (13.19) is valid for $n \times n$ matrices. Moreover, the converse of the latter property holds, so circulants can be regarded as the set of all square matrices which commute with Π. Notice also that because of the simple structure of Π we can write C in (13.18) as

$$C = c_1 I + c_2 \Pi + c_3 \Pi^2 + c_4 \Pi^3.$$

Indeed, in general an alternative way of writing the $n \times n$ circulant matrix (13.15) is

$$C = c_1 I + c_2 \Pi + c_3 \Pi^2 + \cdots + c_n \Pi^{n-1}$$
$$= c(\Pi), \tag{13.20}$$

where

$$c(\lambda) = c_1 + c_2 \lambda + c_3 \lambda^2 + \cdots + c_n \lambda^{n-1}. \tag{13.21}$$

Exercise 13.10 Prove using (13.19) that A is a circulant matrix if and only if A^* is a circulant matrix.

Exercise 13.11 Prove using (13.19) that (a) if C_1 and C_2 are circulant matrices then so is $C_1 C_2$; (b) if C is non-singular then C^{-1} is also circulant.

It is straightforward to show (Problem 13.4) that all circulants of the same order commute. In particular, since C^* is also a circulant matrix (Exercise 13.10) it follows that $CC^* = C^*C$, showing that any circulant matrix C is normal. We can therefore deduce (see Problem 8.5) that every circulant matrix is unitarily similar to a diagonal matrix, and it is instructive to derive an explicit expression for the similarity transformation. Firstly, define the $n \times n$ *Fourier matrix* $F = [f_{rs}]$ where

$$f_{rs} = \frac{1}{\sqrt{n}} \omega^{(r-1)(s-1)}, \qquad r, s = 1, 2, \ldots, n, \tag{13.22}$$

and $\omega = \bar{\theta} = \exp(-2\pi i/n)$. For example, when $n = 4$ we have

$$F = \frac{1}{2} \begin{bmatrix} 1 & 1 & 1 & 1 \\ 1 & \omega & \omega^2 & \omega^3 \\ 1 & \omega^2 & \omega^4 & \omega^6 \\ 1 & \omega^3 & \omega^6 & \omega^9 \end{bmatrix}, \tag{13.23}$$

where $\omega = \exp(-\pi i/2)$ and $\omega^4 = 1$. The name arises from the occurrence of F in discrete Fourier analysis and fast Fourier transforms (Brigham 1988). Notice that the matrix in (13.23) is a particular case of the Vandermonde matrix defined in Problem 4.16. Also, (13.23) illustrates the fact that F defined by (13.22) is symmetric. Furthermore, since $\omega^n = 1$ and the sequence $1, \omega,$

$\omega^2, \ldots$ is periodic, it follows that there are only n distinct elements in F. For example, (13.23) can be written as

$$F = \frac{1}{2} \begin{bmatrix} 1 & 1 & 1 & 1 \\ 1 & \omega & \omega^2 & \omega^3 \\ 1 & \omega^2 & 1 & \omega^2 \\ 1 & \omega^3 & \omega^2 & \omega \end{bmatrix}. \tag{13.24}$$

We are going to show that F defined by (13.22) is the matrix in the unitary similarity transformation which diagonalizes C. The first step is for the reader to establish the following:

Exercise 13.12 Show that $\omega\bar{\omega} = 1$; $\bar{\omega}^p = \omega^{-p}$ for any positive integer p; and

$$1 + \omega + \omega^2 + \cdots + \omega^{n-1} = 0.$$

Hence prove that F defined by (13.22) is unitary.

The next step is to realise that F diagonalizes the circulant matrix Π defined in (13.17), i.e.

$$F\Pi F^* = \Omega = \text{diag}[1, \theta, \theta^2, \ldots, \theta^{n-1}]. \tag{13.25}$$

Exercise 13.13 Check by direct evauation of $F\Pi F^*$ using F in (13.24) that (13.25) holds when $n = 4$.

Exercise 13.14 Show from (13.25) that

$$F\Pi^k F^* = \Omega^k, \qquad k = 2, 3, \ldots. \tag{13.26}$$

Finally, in view of (13.20), (13.25) and (13.26) we have

$$\begin{aligned} FCF^* &= c_1 FF^* + c_2 F\Pi F^* + \cdots + c_n F\Pi^{n-1} F^* \\ &= c_1 I + c_2 \Omega + c_3 \Omega^2 + \cdots + c_n \Omega^{n-1} \\ &= \text{diag}[c(1), c(\theta), c(\theta^2), \ldots, c(\theta^{n-1})], \end{aligned} \tag{13.27}$$

where $c(\lambda)$ is the polynomial defined in (13.21). Equation (13.27) is the promised diagonalization of C, and shows that the eigenvalues of C are $c(1)$, $c(\theta), \ldots, c(\theta^{n-1})$. However, although $1, \theta, \ldots, \theta^{n-1}$ are the n distinct roots of unity the eigenvalues themselves are not necessarily all distinct. It is interesting that the matrix F in (13.27) is independent of the coefficients c_i—that is, the *same* matrix diagonalizes all circulants C of order n. In particular, the columns of F^* are a set of right eigenvectors of C (see Problem 6.24). In other words, a right eigenvector corresponding to the eigenvalue $c(\theta^k)$ is

$$[1, \ \theta^k, \ \theta^{2k}, \dots, \theta^{(n-1)k}]^\mathsf{T}.$$

Furthermore, since by (6.20) det C is equal to the product of its eigenvalues, we have

$$\det C = c(1)c(\theta)c(\theta^2)\cdots c(\theta^{n-1}). \tag{13.28}$$

Exercise 13.15   When $C = \mathrm{circ}(1, 1, 1)$ evaluate FCF^* and verify that (13.27) holds.

Exercise 13.16   Evaluate the determinant of a general circulant matrix by direct expansion when $n = 3$ and confirm that your result agrees with (13.28).

Exercise 13.17   Let Λ denote the diagonal matrix in (13.27). Use (10.24) to show that $C^+ = F^*\Lambda^+F$, where the *MP*-inverse of Λ is given in Exercise 10.6.

There are a number of generalizations of the circulant matrix, including the following:

(a) A *skew circulant matrix* is obtained from a circulant matrix by reversing the signs of all the elements below the principal diagonal. For example, when $n = 4$ we see from (13.18) that the general skew circulant matrix has the form

$$\begin{bmatrix} c_1 & c_2 & c_3 & c_4 \\ -c_4 & c_1 & c_2 & c_3 \\ -c_3 & -c_4 & c_1 & c_2 \\ -c_2 & -c_3 & -c_4 & c_1 \end{bmatrix}.$$

(b) An $n \times n$ *g-circulant matrix* $(g \geqslant 0)$ is a matrix in which each row consists of the preceding row shifted g places to the right, with the 'overflow' being moved to the first g positions. For example, when $n = 5$ and $g = 2$ we have

$$\begin{bmatrix} c_1 & c_2 & c_3 & c_4 & c_5 \\ c_4 & c_5 & c_1 & c_2 & c_3 \\ c_2 & c_3 & c_4 & c_5 & c_1 \\ c_5 & c_1 & c_2 & c_3 & c_4 \\ c_3 & c_4 & c_5 & c_1 & c_2 \end{bmatrix}.$$

In particular, a 1-circulant matrix is an ordinary circulant matrix as originally defined in (13.15); and when $g = 0$ all the rows of the matrix are identical.

(c) If the c_i in (13.15) are $m \times m$ matrices instead of scalars, then C is a *block circulant matrix* of order mn and type (n, m). For example, when $n = 2$, $m = 2$ we have

$$\begin{bmatrix} c_1 & c_2 \\ c_2 & c_1 \end{bmatrix} = \left[\begin{array}{cc:cc} a_1 & a_2 & b_1 & b_2 \\ a_3 & a_4 & b_3 & b_4 \\ \hdashline b_1 & b_2 & a_1 & a_2 \\ b_3 & b_4 & a_3 & a_4 \end{array}\right].$$

It is clear from this example that a block circulant matrix need not be an ordinary circulant matrix when $m > 1$.

Some properties of these generalizations are explored in Problems 13.11 to 13.15, and a comprehensive treatment of these and other aspects of circulant matrices is given by Davis (1979).

13.3 Toeplitz and Hankel matrices

Matrices named after Hankel have already been encountered in Chapter 11. We saw from the examples in (11.5) and (11.53) that an $n \times n$ *Hankel matrix* $H = [h_{ij}]$ is one having equal elements along each diagonal line parallel to the *secondary* diagonal; the matrices $T = [t_{ij}]$ introduced by *Toeplitz* have equal elements along diagonals parallel to the *principal* diagonal. In particular, it was noted in the preceding section that circulant matrices have Toeplitz form.

Example 13.7 When $n = 4$ an arbitrary Hankel or Toeplitz matrix is respectively

$$H = \begin{bmatrix} h_1 & h_2 & h_3 & h_4 \\ h_2 & h_3 & h_4 & h_5 \\ h_3 & h_4 & h_5 & h_6 \\ h_4 & h_5 & h_6 & h_7 \end{bmatrix}, \quad T = \begin{bmatrix} t_0 & t_{-1} & t_{-2} & t_{-3} \\ t_1 & t_0 & t_{-1} & t_{-2} \\ t_2 & t_1 & t_0 & t_{-1} \\ t_3 & t_2 & t_1 & t_0 \end{bmatrix}. \quad (13.29)$$

In general an $n \times n$ Toeplitz matrix can be defined by

$$t_{ij} = t_{i-j}, \quad i, j = 1, 2, \ldots, n, \quad (13.30)$$

so in particular $t_{i+1, j+1} = t_{ij}$. The matrix contains $2n - 1$ independent elements, which occur in the first row and column.

Similarly, Hankel matrices satisfy

$$h_{ij} = h_{i+j-1}, \qquad i, j = 1, 2, \ldots, n \qquad (13.31)$$

and $h_{i+1,j-1} = h_{ij}$, with the $2n - 1$ defining elements in the first row and last column. The term 'orthosymmetric' has also been used to describe Hankel matrices, since the elements are equal along lines perpendicular to the principal diagonal. It is easy to see that Toeplitz and Hankel matrices are closely related: in (13.29) if the rows of H are reversed in order it becomes a Toeplitz matrix; the same thing happens if its columns are reversed in order. Conversely, if the rows or columns of T in (13.29) are reversed in order then it becomes a Hankel matrix. Using the reverse unit matrix K defined in Problem 2.4, these facts can be stated succinctly as follows: if H is Hankel then KH and HK are Toeplitz, and if T is Toeplitz then KT and TK are Hankel. Thus many results obtained for one form can be immediately applied to the other. It is sometimes convenient to use the generic term *striped matrix* to refer to either type. Notice also, as illustrated in (13.29), that a Hankel matrix is symmetric, and that a Toeplitz matrix is symmetric relative to the secondary diagonal—the latter property was used in Section 13.2 to define *persymmetry* (confusingly, one or two authors have applied this term to describe what we have called above 'orthosymmetry').

Exercise 13.18 Show that if H is a Hankel matrix then $(KH)^{\mathsf{T}} = HK$.

Striped matrices arise in a wide variety of applications, including stochastic processes, time series analysis and digital filtering, and frequently occur in the form of a coefficient matrix for a set of linear equations.

Example 13.8 Consider the problem from Padé approximation theory of expressing a polynomial of fourth degree as a rational function:

$$a_0 + a_1\lambda + a_2\lambda^2 + a_3\lambda^3 + a_4\lambda^4 = \frac{c_0 + c_1\lambda + c_2\lambda^2}{1 + d_1\lambda + d_2\lambda^2}.$$

On multiplying both sides by the denominator and equating coefficients of powers of λ up to λ^4 we obtain a system of linear equations

$$Ad = c,$$

where $d = [1, d_1, d_2, 0, 0]^T$, $c = [c_0, c_1, c_2, 0, 0]^T$ and

$$A = \begin{bmatrix} a_0 & 0 & 0 & 0 & 0 \\ a_1 & a_0 & 0 & 0 & 0 \\ a_2 & a_1 & a_0 & 0 & 0 \\ a_3 & a_2 & a_1 & a_0 & 0 \\ a_4 & a_3 & a_2 & a_1 & a_0 \end{bmatrix}.$$

The lower triangular matrix A has Toeplitz form.

The crucial fact about striped matrices is that algorithms can be derived for their inversion which require only $O(n^2)$ operations, compared to $O(n^3)$ operations to invert an arbitrary matrix by gaussian elimination (the same applies to sets of linear equations with striped coefficient matrices). This is especially useful when n is large, and is not entirely unexpected as there are only $2n - 1$ independent elements in a striped matrix, compared with n^2 in general. Before giving details we first develop for striped matrices characterizations which generalize the eqn (13.19) for circulant matrices; recall that this stated that C is circulant if and only if it commutes with the special companion matrix denoted by Π.

Example 13.9 Consider the equation

$$AX = XB^T \tag{13.32}$$

where A and B are arbitrary companion matrices in the standard form (11.2). Let all the matrices be 3×3, and write out (13.32) in full as

$$\begin{bmatrix} 0 & 1 & 0 \\ 0 & 0 & 1 \\ -a_3 & -a_2 & -a_1 \end{bmatrix} \begin{bmatrix} x_1 & x_2 & x_3 \\ x_4 & x_5 & x_6 \\ x_7 & x_8 & x_9 \end{bmatrix} = \begin{bmatrix} x_1 & x_2 & x_3 \\ x_4 & x_5 & x_6 \\ x_7 & x_8 & x_9 \end{bmatrix} \begin{bmatrix} 0 & 0 & -b_3 \\ 1 & 0 & -b_2 \\ 0 & 1 & -b_1 \end{bmatrix}.$$

After performing the multiplications this becomes

$$\begin{bmatrix} x_4 & x_5 & x_6 \\ x_7 & x_8 & x_9 \\ y & y & y \end{bmatrix} = \begin{bmatrix} x_2 & x_3 & y \\ x_5 & x_6 & y \\ x_8 & x_9 & y \end{bmatrix}, \tag{13.33}$$

where the y's denote entries of no special significance. Comparing elements in (13.33) produces

$$x_2 = x_4, \qquad x_3 = x_5 = x_7, \qquad x_6 = x_8$$

showing that X has Hankel form.

The same result holds in general: provided a solution X of (13.32) exists then it has Hankel form. It is interesting to digress a little and investigate when non-trivial solutions of the linear matrix eqn (13.32) do exist. We begin by noting that it is a special case of (5.51), and so can be written as

$$(A \otimes I_n - I_n \otimes B)x = 0, \tag{13.34}$$

where x is a column vector constructed from the elements of X according to (5.48). It follows from Section 5.4.1 that the set of n^2 equations (13.34) has a non-trivial solution if and only if $A \otimes I_n - I_n \otimes B$ is singular. However, we saw in Section 6.3.5 that this matrix has eigenvalues $\lambda_i - \mu_j$, where the λ's and μ's are the eigenvalues of A and B respectively. Hence the condition for there to be a non-trivial solution of (13.32) is that A and B have at least one eigenvalue in common—in other words, that their characteristic polynomials are *not* relatively prime. Clearly this result still holds when A and B have arbitrary, rather than companion, form.

Exercise 13.19 Find the general solution of (13.32) when A and B are companion matrices associated with the polynomials $\lambda^3 + 2\lambda^2 + 4\lambda + 3$, $\lambda^3 - 3\lambda^2 - 5\lambda - 1$ respectively.

Exercise 13.20 By replacing X by XK in (13.32), show that if a solution X exists of the equation

$$AX = XKB^\mathsf{T}K \tag{13.35}$$

where A and B are companion matrices in the form (11.2), then it is Toeplitz.

Notice that if we set $B = A$ in (13.35) and take the characteristic polynomial of A to be $\lambda^n - 1$, then both A and $KB^\mathsf{T}K$ become the matrix Π in (13.17). Hence (13.35) in this case reduces to (13.19), the equation which characterizes circulant matrices.

If we restrict attention to *non-singular* striped matrices, which is often the case in applications, then A and B can be set equal to each other in (13.32) and (13.35) to produce the following result:

A non-singular matrix X is (a) Hankel; (b) Toeplitz if and only if there exists a companion matrix C (in the standard form (11.2)) such that

$$\text{(a) } CX = XC^\mathsf{T}; \quad \text{(b) } CX = XKC^\mathsf{T}K. \tag{13.36}$$

Example 13.10 The matrix

$$X = \begin{bmatrix} 1 & 1 & 1 \\ 1 & 1 & 1 \\ 0 & 1 & 1 \end{bmatrix}$$

is Toeplitz and singular. Equation (13.36(b)) becomes

$$\begin{bmatrix} 0 & 1 & 0 \\ 0 & 0 & 1 \\ -a_3 & -a_2 & -a_1 \end{bmatrix} X = X \begin{bmatrix} -a_1 & 1 & 0 \\ -a_2 & 0 & 1 \\ -a_3 & 0 & 0 \end{bmatrix}. \tag{13.37}$$

Equating the $3, 2$ and $3, 3$ elements in the products on either side of (13.37) gives $a_1 + a_2 + a_3 = 0$, $a_1 + a_2 + a_3 = -1$, showing that there is no solution C of (13.36(b)) in this case. This illustrates that it is essential for X to be non-singular for (13.36) to hold in general.

We have already encountered (13.36(a)) in Problem 11.1, and we used it in Section 11.2.3 to show that the inverse of any non-singular bezoutian matrix has Hankel form. In fact the converse of this latter result also holds: given any non-singular Hankel matrix H, there exist polynomials $a(\lambda)$ and $b(\lambda)$ such that H^{-1} is a bezoutian matrix Z defined by (11.14) (note that in contrast to the circulant case—see Exercise 13.11(b)—H^{-1} is not Hankel). Recalling that in (11.20) we expressed a bezoutian in terms of four triangular matrices, it follows that the inverse of H can always be expressed in a similar way.

Example 13.11 The Hankel matrix

$$H = \begin{bmatrix} -33 & -13 & -5 \\ -13 & -5 & -2 \\ -5 & -2 & -1 \end{bmatrix} \tag{13.38}$$

is non-singular, and its inverse can be expressed as

$$H^{-1} = \underbrace{\begin{bmatrix} 1 & 1 & 2 \\ 0 & 1 & 1 \\ 0 & 0 & 1 \end{bmatrix}}_{H_1} \underbrace{\begin{bmatrix} 0 & 0 & -2 \\ 0 & -2 & 5 \\ -2 & 5 & -1 \end{bmatrix}}_{H_2} - \underbrace{\begin{bmatrix} 3 & -1 & 5 \\ 0 & 3 & -1 \\ 0 & 0 & 3 \end{bmatrix}}_{H_3} \underbrace{\begin{bmatrix} 0 & 0 & -1 \\ 0 & -1 & 2 \\ -1 & 2 & 1 \end{bmatrix}}_{H_4}. \tag{13.39}$$

The coefficients of the polynomials $a(\lambda)$, $b(\lambda)$ for which $H^{-1} = Z$ can be written down by comparing (13.39) with (11.21) to give

$$a(\lambda) = \lambda^3 + \lambda^2 + 2\lambda - 1, \qquad b(\lambda) = 3\lambda^3 - \lambda^2 + 5\lambda - 2.$$

For the remainder of this section we shall prefer to work with Toeplitz matrices. Postmultiply H in (13.38) by K (i.e. reverse the order of its columns) to produce the Toeplitz matrix

$$T = HK = \begin{bmatrix} -5 & -13 & -33 \\ -2 & -5 & -13 \\ -1 & -2 & -5 \end{bmatrix}. \tag{13.40}$$

Inverting, we have $T^{-1} = KH^{-1}$, so from (13.39)

$$T^{-1} = KH_1H_2 - KH_3H_4$$

$$= (KH_1K)(KH_2) - (KH_3K)(KH_4)$$

$$= \begin{bmatrix} 1 & 0 & 0 \\ 1 & 1 & 0 \\ 2 & 1 & 1 \end{bmatrix} \begin{bmatrix} -2 & 5 & -1 \\ 0 & -2 & 5 \\ 0 & 0 & -2 \end{bmatrix} - \begin{bmatrix} 3 & 0 & 0 \\ -1 & 3 & 0 \\ 5 & -1 & 3 \end{bmatrix} \begin{bmatrix} -1 & 2 & 1 \\ 0 & -1 & 2 \\ 0 & 0 & -1 \end{bmatrix}$$

$$\tag{13.41}$$

$$= \begin{bmatrix} 1 & -1 & -4 \\ -3 & 8 & -1 \\ 1 & -3 & 1 \end{bmatrix}. \tag{13.42}$$

Notice that we used the property $K^2 = I$ in obtaining (13.41), which expresses the inverse of the Toeplitz matrix in (13.40) in terms of four triangular Toeplitz matrices.

It can be shown that a result similar to (13.41) holds in general. If the two sets of equations

$$yT = e_1, \qquad Tx = e_1^{\mathsf{T}}/y_0, \tag{13.43}$$

where

$$x = [1, x_1, \ldots, x_{n-1}]^{\mathsf{T}}, \qquad y = [y_0, y_{-1}, \ldots, y_{-n+1}]$$

and e_1 is the first row of I_n, have a solution with $y_0 \neq 0$ then the $n \times n$ Toeplitz matrix T is non-singular and the expression for T^{-1} in terms of Toeplitz matrices is

$$T^{-1} = \begin{bmatrix} 1 & & & \\ x_1 & 1 & & \mathbf{0} \\ \vdots & & \ddots & \\ x_{n-1} & \cdots & x_1 & 1 \end{bmatrix} \begin{bmatrix} y_0 & y_{-1} & \cdots & y_{-n+1} \\ & y_0 & & \vdots \\ & & \ddots & y_{-1} \\ \mathbf{0} & & & y_0 \end{bmatrix}$$

$$- \begin{bmatrix} 0 & & & \\ y_{-n+1} & 0 & & \mathbf{0} \\ \vdots & & \ddots & \\ y_{-1} & y_{-2} & \cdots & y_{-n+1} & 0 \end{bmatrix} \begin{bmatrix} 0 & x_{n-1} & \cdots & x_1 \\ 0 & 0 & x_{n-1} & \vdots \\ \mathbf{0} & & \ddots & x_{n-1} \\ & & & 0 \end{bmatrix}.$$

$$\tag{13.44}$$

Example 13.12 Return to the Toeplitz matrix in (13.40). The first set of equations in (13.43) is

$$-5y_0 - 2y_{-1} - y_{-2} = 1$$
$$-13y_0 - 5y_{-1} - 2y_{-2} = 0$$
$$-33y_0 - 13y_{-1} - 5y_{-2} = 0$$

and the reader can easily check that the unique solution is $y_0 = 1$, $y_{-1} = -1$, $y_{-2} = -4$. Similarly, the solution of the second set of equations in (13.43) is $x_1 = -3$, $x_2 = 1$. Hence (13.44) gives

$$T^{-1} = \begin{bmatrix} 1 & 0 & 0 \\ -3 & 1 & 0 \\ 1 & -3 & 1 \end{bmatrix} \begin{bmatrix} 1 & -1 & -4 \\ 0 & 1 & -1 \\ 0 & 0 & 1 \end{bmatrix} - \begin{bmatrix} 0 & 0 & 0 \\ -4 & 0 & 0 \\ -1 & -4 & 0 \end{bmatrix} \begin{bmatrix} 0 & 1 & -3 \\ 0 & 0 & 1 \\ 0 & 0 & 0 \end{bmatrix},$$

$$(13.45)$$

which agrees with (13.42), although the actual triangular Toeplitz matrices in (13.45) are different from those in (13.41).

Exercise 13.21 When

$$T = \begin{bmatrix} 18 & 13 & 58 \\ -2 & 18 & 13 \\ 78 & -2 & 18 \end{bmatrix} \tag{13.46}$$

solve the equations (13.43), and hence express T^{-1} in the form (13.44).

We now develop a crucial property of the inverse of a Toeplitz matrix, thereby producing a key step in the construction of an efficient algorithm for inversion. One possible starting point is the formula (13.44), but we choose instead to use the characterization involving a companion matrix in (13.36). Suppose T is a non-singular Toeplitz matrix, and denote its inverse by $Y = [y_{ij}]$, $i, j = 1, 2, \ldots, n$. We know that T satisfies (13.36(b)), i.e.

$$CT = TKC^TK,$$

where C is a companion matrix, and multiplying this on the left and right by Y produces

$$YC = KC^TKY \tag{13.47}$$

Also, since T is persymmetric, so is Y (see Exercise 13.9(b)).

Example 13.13 When $n = 3$, (13.47) becomes

$$\underbrace{\begin{bmatrix} y_1 & y_2 & y_3 \\ y_4 & y_5 & y_2 \\ y_6 & y_4 & y_1 \end{bmatrix}}_{Y} \underbrace{\begin{bmatrix} 0 & 1 & 0 \\ 0 & 0 & 1 \\ -a_3 & -a_2 & -a_1 \end{bmatrix}}_{C} = \underbrace{\begin{bmatrix} -a_1 & 1 & 0 \\ -a_2 & 0 & 1 \\ -a_3 & 0 & 0 \end{bmatrix}}_{KC^TK} \underbrace{\begin{bmatrix} y_1 & y_2 & y_3 \\ y_4 & y_5 & y_6 \\ y_6 & y_4 & y_1 \end{bmatrix}}_{Y}, \tag{13.48}$$

where the persymmetry of Y has been included. On expanding (13.48) and equating elements the reader can check that only the following three independent equations are obtained:

$$y_5 = y_1 + a_1 y_2 - a_2 y_3, \qquad \text{from elements 1, 2 or 2, 3,}$$

$$y_4 - a_1 y_1 = -a_3 y_3, \qquad \text{from elements 1, 1 or 3, 3,}$$

$$y_6 - a_2 y_1 = -a_3 y_2, \qquad \text{from elements 2, 1 or 3, 2.}$$

Substituting for a_1 and a_2 from the second and third of these into the first, assuming $y_1 \neq 0$, produces

$$y_5 = y_1 + \frac{1}{y_1}(y_2 y_4 - y_3 y_6). \qquad (13.49)$$

Equation (13.49) shows that once y_1, y_2, y_3, y_4 and y_6 are known then the remaining unknown element y_5 in the inverse matrix Y can be obtained directly.

By extending the argument in Example 13.13 to the general case it is found that provided $y_{11} \neq 0$ then

$$y_{i+1,j+1} = y_{ij} + \frac{1}{y_{11}}(y_{i+1,1}y_{1,j+1} - y_{1,n-i+1}y_{n-j+1,1}),$$
$$i, j = 1, 2, \ldots, n-1. \quad (13.50)$$

This shows that *all* the elements of the inverse Y can be determined sequentially from (13.50) and the persymmetry condition $y_{ij} = y_{n+1-j,n+1-i}$ (see (13.16)) once those in the first row and first column of Y (a total of $2n - 1$) are known. The number of elements which determines T^{-1} is thus the same as the number of independent elements in T.

Example 13.14 For the 3×3 matrix T in (13.40), the 2, 2 element of its inverse is given by (13.50) as

$$y_{22} = y_{11} + \frac{1}{y_{11}}(y_{21}y_{12} - y_{13}y_{31}).$$

Using the elements in the first row and column of T^{-1} in (13.42) gives

$$y_{22} = 1 + (-3)(-1) - (-4)(1) = 8,$$

agreeing with the value in (13.42).

Exercise 13.22 The first row and column of the inverse of a 5×5 Toeplitz matrix are respectively

$$[1, 1, 1, -4, 6], \qquad \left[1, -\frac{2}{5}, 0, \frac{1}{5}, \frac{1}{5}\right]^{\mathrm{T}}.$$

Determine all the remaining elements of the inverse.

To complete our algorithm for efficient determination of the inverse of an $n \times n$ Toeplitz matrix T_n, it remains to describe the procedure for computing the first row and column of the inverse. Write T_n in the partitioned form

$$T_n = \begin{bmatrix} t_0 & r_{n-1}^T \\ s_{n-1} & T_{n-1} \end{bmatrix} \begin{matrix} 1 \\ n-1 \end{matrix}. \qquad (13.51)$$
$$\begin{matrix} 1 & n-1 \end{matrix}$$

Because T_{n-1} has the same Toeplitz structure, (13.51) still holds with n replaced by k, for $k = 2, 3, \ldots, n-1$. It is necessary to assume that T_n is *strongly non-singular*—that is, $\det T_k \neq 0$, $k = 1, 2, \ldots, n$ (notice that this includes $t_0 \neq 0$). Positive definite matrices are an important class of strongly non-singular matrices (see the remark at the end of Section 7.4). Write T_n^{-1} in the same partitioned form as (13.51), namely

$$Y_n = \begin{bmatrix} y_{n-1} & f_{n-1}^T \\ g_{n-1} & Y_{n-1} \end{bmatrix} \begin{matrix} 1 \\ n-1 \end{matrix}. \qquad (13.52)$$
$$\begin{matrix} 1 & n-1 \end{matrix}$$

At this point we can remark that from the formula (4.46) we have

$$T_n^{-1} = \left(\frac{1}{\det T_n} \right) \text{adj } T_n$$

and, since the cofactor of t_0 in (13.51) (obtained by deleting the first row and column) is $\det T_{n-1}$, it follows that

$$y_{n-1} = \frac{\det T_{n-1}}{\det T_n} \neq 0.$$

The first row and column of the inverse matrix Y_n are obtained recursively from the formulae

$$\gamma_k = y_k r'_{k+1} + f_k^T K r_k \qquad (13.53)$$

$$\delta_k = y_k s'_{k+1} + g_k^T K s_k \qquad (13.54)$$

$$y_{k+1} = \frac{y_k}{1 - \gamma_k \delta_k} \qquad (13.55)$$

for $k = 0, 1, 2, \ldots, n-2$, starting with $y_0 = 1/t_0$, $f_0 = 0$, $g_0 = 0$, and

$$f_{k+1} = \frac{y_{k+1}}{y_k} \begin{bmatrix} f_k - \gamma_k K g_k \\ -\gamma_k y_k \end{bmatrix} \tag{13.56}$$

$$g_{k+1} = \frac{y_{k+1}}{y_k} \begin{bmatrix} g_k - \delta_k K f_k \\ -\delta_k y_k \end{bmatrix} \tag{13.57}$$

for $k = 1, 2, \ldots, n-2$ with $f_1 = [-\gamma_0 y_1]$, $g_1 = [-\delta_0 y_1]$. In (13.53)–(13.57) each of r_k, s_k, f_k, g_k is a column k-vector, and K is as usual the reverse unit matrix of appropriate order. The notation r'_{k+1} denotes the last element of r_{k+1}, so that

$$r_{k+1} = \begin{bmatrix} r_k \\ r'_{k+1} \end{bmatrix}, \qquad k = 1, 2, \ldots, n-2 \tag{13.58}$$

with $r'_1 = r_1$, and similar expressions for s_{k+1} and s'_1.

In addition, it turns out that

$$\det T_n = (y_0 y_1 \cdots y_{n-1})^{-1}. \tag{13.59}$$

When T_n is symmetric the computation is reduced by about half, since (13.54) becomes equivalent to (13.53), and (13.57) to (13.56). In general, the efficiency of the algorithm is worthwhile for $n > 4$, but we present an illustrative example for $n = 3$.

Example 13.15 We apply the procedure to the 3×3 Toeplitz matrix in (13.40), written here using the notation in (13.51) as

$$T_3 = \begin{bmatrix} -5 & -13 & -33 \\ -2 & -5 & -13 \\ -1 & -2 & -5 \end{bmatrix} = \begin{bmatrix} t_0 & r_2^T \\ s_2 & T_2 \end{bmatrix}, \tag{13.60}$$

where

$$T_2 = \begin{bmatrix} -5 & -13 \\ -2 & -5 \end{bmatrix} = \begin{bmatrix} -5 & r_1^T \\ s_1 & T_1 \end{bmatrix},$$

so that $t_0 = -5$, $r'_1 = -13$, $s'_1 = -2$. Hence $y_0 = -1/5$, and setting $k = 0$ in (13.53)–(13.55) gives

$$\gamma_0 = y_0 r'_1 = \frac{13}{5}, \qquad \delta_0 = y_0 s'_1 = \frac{2}{5},$$

$$y_1 = \frac{y_0}{1 - \gamma_0 \delta_0} = 5,$$

so that

$$f_1 = [-\gamma_0 y_1] = [-13], \qquad g_1 = [-\delta_0 y_1] = [-2].$$

In (13.60), $r_2^T = [-13, -33]$ so from (13.58) $r'_2 = -33$, and similarly

$s_2' = -1$. Repeating the above procedure with $k = 1$ gives

$$\gamma_1 = y_1 r_2' + f_1^T K r_1, \qquad \text{by (13.53),}$$
$$= 5(-33) + (-13)(-13) = 4,$$
$$\delta_1 = y_1 s_2' + g_1^T K s_1, \qquad \text{by (13.54),}$$
$$= 5(-1) + (-2)(-2) = -1$$

$$y_2 = \frac{y_1}{1 - \gamma_1 \delta_1} = 1, \qquad \text{by (13.55),} \tag{13.61}$$

and from (13.56) and (13.57) with $k = 1$

$$f_2 = \frac{y_2}{y_1} \begin{bmatrix} f_1 - \gamma_1 K g_1 \\ -\gamma_1 y_1 \end{bmatrix} = \begin{bmatrix} -1 \\ -4 \end{bmatrix} \tag{13.62}$$

$$g_2 = \frac{y_2}{y_1} \begin{bmatrix} g_1 - \delta_1 K f_1 \\ -\delta_1 y_1 \end{bmatrix} = \begin{bmatrix} -3 \\ 1 \end{bmatrix}. \tag{13.63}$$

The first row and column of $Y_3 = T_3^{-1}$ are therefore given by substituting (13.61), (13.62) and (13.63) into (13.52) with $n = 3$, producing

$$Y_3 = \begin{bmatrix} 1 & -1 & & -4 \\ -3 & & & \\ & & Y_2 & \\ 1 & & & \end{bmatrix}.$$

This agrees with the result found previously in (13.42). The remaining elements of Y_3 are then found using (13.50), as indicated in Example 13.14, together with the persymmetry property of the inverse. In addition, from (13.59) we have

$$\det T_3 = 1 \bigg/ \left(-\frac{1}{5}\right)(5)(1) = -1.$$

Exercise 13.23 Use the algorithm described above to compute the inverse and determinant of the matrix in (13.46).

Exercise 13.24 The first row and column of a 5×5 Toeplitz matrix are respectively

$$[2, 3, -1, 0, 1], \qquad [2, 1, 0, -1, -1]^T.$$

Determine the first row and column of the inverse matrix.

Check that these agree with the expressions in Exercise 13.22, where all the remaining elements of the inverse were computed.

The solution of the set of n equations

$$T_n x_n = b_n \tag{13.64}$$

with a Toeplitz coefficient matrix is obtained by performing the recursion

$$x_{k+1} = \begin{bmatrix} x_k + \theta_k K f_k \\ y_k \theta_k \end{bmatrix}, \qquad k = 0, 1, 2, \ldots, n-1, \quad (13.65)$$

where $x_0 = 0$,

$$\theta_k = \frac{1}{y_k} g_k^T K b_k + b_{k+1}', \qquad (13.66)$$

f_k, g_k are given by (13.56) and (13.57) respectively, and b_{k+1} is defined in an analogous fashion to (13.58). Notice that it is not necessary to use the formula (13.50) when solving the equations (13.64).

Example 13.15 (continued)   Consider the equations

$$T_3 x_3 = \begin{bmatrix} 116 \\ 46 \\ 17 \end{bmatrix}, \qquad (13.67)$$

where T_3 is the matrix in (13.60). By analogy with (13.58)

$$b_3 = \begin{bmatrix} 116 \\ 46 \\ 17 \end{bmatrix} = \begin{bmatrix} b_2 \\ b_3' \end{bmatrix}$$

$$b_2 = \begin{bmatrix} 116 \\ 46 \end{bmatrix} = \begin{bmatrix} b_1 \\ b_2' \end{bmatrix}$$

and $b_1' = b_1 = 116$. From (13.66) and (13.65) we obtain successively, using the previously computed y_k, f_k and g_k:

$k = 0$ $\theta_0 = b_1' = 116,$ $x_1 = [y_0 \theta_0] = -\dfrac{116}{5};$

$k = 1$ $\theta_1 = \dfrac{1}{y_1} g_1^T K b_1 + b_2'$

$\qquad\qquad = \dfrac{1}{5}(-2)(116) + 46 = -\dfrac{2}{5},$

$$x_2 = \begin{bmatrix} x_1 + \theta_1 K f_1 \\ y_1 \theta_1 \end{bmatrix} = \begin{bmatrix} -\dfrac{116}{5} & -\dfrac{2}{5}(-13) \\ & 5\left(-\dfrac{2}{5}\right) \end{bmatrix} = \begin{bmatrix} -18 \\ -2 \end{bmatrix};$$

$k = 2$ $\theta_2 = \dfrac{1}{y_2} g_2^T K b_2 + b_3'$

$\qquad\qquad = [-3, 1]\begin{bmatrix} 0 & 1 \\ 1 & 0 \end{bmatrix}\begin{bmatrix} 116 \\ 46 \end{bmatrix} + 17 = -5;$

$$x_3 = \begin{bmatrix} x_2 + \theta_2 K f_2 \\ y_2 \theta_2 \end{bmatrix}$$

$$= \begin{bmatrix} \begin{pmatrix} -18 \\ -2 \end{pmatrix} - 5 \begin{pmatrix} 0 & 1 \\ 1 & 0 \end{pmatrix} \begin{pmatrix} -1 \\ -4 \end{pmatrix} \\ -5 \end{bmatrix} = \begin{bmatrix} 2 \\ 3 \\ -5 \end{bmatrix};$$

and this last vector is the solution of the eqns (13.67).

Exercise 13.25 Determine the solution of the eqns (13.64) when T_3 is the matrix in (13.46) and $b_3 = [-29, 11, -44]^\mathsf{T}$ (use the computations in Exercise 13.23).

We close this section with an explicit formula for the inverse of a tridiagonal Toeplitz matrix $T = [t_{i-j}]$ with $t_0 = 1$, $t_{i-j} = 0$ for $|i - j| > 1$ and t_{-1}, t_1 having arbitrary values, subject only to T being non-singular. If $T^{-1} = [y_{ij}]$, first define

$$\alpha_0 = 1, \quad \alpha_1 = 1 - t_1 t_{-1}, \quad \alpha_k = \alpha_{k-1} - t_1 t_{-1} \alpha_{k-2}, \qquad k = 2, \ldots, n$$

and

$$\beta_k = \begin{cases} \alpha_{k-1}/(t_{-1})^k, & i \leqslant j \\ \alpha_{k-1}/(t_1)^k, & i > j, \quad k = 1, 2, \ldots, n \end{cases}$$

with $\beta_0 = 1$. Then

$$y_{ij} = \begin{cases} (-1)^{i+j} \beta_{i-1} \beta_{n-j}/t_{-1} \beta_n, & i \leqslant j \\ (-1)^{i+j} \beta_{j-1} \beta_{n-i}/t_1 \beta_n, & i > j \end{cases} \qquad (13.68)$$

and T^{-1} is persymmetric (Exercise 13.9(b)).

Exercise 13.26 Use (13.68) to determine the inverse of the 4×4 tridiagonal Toeplitz matrix with $t_0 = 1$, $t_1 = 2$, $t_{-1} = 3$.

For further results on Toeplitz matrices, see Heinig and Rost (1984).

13.4 Other forms

In this section we record definitions and properties of a few other useful special forms of matrix.

13.4.1 Brownian matrices

An $n \times n$ matrix $B = [b_{ij}]$ is *brownian* if

$$b_{i,j+1} = b_{ij}, \qquad j > i, \qquad \text{and} \quad b_{i+1,j} = b_{ij}, \qquad i > j,$$

for all i and j. For example, when $n = 4$

$$B = \begin{bmatrix} b_1 & b_5 & b_5 & b_5 \\ b_8 & b_2 & b_6 & b_6 \\ b_8 & b_9 & b_3 & b_7 \\ b_8 & b_9 & b_{10} & b_4 \end{bmatrix},$$

and in general B is completely determined by the $3n - 2$ elements on the tridiagonal band which includes the principal diagonal. The name arises from brownian motion, and there are also other applications. Any brownian matrix is congruent to a tridiagonal matrix A according to $A = PBP^{\mathsf{T}}$ where $P = [p_{ij}]$ is a bidiagonal Toeplitz matrix with

$$p_{ij} = \begin{cases} 1, & j = i, & i = 1, 2, \ldots, n \\ -1, & j = i - 1, & i = 2, 3, \ldots, n. \\ 0, & \text{otherwise} \end{cases}$$

Like Toeplitz matrices, algorithms have been devised for inversion in only $O(n^2)$ operations.

13.4.2 Centrosymmetric matrices

An $n \times n$ matrix $C = [c_{ij}]$ is *centrosymmetric* if

$$c_{ij} = c_{n+1-i, n+1-j}, \qquad i, j = 1, 2, \ldots, n.$$

For example, when $n = 4$

$$C = \begin{bmatrix} c_1 & c_2 & c_3 & c_4 \\ c_5 & c_6 & c_7 & c_8 \\ c_8 & c_7 & c_6 & c_5 \\ c_4 & c_3 & c_2 & c_1 \end{bmatrix},$$

and in general C has $\frac{1}{2}n^2$ independent elements when n is even, and $\frac{1}{2}(n^2 + 1)$ when n is odd. The rth row from the bottom consists of the rth row from the top, with the order of the elements reversed. An alternative way of expressing this is

$$C = KCK,$$

where, as usual, K is the reverse unit matrix defined in Problem 2.4. Any centrosymmetric matrix of even order $n = 2m$ can be expressed in the partitioned form

$$C = \begin{matrix} m & & m \\ \begin{bmatrix} C_1 & C_2 K \\ K C_2 & K C_1 K \end{bmatrix} & \begin{matrix} m \\ m \end{matrix} \end{matrix} = \begin{bmatrix} I & 0 \\ 0 & K \end{bmatrix} \begin{bmatrix} C_1 & C_2 \\ C_2 & C_1 \end{bmatrix} \begin{bmatrix} I & 0 \\ 0 & K \end{bmatrix} \qquad (13.69)$$

and when the order $n = 2m + 1$ is odd in the form

$$
\begin{array}{ccc}
m & 1 & m
\end{array}
$$
$$
\begin{bmatrix} C_1 & a & C_2K \\ b & c & bK \\ KC_2 & Ka & KC_1K \end{bmatrix}\begin{matrix} m \\ 1 \\ m \end{matrix} = \begin{bmatrix} I & 0 & 0 \\ 0 & 1 & 0 \\ 0 & 0 & K \end{bmatrix}\begin{bmatrix} C_1 & a & C_2 \\ b & c & b \\ C_2 & a & C_1 \end{bmatrix}\begin{bmatrix} I & 0 & 0 \\ 0 & 1 & 0 \\ 0 & 0 & K \end{bmatrix},
$$

where C_1 and C_2 are arbitrary $m \times m$ matrices, a and b are arbitrary vectors and c is an arbitrary scalar. Some properties of C in (13.69) can be expressed in terms of matrices of order m—see Problem 13.23.

Finally, if C is $m \times n$ then the definitions of centrosymmetry become $c_{ij} = c_{m+1-i,n+1-j}$ and $K_m C K_n = C$.

Exercise 13.27 Using the first part of Exercise 13.9, deduce that a matrix which is symmetric about both its principal and secondary diagonals is centrosymmetric.

Exercise 13.28 Verify that each of the two standard partitioned forms given above for a centrosymmetric matrix satisfies $KCK = C$, where K has appropriate order.

13.4.3 Comrade matrix

The $n \times n$ comrade matrix

$$
A = \begin{bmatrix}
-\dfrac{\beta_1}{\alpha_1} & \dfrac{1}{\alpha_1} & 0 & 0 & & \\
\dfrac{\gamma_2}{\alpha_2} & -\dfrac{\beta_2}{\alpha_2} & \dfrac{1}{\alpha_2} & 0 & & \Large 0 \\
0 & \dfrac{\gamma_3}{\alpha_3} & -\dfrac{\beta_3}{\alpha_3} & \dfrac{1}{\alpha_3} & & \\
& \cdot & \cdot & \cdot & \cdot & \cdot \\
& \cdot & \cdot & \cdot & \cdot & \cdot \\
\Large 0 & & & \dfrac{\gamma_{n-1}}{\alpha_{n-1}} & -\dfrac{\beta_{n-1}}{\alpha_{n-1}} & \dfrac{1}{\alpha_{n-1}} \\
-\dfrac{a_n}{\alpha_n} & \cdot & \cdot & \cdot & -\dfrac{a_3}{\alpha_n} & \dfrac{\gamma_n - a_2}{\alpha_n} & \dfrac{-a_1 - \beta_n}{\alpha_n}
\end{bmatrix} \qquad (13.70)
$$

where $\alpha_i > 0$, for all i, consists of a tridiagonal band together with a last row which enables the characteristic polynomial to be written down in the following way. From Problem 13.2 we have

$$\det(\lambda I_n - A) = a(\lambda)/(\alpha_1 \alpha_2 \cdots \alpha_n), \qquad (13.71)$$

where

$$a(\lambda) = p_n(\lambda) + a_1 p_{n-1}(\lambda) + \cdots + a_{n-1} p_1(\lambda) + a_n p_0(\lambda) \qquad (13.72)$$

and the polynomials $p_i(\lambda)$ in (13.72) are defined recursively by

$$p_0(\lambda) = 1, \qquad p_1(\lambda) = \alpha_1 \lambda + \beta_1,$$
$$p_i(\lambda) = (\alpha_i \lambda + \beta_i)p_{i-1}(\lambda) - \gamma_i p_{i-2}(\lambda), \qquad i = 2, 3, \ldots, n. \qquad (13.73)$$

The name 'comrade' arises by analogy with 'companion', used many times in this book, and given special attention in Section 11.1. In both cases the matrix is a 'partner' to a specific polynomial. It can be seen that setting $\alpha_i = 1$, $\beta_i = 0$, $\gamma_i = 0$ for all i reduces A to the standard companion form (11.2), and in (13.73) $p_i(\lambda)$ becomes simply λ^i. In general, if $\alpha_i > 0$, $\gamma_i > 0$ for all i then the polynomials $p_i(\lambda)$ form an *orthogonal set*—that is, there exists a weight function $w(\lambda) \geq 0$ and an interval $[c, d]$ such that

$$\int_c^d w(\lambda) \, d\lambda > 0$$

and

$$\int_c^d p_i(\lambda)p_j(\lambda)w(\lambda) \, d\lambda = 0, \qquad i \neq j$$
$$\neq 0, \qquad i = j.$$

Some useful special cases of (13.70) are:

(i)   When $\alpha_i = 1$, $\beta_i = 0$, $\gamma_i = 1$, for all i then A takes the *colleague form*, which is just like the companion form (11.2) but with an extra line of 1's on the subdiagonal. The polynomials defined by (13.73) are a set of Chebyshev polynomials $S_i(\lambda)$ defined by

$$S_0 = 1, \qquad S_1 = \lambda, \qquad S_i = \lambda S_{i-1} - S_{i-2}, \qquad i \geq 2.$$

(ii) The *Leslie matrix*

$$L = \begin{bmatrix} 0 & f_1 & 0 & & & \\ 0 & 0 & f_2 & & \mathbf{0} & \\ & \mathbf{0} & & \cdot & 0 & f_{n-1} \\ l_1 & l_2 & l_3 & \cdots & l_{n-1} & l_n \end{bmatrix},$$

with all $f_i \neq 0$, arises in population models, simple cases being given in Example 1.8 and Problem 1.6.

(iii) The *Schwarz matrix* defined in (11.101) arises in stability theory and network theory. It is tridiagonal, and the location of its eigenvalues relative to the imaginary axis can be determined directly (see Problem 11.10).

(iv) The *Routh matrix* $R = [r_{ij}]$ is tridiagonal, with

$$r_{11} = -b_1, \qquad r_{ii} = 0, \qquad i \geqslant 2,$$
$$r_{i-1,i} = b_i^{1/2}, \qquad r_{i,i-1} = -b_i^{1/2}, \qquad i = 2, 3, \ldots, n, \qquad (13.74)$$

where the b_i are real and non-zero, and is similar to (11.101) (see Problem 13.21). It is encountered in linear control theory.

A polynomial expressed in the form (13.72) in terms of a basis of orthogonal polynomials is said to be in generalized form. The comrade matrix plays the same role in investigating algebraic properties of generalized polynomials as the companion matrix does for ordinary polynomials (see Chapter 11). Details are outside the scope of this book, but it can be mentioned that a further extension of (13.70) in lower Hessenberg form has been termed the *confederate matrix*, the whole class of such matrices associated with polynomials being called *congenial* (Barnett 1983).

Exercise 13.29 Prove that the Leslie matrix L is similar to a companion matrix by considering DLD^{-1}, where

$$D = \text{diag}[1, f_1, f_1 f_2, \ldots, f_1 f_2 f_3 \cdots f_{n-1}].$$

Exercise 13.30 Deduce from Problem 11.10 that the Schwarz matrix defined in (11.101) is a stability matrix (i.e. all its eigenvalues have negative real parts) if and only if all the b_i are positive.

Exercise 13.31 For each of the following cases, write down the appropriate comrade matrix.

(a) The basis (13.73) consisting of the *Chebyshev polynomials* $T_i(\lambda)$ of the first kind, with $T_0 = 1$, $T_1 = \lambda$, $T_i = 2\lambda T_{i-1} - T_{i-2}$, $i \geqslant 2$.

(b) The basis (13.73) consisting of the *Hermite polynomials* $H_i(\lambda)$ with $H_0 = 1$, $H_1 = 2\lambda$, $H_i = 2\lambda H_{i-1} - 2(i-1)H_{i-2}$, $i \geqslant 2$.

(c) The basis (13.73) consisting of the *Legendre polynomials* $P_i(\lambda)$, with $P_0 = 1$, $P_1 = \lambda$,

$$P_i = \frac{2i-1}{i} P_{i-1} - \frac{i-1}{i} P_{i-2}, \qquad i \geqslant 2.$$

Finally, if the comrade matrix A in (13.70) has distinct eigenvalues $\lambda_1, \lambda_2, \ldots, \lambda_n$ then corresponding right eigenvectors are

$$u(\lambda_j) = [1, p_1(\lambda_j), p_2(\lambda_j), \ldots, p_{n-1}(\lambda_j)]^T, \qquad j = 1, 2, \ldots, n$$

and

$$M^{-1}AM = \text{diag}[\lambda_1, \lambda_2, \ldots, \lambda_n], \tag{13.75}$$

where

$$M = [u(\lambda_1), u(\lambda_2), \ldots, u(\lambda_n)]. \tag{13.76}$$

Notice that when $p_i(\lambda_j) = \lambda_j^i$ so that A reduces to a companion matrix, then M in (13.76) becomes the Vandermonde matrix defined in (4.80). The diagonalization formula (13.75) is therefore a generalization of that for the companion matrix in (11.3).

13.4.4 Loewner matrix

The *Loewner matrix* $L = [l_{ij}]$ does not have a visually recognizable pattern, but is mentioned here because of its relationship to Hankel matrices. The elements are defined by

$$l_{ij} = \frac{c_i - d_j}{y_i - z_j}, \qquad i, j = 1, 2, \ldots, n, \tag{13.77}$$

for any scalars $c_1, \ldots, c_n$, $d_1, \ldots, d_n$, and any *distinct* scalars $y_1, \ldots, y_n$, $z_1, \ldots, z_n$. There are applications to rational interpolation. Let $W(t) = W(t_1, t_2, \ldots, t_n)$ be the $n \times n$ matrix whose ith row consists of the coefficients of the $(n-1)$th degree polynomial

$$f_i(\lambda) = \prod_{j=1, j \neq i}^{n} (\lambda - t_j) \tag{13.78}$$

in ascending order of powers of λ. Thus, for example, when $n = 3$

$$W(t) = \begin{bmatrix} t_2 t_3 & -(t_2 + t_3) & 1 \\ t_1 t_3 & -(t_1 + t_3) & 1 \\ t_1 t_2 & -(t_1 + t_2) & 1 \end{bmatrix}.$$

The relationship between Hankel and Loewner matrices is that for any $n \times n$ Hankel matrix H the matrix

$$L = W(y)HW^\mathsf{T}(z) \tag{13.79}$$

is a Loewner matrix, where $y = (y_1, y_2, \ldots, y_n)$, $z = (z_1, z_2, \ldots, z_n)$. Conversely, if L in (13.79) is Loewner then H is Hankel and

$$H = V(y)D^{-1}(y)LD^{-1}(z)V^\mathsf{T}(z),$$

where

$$D(t) = \mathrm{diag}[f_1(t_1), f_2(t_2), \ldots, f_n(t_n)]$$

and $V(t)$ is the Vandermonde matrix defined in (4.80), having ith column,

$$[1, t_i, t_i^2, \ldots, t_i^{n-1}]^\mathsf{T}, \qquad i = 1, 2, \ldots, n. \tag{13.80}$$

Exercise 13.32 Express $f_i(\lambda)$ in (13.78) as a finite McLaurin series, and hence show that premultiplication of the vector in (13.80) by $W(t)$ produces $[f_1(t_i), f_2(t_i), \ldots, f_n(t_i)]^\mathsf{T}$. Deduce from this that $W(t)V(t) = D(t)$.

13.4.5 Permutation matrices

An $n \times n$ *permutation matrix* P is one in which there is a single element equal to 1 in every row and column, all other elements being zero. In other words (see Problem 5.12), the rows of P consist of a permutation of the rows of I_n. For example, the special circulant matrix Π defined in (13.17) and the reverse unit matrix K_n defined in Problem 2.4 are permutation matrices. Premultiplying any $n \times m$ matrix A by P has the effect of applying the same permutation to the rows of A. Similarly, regarding the columns of P as a (different) permutation of the columns of I_n, then postmultiplying an $m \times n$ matrix by P performs this permutation on its columns.

Example 13.16 When $n = 3$ the permutation $(1, 2, 3) \rightarrow (3, 1, 2)$ on the rows or columns of I_3 produces respectively the permutation matrices

$$P_1 = \begin{bmatrix} 0 & 1 & 0 \\ 0 & 0 & 1 \\ 1 & 0 & 0 \end{bmatrix}, \qquad P_2 = \begin{bmatrix} 0 & 0 & 1 \\ 1 & 0 & 0 \\ 0 & 1 & 0 \end{bmatrix}. \tag{13.81}$$

The reader will find it instructive to write out the products P_1A and BP_2 for arbitrary $3 \times m$ and $m \times 3$ matrices A and B respectively.

Notice that in (13.81) $P_2 = P_1^T$ and $P_2 P_1 = I$ and this is easily seen to be true in general: for if P has rows $e_{\pi_1}, e_{\pi_2}, \ldots, e_{\pi_n}$, where $\pi_1, \pi_2 \cdots \pi_n$ is some permutation of $1\,2 \cdots n$, then P^T has columns $e_{\pi_1}^T, e_{\pi_2}^T, \ldots, e_{\pi_n}^T$. It then follows that $P^T P = I$, so that $P^{-1} = P^T$, i.e. P is orthogonal.

An $n \times n$ matrix A is *reducible* (or *decomposable*) if there exists a permutation matrix P such that

$$P^T A P = \begin{matrix} k & n-k \\ \begin{bmatrix} A_1 & 0 \\ A_2 & A_3 \end{bmatrix} & \begin{matrix} k \\ n-k \end{matrix} \end{matrix}, \tag{13.82}$$

where A_1 and A_3 are square. The terminology arises because a set of linear equations

$$Ax = b \tag{13.83}$$

can be written as

$$P^T A P y = c, \tag{13.84}$$

where

$$c = P^T b = \begin{bmatrix} c_1 \\ c_2 \end{bmatrix} \begin{matrix} k \\ n-k \end{matrix}, \qquad y = P^T x = \begin{bmatrix} y_1 \\ y_2 \end{bmatrix} \begin{matrix} k \\ n-k \end{matrix}.$$

Substituting (13.82) into (13.84) produces

$$A_1 y_1 = c_1, \qquad A_2 y_1 + A_3 y_2 = c_2. \tag{13.85}$$

This shows that the original eqns (13.83) in n variables can be reduced (or 'decomposed') to the pair of eqns (13.85), the first of which can be solved for the k variables y_1 and then the second can be solved for the remaining $n - k$ variables y_2. If A is not reducible it is called *irreducible* (or *indecomposable*).

There is an interesting application of permutation matrices to the Kronecker product of a $p \times m$ matrix A and a $q \times n$ matrix B. Recall from Section 5.6 that if $X = [x_{ij}]$ is an $m \times n$ matrix then by 'stacking up' the rows of X, we can construct a column mn-vector denoted by

$$v(X) = [x_{11}, x_{12}, \ldots, x_{1m}, x_{21}, \ldots, x_{2m}, \ldots, x_{n1}, \ldots, x_{nm}]^T. \tag{13.86}$$

Since the elements of $v(X^T)$ are simply a rearrangement of those of $v(X)$, we can define an $mn \times mn$ permutation·matrix Q which satisfies the relationship

$$v(X) = Qv(X^\mathsf{T}).$$ (13.87)

Example 13.17 If $m = 2$, $n = 3$ then from (13.86)

$$v(X) = [x_{11}, x_{12}, x_{13}, x_{21}, x_{22}, x_{23}]^\mathsf{T}$$

and

$$v(X^\mathsf{T}) = [x_{11}, x_{21}, x_{12}, x_{22}, x_{13}, x_{23}]^\mathsf{T}.$$

It is easy to determine that the matrix in (13.87) is

$$Q = \begin{bmatrix} 1 & 0 & 0 & 0 & 0 & 0 \\ 0 & 0 & 1 & 0 & 0 & 0 \\ 0 & 0 & 0 & 0 & 1 & 0 \\ 0 & 1 & 0 & 0 & 0 & 0 \\ 0 & 0 & 0 & 1 & 0 & 0 \\ 0 & 0 & 0 & 0 & 0 & 1 \end{bmatrix}.$$ (13.88)

Now consider the equation

$$AXB^\mathsf{T} = C,$$ (13.89)

where C is $p \times q$. As in Problem 5.11, applying the operator $v(\ )$ to both sides of (13.89) results in

$$(A \otimes B)v(X) = v(C).$$ (13.90)

Now transpose both sides of (13.89), producing

$$BX^\mathsf{T}A^\mathsf{T} = C^\mathsf{T},$$

and again apply $v(\ )$ to obtain

$$(B \otimes A)v(X^\mathsf{T}) = v(C^\mathsf{T}).$$ (13.91)

As in (13.87) we can write

$$v(C^\mathsf{T}) = Pv(C),$$ (13.92)

where P is a permutation matrix depending only upon the dimensions p and q of C. Substituting (13.87) and (13.92) into (13.90) gives

$$(A \otimes B)Qv(X^\mathsf{T}) = P^{-1}v(C^\mathsf{T}).$$ (13.93)

Comparing (13.91) and (13.93) reveals that

$$B \otimes A = P(A \otimes B)Q,$$ (13.94)

where P, Q are permutation matrices depending only on p, q and m, n respectively. Equation (13.94) shows that the elements of the Kronecker product $A \otimes B$ are simply a rearrangement of those of $B \otimes A$. This fact was noted without proof in Section 2.5, and Example 2.18 illustrated a simple case. The reader was asked in Problems 5.11 and 5.12 to establish (13.94) when A and B are square.

Exercise 13.33 Deduce that when A and B are both $n \times n$ then in (13.94) $Q = P^T$.

The permutation matrix Q defined by (13.87) is also useful in handling a partitioned $mn \times mn$ matrix A where each $n \times n$ block A_{ij} possesses the same particular pattern. It is helpful to consider a simple specific case.

Example 13.18 Consider the case $m = 2$, $n = 3$ where

$$A = \begin{bmatrix} A_{11} & A_{12} \\ A_{21} & A_{22} \end{bmatrix}$$

and each A_{ij} is a 3×3 Toeplitz matrix, so we can write

$$A_{11} = \begin{bmatrix} a_0 & a_{-1} & a_{-2} \\ a_1 & a_0 & a_{-1} \\ a_2 & a_1 & a_0 \end{bmatrix}$$

and similarly $[A_{12}]_{ij} = [b_{i-j}]$, $[A_{21}]_{ij} = [c_{i-j}]$, $[A_{22}]_{ij} = [d_{i-j}]$. Using the permutation matrix Q given in (13.88) we obtain

$$Q^T A Q = D, \tag{13.95}$$

where D is a *block-Toeplitz matrix* in the form

$$D = \begin{bmatrix} D_0 & D_{-1} & D_{-2} \\ D_1 & D_0 & D_{-1} \\ D_2 & D_1 & D_0 \end{bmatrix}, \qquad D_i = \begin{bmatrix} a_i & b_i \\ c_i & d_i \end{bmatrix}. \tag{13.96}$$

When A has order mn the result (13.95) still holds, with Q defined by (13.87) and the blocks in the $mn \times mn$ matrix D each having order m. In fact the result also holds *whatever* the pattern of the elements in the blocks of A. Thus if all the A_{ij} are Hankel, then D is *block-Hankel*; and similarly D is block-circulant (see Section 13.2) when all the A_{ij} are circulant matrices. The importance of (13.95) lies in the fact that inversion of A can be achieved by inverting D, for which algorithms paralleling the patterned cases with $m = 1$ have been devised.

Exercise 13.34 Evaluate $Q^T A Q$ in (13.95) and hence verify that D has the stated form (13.96).

13.4.6 Sequence Hankel matrices

In this section we quote some results for $n \times n$ Hankel matrices $H = [h_{i+j-1}]$, as defined in (13.31), where the elements belong to a simple sequence.

(a) *Arithmetic sequence*
   When

$$h_i = a + (i-1)d, \qquad i = 1, 2, \ldots, 2n-1,$$

then H is an *arithmetic matrix*, and $R(H) = 2$ provided $d \neq 0$. A general result is that for *any* $n \times n$ matrix H of rank 2, provided $I_n + H$ is non-singular, then

$$(I + H)^{-1} = I + \frac{H^2 - (1 + \sigma_1)H}{1 + \sigma_1 + \sigma_2},$$

where

$$\sigma_1 = \operatorname{tr}(H), \qquad \sigma_2 = \frac{1}{2}[(\operatorname{tr} H)^2 - \operatorname{tr}(H^2)]. \qquad (13.97)$$

When H is the arithmetic matrix we obtain

$$\sigma_1 = na + n(n-1)d, \qquad \sigma_2 = -n^2(n^2-1)d^2/12.$$

(b) *Geometric sequence*
   When

$$h_1 = 1, \qquad h_i = \lambda^{i-1}, \qquad i = 2, 3, \ldots, 2n-1,$$

then H is a *geometric matrix*, $R(H) = 1$ and

$$(I + H)^{-1} = I - \frac{(1 - \lambda^2)H}{2 - \lambda^2 - \lambda^{2n}}, \qquad \lambda \neq \pm 1. \qquad (13.98)$$

Exercise 13.35 Show that the only non-zero eigenvalue of a geometric matrix is equal to $(1 - \lambda^{2n})/(1 - \lambda^2)$, if $\lambda \neq \pm 1$; and n if $\lambda = \pm 1$. Also, obtain (13.98) using (5.81).

(c) *Harmonic sequence*
   When

$$h_i = 1/[a + (i-1)d], \qquad i = 1, 2, \ldots, 2n-1,$$

then H is a *harmonic matrix*. It is assumed that a and d are such that no denominator is zero. If

$$D_k = \prod_{j=k-1}^{n+k-2} (a + jd)$$

denotes the product of the denominators in the kth row (or column) of H then

$$\det H = \frac{\prod\limits_{k=1}^{n} [(n-k)!\, d^{n-k}]^2}{\prod\limits_{k=1}^{n} D_k}.$$

In the special case when $a = 1$, $d = 1$, H is the *Hilbert matrix*, and an explicit formula for the elements H_{ij} of H^{-1} is

$$H_{ij} = \frac{(-1)^{i+j}(n+i-1)!\,(n+j-1)!}{[(i-1)!\,(j-1)!]^2(n-i)!\,(n-j)!\,(i+j-1)}. \quad (13.99)$$

Exercise 13.36 Compute the determinant and inverse of the 4×4 Hilbert matrix.

Exercise 13.7 Show that in (13.99) $H_{11} = n^2$.

Problems

13.1   The set of orthogonal polynomials in (13.72) can be defined by

$$p_i(\lambda) = \det \begin{bmatrix} \alpha_1\lambda + \beta_1 & -1 & 0 & & \\ -\gamma_2 & \alpha_2\lambda + \beta_2 & -1 & & \\ 0 & -\gamma_3 & \alpha_3\lambda + \beta_3 & & \\ & & & \ddots & -1 \\ & & & -\gamma_i & \alpha_i\lambda + \beta_i \end{bmatrix}.$$

Use (13.12) to prove (13.73).
   Notice that in the special case

$$\alpha_1 = 1; \qquad \alpha_i = 2, \quad i \geqslant 2; \qquad \beta_i = 0, \quad \gamma_i = 1$$

the polynomials become a set of Chebyshev polynomials, and the result in this problem therefore generalizes that in Problem 6.8.

13.2   Let A be the comrade matrix in (13.70) and define

$$D = \text{diag}[\alpha_1, \alpha_2, \ldots, \alpha_n].$$

Show that

$$\det(\lambda D - DA) = \alpha_1\alpha_2\cdots\alpha_n \det(\lambda I - A).$$

By expanding $\det(\lambda D - DA)$ by its last row show also that

$$\det(\lambda D - DA) = a_n + a_{n-1}p_1(\lambda) + a_{n-2}p_2(\lambda) +$$
$$\cdots + (a_2 - \gamma_n)p_{n-2}(\lambda) + (a_1 + a_n\lambda + \beta_n)p_{n-1}(\lambda),$$

where the $p_i(\lambda)$ are defined by (13.73). Hence establish (13.71).

13.3 Apply (13.12) to the $k \times k$ tridiagonal determinant

$$L_k = \begin{vmatrix} 3 & i & 0 & 0 & & \\ i & 1 & i & 0 & & \\ 0 & i & 1 & i & & \\ & & \ddots & \ddots & \ddots & \\ & & & \cdot & 1 & i \\ & & & & i & 1 \end{vmatrix},$$

where $i^2 = -1$, to prove that $L_k = L_{k-1} + L_{k-2}$, $k \geqslant 3$. With $L_1 = 3$, $L_2 = 4$, this defines the sequence of *Lucas numbers*. These are related to the Fibonacci numbers x_k (see Problems 2.11 and 4.26) by $L_x = x_k + x_{k+2}$, $k \geqslant 1$.

13.4 Use the diagonalization formula (13.27) to show that any two circulant matrices of the same order commute.

13.5 Let Π be the matrix defined in (13.17).

(a) Prove that if A is any square matrix satisfying $A\Pi = \Pi A$, then A is a circulant matrix.

(b) Prove that $\Pi^n = I$, $\Pi^{n-1} = \Pi^T = \Pi^{-1}$.

13.6 It was seen in Problem 4.17 that there is a unique polynomial of degree $n - 1$ passing through n given distinct points. Apply this result to the points $(1, \lambda_1)$, (θ, λ_2), $(\theta^2, \lambda_3), \ldots, (\theta^{n-1}, \lambda_n)$, where $\theta = \exp(2\pi i/n)$. Hence deduce from (13.27) that $F^*\Lambda F$ is a circulant matrix, where F is the Fourier matrix defined by (13.22) and $\Lambda = \text{diag}[\lambda_1, \lambda_2, \ldots, \lambda_n]$.

Deduce using Exercise 13.17 that the MP-inverse of a circulant matrix is also circulant.

13.7 Show that the Fourier matrix F defined in (13.22) satisfies

$$F^2 = (F^*)^2, \qquad F^3 = F^*, \qquad F^4 = I.$$

Hence deduce that the eigenvalues of F are ± 1, $\pm i$, with appropriate multiplicities.

13.8 Use (13.28) to show that the resultant (defined in Section 11.2) of $\lambda^n - 1$ and $c(\lambda)$ in (13.21) is equal to $\det \text{circ}(c_1, c_2, \ldots, c_n)$.

13.9   Interchange rows 2 and 3 of the matrix in (13.24), and verify that it can then be expressed as the product

$$\begin{bmatrix} F_1 & 0 \\ 0 & F_2 \end{bmatrix}\begin{bmatrix} I & I \\ I & F_3 \end{bmatrix},$$

where

$$F_1 = \begin{bmatrix} 1 & 1 \\ 1 & \omega^2 \end{bmatrix}, \qquad F_2 = \begin{bmatrix} 1 & \omega \\ 1 & \omega^3 \end{bmatrix}, \qquad F_3 = \omega^2 I.$$

This type of factorization is the key to the fast Fourier transform (Brigham 1988).

13.10 Let A and B be two $n \times n$ circulant matrices having eigenvalues $\lambda_1, \lambda_2, \ldots, \lambda_n$ and $\mu_1, \mu_2, \ldots, \mu_n$ respectively. Use (13.27) to prove that the eigenvalues of $A + B$, $A - B$, AB are respectively $\lambda_i + \mu_i$, $\lambda_i - \mu_i$, $\lambda_i\mu_i$, $i = 1, 2, \ldots, n$.

13.11 Define an $n \times n$ matrix η by replacing the $n, 1$ element of Π in (13.17) by -1. This matrix plays the same role for skew circulant matrices as does Π for ordinary circulant matrices.
      Prove that (a) $\eta^n = -I$, $\eta^{\mathsf{T}} = -\eta^{n-1}$; (b) an $n \times n$ matrix A is skew circulant if and only if $A\eta = \eta A$.

13.12 Prove that an $n \times n$ matrix A is a g-circulant if and only if

$$\Pi A = A\Pi^g,$$

where Π is defined in (13.17) and $g > 0$.

13.13 Show that if A is a g-circulant with $g > 0$ then its rows are all distinct if and only if g and n are relatively prime. Hence deduce that if A is non-singular then g and n are relatively prime.

13.14 If g and n are relatively prime positive integers then there exists a unique positive integer x such that $gx = 1 \pmod n$, so $x = g^{-1}$.
      If A is a non-singular g-circulant matrix of order n, use the results in Problems 13.12 and 13.13 to show that $\Pi A^{-1} = A^{-1}\Pi^x$, i.e. A^{-1} is a g^{-1}-circulant matrix.

13.15 Show that an $mn \times mn$ matrix A is a block circulant matrix of type (n, m) if and only if it commutes with $\Pi \otimes I_m$, where Π is the $n \times n$ matrix in (13.17).

13.16 Let A be the $n \times n$ tridiagonal Toeplitz matrix in Example 13.5(a). Let B be an $m \times m$ matrix having the same tridiagonal Toeplitz form but with $b_{ii} = c$, $b_{i,i+1} = b_{i,i-1} = d$. Prove, using an appropriate result from Section 6.3.5, that $A \otimes I_m + I_n \otimes B$ has eigenvalues

$$a + c + 2b \cos \frac{k\pi}{n+1} + 2d \cos \frac{l\pi}{m+1}, \qquad k = 1, 2, \ldots, n$$
$$l = 1, 2, \ldots, m.$$

13.17 Let $T = [t_{ij}]$ be an $n \times n$ bidiagonal Toeplitz matrix with all $t_{ii} = 1$ and $t_{i+1,i} = x$, $i = 1, 2, \ldots, n - 1$. Show that its inverse $Y = [y_{ij}]$ is a lower triangular Toeplitz matrix with $y_{ii} = 1$, $y_{ij} = (-x)^{i-j}$, $i > j$.

13.18 Let $T = [t_{i-j}]$ be an $n \times n$ lower triangular Toeplitz matrix with $t_0 = 1$ and $t_k = 0$ for $k < 0$. Show that its inverse is a lower triangular Toeplitz matrix $Y = [y_{i-j}]$ with $y_0 = 1$ and

$$y_{i-j} = -\sum_{k=1}^{i-j} t_k y_{i-j-k}, \qquad i - j \geq 1.$$

13.19 Let T be an $n \times n$ symmetric pentadiagonal Toeplitz matrix whose first row is $[a, b, 1, 0, \ldots, 0]$. It can be factorized as $T = PDP^\mathsf{T}$ where $P = [p_{ij}]$ is bidiagonal and $D = [d_{ij}]$ is symmetric and tridiagonal. The elements of D are defined by

$$d_{i,i+1} = b_i, \qquad i = 1, 2, \ldots, n - 1,$$
$$d_{11} = a, \qquad d_{22} = (b - b_2)b_1,$$
$$d_{ii} = \left(b - b_i - \frac{1}{b_{i-2}} \right) b_{i-1}, \qquad i = 3, 4, \ldots, n,$$

where

$$b_1 = b, \qquad b_2 = b - \frac{a}{b}, \qquad b_3 = b - \frac{a}{b_2} + \frac{b}{b_2 b_1},$$
$$b_{i+1} = b - \frac{a}{b_i} + \frac{b}{b_i b_{i-1}} - \frac{1}{b_i b_{i-1} b_{i-2}}, \qquad i = 3, 4, \ldots, n - 1.$$

The factorization exists provided no b_i is zero, and P has 1's along the principal diagonal and subdiagonal elements $p_{21} = 0$, $p_{i+1,i} = 1/b_{i-1}$, $i = 2, \ldots, n - 1$. Hence inversion of T reduces to the inversion of a bidiagonal and a tridiagonal matrix.

Determine P and D when $n = 5$, $a = 6$, $b = 2$.

13.20 It can be shown that any strongly non-singular Hankel matrix H can be uniquely expressed as $H = R^* DR$, where D is diagonal and R is upper triangular with 1's along its principal diagonal. Perform this decomposition when

$$H = \begin{bmatrix} 1 & 2 & 3 \\ 2 & 3 & -1 \\ 3 & -1 & 10 \end{bmatrix}.$$

A general algorithm has been devised for determining D and R in $O(n^2)$ operations.

13.21 Assume that in the definition (11.101) of the Schwarz matrix A we have $b_i > 0$, all i, so that by Exercise 13.30 A is a stability matrix. Show that the Routh matrix R defined in (13.74) satisfies $YAY^{-1} = R$, where

$$Y = K \operatorname{diag}[(-1)^{n+1}(b_1 b_2 \cdots b_n)^{1/2}, (-1)^n (b_1 b_2 \cdots b_{n-1})^{1/2}, \ldots,$$
$$-(b_1 b_2)^{1/2}, b_1^{1/2}].$$

13.22 Apply Gershgorin's theorem (quoted in Problem 6.4) to the comrade matrix A in (13.70), and hence deduce that the roots of the polynomial $a(\lambda)$ in (13.72) lie in the region of the complex z-plane consisting of the discs

$$|z + (\beta_1/\alpha_1)| \le 1/\alpha_1, \qquad |z + \beta_i/\alpha_i| \le (1 + \gamma_i)/\alpha_i,$$
$$i = 2, 3, \ldots, n-1,$$

$$|z + (\beta_n + a_1)/\alpha_n| \le \left(\sum_{j=3}^n |a_j| + |\gamma_n - a_2| \right) \Big/ \alpha_n.$$

Write down the corresponding result obtained using A^T.

13.23 By considering the product $T^{-1}CT$ where C is the centrosymmetric matrix in (13.69) and

$$T = \begin{bmatrix} I & -I \\ K & K \end{bmatrix} \begin{matrix} m \\ m \end{matrix},$$

with m and m labelling the columns,

show that:

(a) $\det C = \det X \det Y$;

(b) the eigenvalues of C are those of X and of Y;

(c) $C^{-1} = \begin{bmatrix} C_3 & C_4 K \\ KC_4 & KC_3 K \end{bmatrix}$

where $X = C_1 + C_2$, $Y = C_1 - C_2$, $C_3 = \frac{1}{2}(X^{-1} + Y^{-1})$, $C_4 = \frac{1}{2}(X^{-1} - Y^{-1})$.

13.24 By analogy with skew symmetric matrices (Section 2.3.2) define S to be *skew centrosymmetric* if $S = -KSK$. When S has order $2m + 1$ it can be written as

$$S = \begin{bmatrix} S_1 & a & -S_2 K \\ b & 0 & -bK \\ KS_2 & -Ka & -KS_1 K \end{bmatrix} \begin{matrix} m \\ 1 \\ m \end{matrix} \qquad (13.100)$$

with columns labelled m, 1, m.

(a) Verify that (13.100) satisfies the defining equation.

(b) By considering the product $V^{-1}SV$ where

$$
\begin{array}{cccc}
 & m & 1 & m \\
V = & \begin{bmatrix} I & 0 & I \\ 0 & 1 & 0 \\ K & 0 & -K \end{bmatrix} & & \begin{matrix} m \\ 1 \\ m \end{matrix}
\end{array} ,
$$

show that $\det S = 0$.

13.25 When C has complex elements it is called *centrohermitian* if $C = KC^*K$ and *skew centrohermitian* if $C = -KC^*K$. Prove:

(a) if C is centrohermitian then so is C^+;

(b) if C is skew centrohermitian then $\pm iC$ is centrohermitian (compare with Exercise 2.26).

13.26 Consider the arithmetic matrix H defined in Section 13.4.6(a). When $d = 0$ let $H = aE$, where E is the $n \times n$ matrix all of whose elements are equal to 1. Deduce that $R(H) = 1$, and use Problem 6.20 to show that

$$
\det(I + H) = 1 + na, \qquad (I + H)^{-1} = I - \frac{H}{1 + na}.
$$

Use these results to obtain the determinant and inverse of the $n \times n$ symmetric Toeplitz matrix having all elements on the principal diagonal equal to α, and all other elements equal to β ($\neq \alpha$).

13.27 Let H be *any* $n \times n$ matrix having rank two and non-zero eigenvalues λ_1 and λ_2. Deduce that σ_1 and σ_2 defined in (13.97) satisfy $\sigma_1 = \lambda_1 + \lambda_2$, $\sigma_2 = \lambda_1\lambda_2$ and hence show that

$$
\det(I + H) = 1 + \sigma_1 + \sigma_2.
$$

13.28 Let V_n be the Vandermonde matrix defined in (4.80), and let $\lambda_1, \lambda_2, \ldots, \lambda_n$ be the roots of an nth degree polynomial $a(\lambda)$. Show that $S = V_n V_n^T$ is a Hankel matrix with i, j element s_{i+j-2} where $s_0 = 1$ and

$$
s_k = \lambda_1^k + \lambda_2^k + \cdots + \lambda_n^k, \qquad k = 1, 2, \ldots, 2n - 2.
$$

Hence deduce using (4.81) that $\det S$ is a *discriminant* for $a(\lambda)$, defined in Problem 6.14 (i.e. $\det S \neq 0$ if and only if $a(\lambda)$ has no repeated roots).

It can be shown that if S_k is the $k \times k$ leading principal submatrix of S then

(a) $a(\lambda)$ has exactly t distinct roots if and only if $\det S_t \neq 0$, $\det S_{t+1} = \det S_{t+2} = \cdots = \det S_{n-1} = \det S = 0$;

(b) when $a(\lambda)$ is real the number of distinct *real* roots is $t - 2v$, where v is the number of variations in sign in the sequence 1, $\det S_2, \det S_3, \ldots, \det S_t$.

For a proof using the *Hankel form* $x^{\mathsf{T}} S x$ see Gantmacher (1959).

References

Barnett (1983), Brigham (1988), Davis (1979), Fiedler (1986), Gant-macher (1959, Vol. 2), Golub and Van Loan (1989), Gregory and Karney (1969), Heinig and Rost (1984), Householder (1964), Miller (1987).

14 Miscellaneous topics

In this chapter we bring together some further material which is of interest for applications. By necessity we shall have to use a rather terser style than in the rest of the book, since many of the topics we discuss have been the subject of entire texts in their own right. To emphasize the reference nature of this chapter, the exercises are collected together at the end.

14.1 Matrix equations

14.1.1 $AX = XB$

Consider the matrix equation

$$AX = XB \qquad (14.1)$$

where A and B are given $n \times n$ and $m \times m$ matrices respectively. Using the same argument as in Section 13.3, employing a Kronecker product representation like (13.34), we find that (14.1) has a non-zero solution X if and only if A and B have at least one eigenvalue in common.

If the Jordan forms of A, B are $\mathcal{J}_A$, $\mathcal{J}_B$ respectively so that

$$T^{-1}AT = \mathcal{J}_A, \qquad P^{-1}BP = \mathcal{J}_B, \qquad (14.2)$$

then substituting (14.2) into (14.1) shows that the general solution of (14.1) is $X = TYP^{-1}$, where Y is the general solution of

$$\mathcal{J}_A Y = Y \mathcal{J}_B. \qquad (14.3)$$

Let the eigenvalues of A, B be λ_i, μ_j respectively, and write the Jordan forms as

$$\mathcal{J}_A = \mathrm{diag}[J_1, J_2, \ldots, J_q], \qquad \mathcal{J}_B = \mathrm{diag}[L_1, L_2, \ldots, L_p], \qquad (14.4)$$

where J_i is $k_i \times k_i$, L_j is $l_j \times l_j$ and each is an upper triangular Jordan block in the form (8.2). The solution Y of (14.3) is non-zero provided $\lambda_i = \mu_j$ for at least one pair of suffices i and j.

The $n \times m$ matrix Y can be expressed in block-partitioned form, with its $k_i \times l_j$ submatrix Y_{ij} given by:

$$0 \qquad \text{if } \lambda_i \neq \mu_j,$$

$$[0, Z] \qquad \text{if } \lambda_i = \mu_j, \quad k_i \leq l_j, \tag{14.5}$$

$$\begin{bmatrix} Z \\ 0 \end{bmatrix} \qquad \text{if } \lambda_i = \mu_j, \quad k_i \geq l_j, \tag{14.6}$$

where Z is an arbitrary upper triangular Toeplitz matrix having appropriate dimensions. The number of independent parameters in Y (and hence X) is

$$\sum_{i=1}^{q} \sum_{j=1}^{p} d_{ij}, \tag{14.7}$$

where $d_{ij} = \min(k_i, l_j)$ if $\lambda_i = \mu_j$, and $d_{ij} = 0$ otherwise.

Example 14.1 Suppose that in (14.3) and (14.4) we have

$$n = 9, \quad \lambda_1 = 2, \quad k_1 = 4, \quad \lambda_2 = 3, \quad k_2 = 2,$$

$$\lambda_3 = 1, \quad k_2 = 2, \quad \lambda_4 = 2, \quad k_4 = 1,$$

$$m = 7, \quad \mu_1 = 4, \quad l_1 = 1, \quad \mu_2 = 2, \quad l_2 = 2, \quad \mu_3 = 3, \quad l_3 = 4.$$

The general solution Y of (14.3) is 9×7, and is partitioned into blocks Y_{ij}, each $k_i \times l_j$, where since $q = 4$ and $p = 3$ we have $i = 1, 2, 3, 4$ and $j = 1, 2, 3$. These blocks are given by (14.5) and (14.6), and are non-zero only when $\lambda_i = \mu_j$, that is for

$$\lambda_1 = \mu_2, \qquad \lambda_2 = \mu_3, \qquad \lambda_4 = \mu_2,$$

so only Y_{12}, Y_{23} and Y_{42} are non-zero, all the other blocks in the partitioned form of Y being identically zero. In (14.6) we have

$$\lambda_1 = \mu_2, \qquad k_1 > l_2, \qquad \text{so } Y_{12} = \begin{bmatrix} y_1 & y_2 \\ 0 & y_1 \\ 0 & 0 \\ 0 & 0 \end{bmatrix},$$

and from (14.5)

$$\lambda_2 = \mu_3, \qquad k_2 < l_3, \qquad \text{so } Y_{23} = \begin{bmatrix} 0 & 0 & y_3 & y_4 \\ 0 & 0 & 0 & y_3 \end{bmatrix}$$

$$\lambda_4 = \mu_2, \qquad k_4 < l_2, \qquad \text{so } Y_{42} = [0 \quad y_5],$$

where $y_1, \ldots, y_5$ are arbitrary parameters. This agrees with (14.7), where the only non-zero terms in the sum are $d_{12} = 2$, $d_{23} = 2$, $d_{42} = 1$, giving a total of five parameters.

14.1.2 Commuting matrices

An important special case of (14.2) is when $A = B$, so that we seek all matrices X which commute with a given matrix A, i.e.

$$AX = XA. \tag{14.8}$$

We see from the previous section that the general solution of (14.8) is $X = TYT^{-1}$, where T is defined in (14.2) and Y is the general solution of

$$\mathcal{J}_A Y = Y \mathcal{J}_A. \tag{14.9}$$

Furthermore, all the diagonal blocks Y_{ii} in the partitioned form of Y are square and non-zero, and off-diagonal blocks Y_{ij} are non-zero only when the Jordan blocks J_i, J_j in the Jordan form of A correspond to equal eigenvalues, i.e. $\lambda_i = \lambda_j$.

Example 14.2 If

$$\mathcal{J}_A = \operatorname{diag}\left\{\begin{bmatrix} 2 & 1 \\ 0 & 2 \end{bmatrix}, 2\right\},$$

then in (14.4) we have

$$\lambda_1 = 2, \qquad k_1 = 2, \qquad \lambda_2 = 2, \qquad k_2 = 1, \qquad q = 2.$$

Using (14.5) and (14.6) the diagonal blocks in $Y = [Y_{ij}]$ are

$$Y_{11} = \begin{bmatrix} y_1 & y_2 \\ 0 & y_1 \end{bmatrix}, \qquad Y_{22} = [y_3],$$

and

$$\lambda_1 = \lambda_2, \qquad k_1 > k_2, \qquad \text{so } Y_{12} = \begin{bmatrix} y_4 \\ 0 \end{bmatrix}$$

$$\lambda_2 = \lambda_1, \qquad k_2 < k_1, \qquad \text{so } Y_{21} = [0 \;\; y_5].$$

The general solution of (14.9) in this case is therefore

$$Y = \begin{bmatrix} Y_{11} & Y_{12} \\ Y_{21} & Y_{22} \end{bmatrix} = \begin{bmatrix} y_1 & y_2 & y_4 \\ 0 & y_1 & 0 \\ 0 & y_5 & y_3 \end{bmatrix}$$

and contains five arbitrary parameters.

A few additional remarks about commuting matrices are worth noting. Obviously, if X is a polynomial in A (with scalar coefficients) then it commutes with A, since powers of A

commute with each other. However, the converse is not true in general: if X commutes with A then it is a polynomial in A only if A is non-derogatory (i.e. its minimum and characteristic polynomials are equal). What can be shown is that if for every X which commutes with A we have $BX = XB$, then B can be expressed as a polynomial in A. An interesting special case of this occurs when B is the MP-inverse A^+ of A: for if $AX = XA$ and we assume $A^+A = AA^+$, then using (10.11) and (10.12) we obtain

$$A^+X = A^+AA^+X = (A^+)^2AX = (A^+)^2XA = (A^+)^2XA^3(A^+)^2$$
$$= (A^+)^2A^3X(A^+)^2 = AX(A^+)^2 = XA(A^+)^2 = XA^+.$$

Hence, if A commutes with A^+ it follows that A^+ can be expressed as a polynomial in A. As above, the converse is obvious, so a necessary and sufficient condition for A^+ to commute with A is that A^+ can be expressed as a polynomial in A; this is equivalent to A being an EP-matrix (see Section 10.1.2).

14.1.3 $f(X) = 0$

We now consider solving the polynomial equation

$$f(X) = 0, \tag{14.10}$$

where X is an $n \times n$ matrix, and $f(\lambda)$ is a given polynomial

$$f(\lambda) = (\lambda - \lambda_1)^{\alpha_1}(\lambda - \lambda_2)^{\alpha_2} \cdots (\lambda - \lambda_r)^{\alpha_r}. \tag{14.11}$$

The general solution of (14.10) can be expressed in the form $X = TYT^{-1}$, where T is non-singular and $n \times n$ but otherwise arbitrary. The matrix Y is

$$Y = \text{diag}\{\lambda_{j_1}I_{\beta_1} + H_{\beta_1}, \lambda_{j_2}I_{\beta_2} + H_{\beta_2}, \ldots, \lambda_{j_k}I_{\beta_k} + H_{\beta_k}\}, \tag{14.12}$$

where $j_i \in \{1, 2, \ldots, r\}$ (some of the j_i may be equal), the integers β_i are such that $1 \le \beta_i \le \alpha_{j_i}$ and $\beta_1 + \cdots + \beta_k = n$, and H_p is the $p \times p$ matrix having ones along the superdiagonal and zeros elsewhere.

Example 14.3 An $n \times n$ matrix X is said to be nilpotent of index 2 if $X^2 = 0$ (see Problem 4.15). In (14.11) we have $f(\lambda) = \lambda^2$, so $\lambda_1 = 0$, $\alpha_1 = 2$ and $r = 1$. The conditions associated with (14.12) give $1 \le \beta_i \le 2$, and $\beta_1 + \beta_2 + \cdots + \beta_k = n$ for some integer k. The only possibilities are $\beta_i = 1$, when the corresponding diagonal block in (14.12) is just $[0]$; and

$\beta_i = 2$, when the block is

$$H_2 = \begin{bmatrix} 0 & 1 \\ 0 & 0 \end{bmatrix}.$$

Hence by (14.12) the general solution of $X^2 = 0$ is

$$T \, \text{diag} \left\{ 0, 0, \ldots, \begin{bmatrix} 0 & 1 \\ 0 & 0 \end{bmatrix}, \ldots \right\} T^{-1}, \tag{14.13}$$

where if there are t_1 entries of the first type and t_2 of the second type then $t_1 + 2t_2 = n$, but the order in which these entries occur is arbitrary. Thus for example when $n = 3$, the only non-trivial forms of (14.13) are

$$T \begin{bmatrix} 0 & 1 & 0 \\ 0 & 0 & 0 \\ \hline 0 & 0 & 0 \end{bmatrix} T^{-1}, \qquad T \begin{bmatrix} 0 & 0 & 0 \\ \hline 0 & 0 & 1 \\ 0 & 0 & 0 \end{bmatrix} T^{-1}.$$

Example 14.4 An $n \times n$ matrix X is said to be idempotent when $X^2 = X$ (see Problem 4.14). In this case the polynomial (14.11) is

$$f(\lambda) = \lambda^2 - \lambda = \lambda(\lambda - 1),$$

so $\lambda_1 = 0$, $\alpha_1 = 1$, $\lambda_2 = 1$, $\alpha_2 = 1$, $r = 2$. The condition on the parameters in (14.12) is $1 \leqslant \beta_i \leqslant 1$, so all the diagonal blocks in (14.12) are scalars. Hence the general expression for an idempotent matrix is

$$X = T \, \text{diag}[0, 0, \ldots, 0, 1, 1, \ldots, 1] T^{-1},$$
$$\longleftarrow t_1 \longrightarrow \longleftarrow t_2 \longrightarrow$$

where $t_1 + t_2 = n$, but the ordering of the entries of each type is arbitrary.

If the polynomial in (14.11) has matrix coefficients, then the equation (14.10) becomes

$$A_0 X^N + A_1 X^{N-1} + \cdots + A_{N-1} X + A_N = 0,$$

where each A_i is $n \times n$, so X is a *right solvent* of

$$A(\lambda) = A_0 \lambda^N + A_1 \lambda^{N-1} + \cdots + A_N$$

(see Problem 12.1). If $g(\lambda) = \det A(\lambda)$ then any right (or left) solvent of $A(\lambda)$ satisfies $g(X) = 0$, which is a polynomial equation with scalar coefficients and so can be solved as indicated above.

14.1.4 Other equations

For completeness we recall here some results on various other matrix equations given in previous chapters. The linear equation

$$AX + XB = C$$

was discussed in Sections 5.6 and 6.3.5; the related equation

$$AXB - X = C$$

was dealt with in Problem 6.27; and special cases of each of these equations when $B = A^T$ were involved in stability considerations in Section 11.5.1. The solution of $AXC = B$ via generalized inverses was discussed in Section 10.3. A linear matrix *differential* equation was considered in Problem 9.12. Finally, the so-called Riccati quadratic matrix equation was studied in Section 11.5.2, with an extension for a more general equation in Problem 11.13.

14.2 Non-negative matrices

14.2.1 *Basic properties*

A real $n \times n$ matrix $A = [a_{ij}]$ is called *non-negative* (*positive*) if all its elements are non-negative (positive), and we write $A \geqslant 0$ ($A > 0$). The same terminology applies to vectors. If $B = [b_{ij}]$ is a second such matrix then the notation $A \geqslant B$ means $a_{ij} \geqslant b_{ij}$ for all i and j, and similarly $A > B$ if $a_{ij} > b_{ij}$. Non-negative matrices have been encountered earlier, in Section 9.1 and Problem 9.15. Of course the eigenvalues λ_i of A need not always be real, but a remarkable result (called in various forms the *Perron–Frobenius theorem*) states that if A is non-negative and irreducible (defined in Section 13.4.5) then it always has a real positive eigenvalue p such that $p \geqslant |\lambda_i|$ for all i. Furthermore, p is a simple root of the characteristic polynomial of A, and there is a positive (right) eigenvector corresponding to p. The eigenvalue p and associated eigenvector are called the *Perron* root and vector of A. In the special case when A is positive (and not necessarily reducible) then p is *strictly* greater than the modulus of any other eigenvalue. However, if $A \geqslant 0$ but is not necessarily reducible, then we have only the weaker result that A has a real eigenvalue $p \geqslant 0$ with a corresponding non-negative eigenvector, and $p \geqslant |\lambda_i|$.

A simple necessary and sufficient condition for a non-negative $n \times n$ matrix A to be irreducible is $(I + A)^{n-1} > 0$.

Example 14.5 The 2×2 matrix A in (6.5) is positive. We saw in Example 6.1 that it has a positive eigenvalue $\lambda_2 = 4 = p$ and a corresponding positive right eigenvector $[1, 1]^T$. The other eigenvalue of A is $\lambda_1 = -1$ and $p > |\gamma_1|$.

In all cases the *spectral radius* $\rho(A)$ of A, which is the radius of the smallest circle in the complex plane containing all the eigenvalues of A (see Problem 9.15), is equal to the *maximal eigenvalue* p. A non-negative irreducible matrix is called *primitive* if there is only a single eigenvalue with modulus p. Thus, in particular, any positive matrix is primitive. Define the *row sums* and *column sums* respectively of A by

$$r_i = \sum_{k=1}^n a_{ik}, \qquad c_i = \sum_{k=1}^n a_{ki}. \tag{14.14}$$

Then bounds on the spectral radius of any non-negative matrix A are provided by

$$\min_i(r_i) \leq p \leq \max_i(r_i), \tag{14.15}$$

together with a corresponding result involving the column sums obtained by applying (14.15) to the non-negative matrix A^T. If A is irreducible then equality can hold on either side of (14.15) if and only if all the r_i are equal. This is illustrated by the 2×2 matrix A in (6.5), used in Example 14.5, where $r_1 = r_2 = 4$ and $p = 4$. A number of bounds which are sharper than (14.15) have been discovered (see Minc 1988). Finally, if $B \geq A \geq 0$ then $\rho(B) \geq \rho(A)$; and if in addition A is irreducible and $B \neq A$ then $\rho(B) > \rho(A)$.

14.2.2 M-matrices

In a number of applications, especially in the fields of economics and operational research, matrices $A = [a_{ij}]$ arise having all off-diagonal elements non-positive, i.e. $a_{ij} \leq 0$ $(i \neq j)$, and all diagonal elements a_{ii} non-negative. An example of such a matrix occurred in the stretched string problems in Examples 6.4 and 13.1. Let Z denote the set of $n \times n$ real matrices all of whose off-diagonal elements are non-positive. If $A \in Z$ is non-singular and $A^{-1} \geq 0$, then A is called an *M-matrix* (named after Minkowski).

Example 14.6 It is easy to check that the 3×3 case of (13.3), namely

$$A = \begin{bmatrix} 2 & -1 & 0 \\ -1 & 2 & -1 \\ 0 & -1 & 2 \end{bmatrix} \tag{14.16}$$

has det $A \neq 0$ and

$$A^{-1} = \frac{1}{4} \begin{bmatrix} 3 & 2 & 1 \\ 2 & 4 & 2 \\ 1 & 2 & 3 \end{bmatrix} > 0,$$

so A in (14.16) is an M-matrix.

The above example is a special case of the fact that a general $n \times n$ *real* tridiagonal matrix A defined in (13.9) is an M-matrix provided

$$a_i > 0, \qquad b_i < 0, \qquad c_i < 0, \qquad \text{for all } i,$$

and the row sums of A satisfy

$$r_1 = a_1 + b_1 > 0, \qquad r_n = a_n + c_{n-1} > 0,$$
$$r_i = a_i + b_i + c_{i-1} \geqslant 0, \qquad i = 2, 3, \ldots, n - 1.$$

There are very many equivalent definitions of an M-matrix, and we list below just a few of the equivalent necessary and sufficient conditions for a nonsingular matrix $A \in Z$ to be an M-matrix:

(M1) Every real eigenvalue of A is positive.

(M2) All principal minors of A are positive.

(M3) There exists a matrix $B \geqslant 0$ and a scalar $c > \rho(B)$ (the spectral radius of B) such that $A = cI - B$.

(M4) There exists a triangular decomposition $A = LU$ (see Section 3.3) such that all the diagonal elements of both L and U are positive.

(M5) The real part of every eigenvalue of A is positive. (By an abuse of terminology some authors say A is 'positive stable', but it is preferable to state that $-A$ is a stability matrix in the sense defined in Section 11.5.1.)

(M6) There exists a diagonal matrix $P > 0$ such that $A^{\mathrm{T}}P + PA$ is positive definite.

Property (M5) follows immediately from (M6) by appealing to the Liapunov theorem in Section 11.5.1, since $(-A^{\mathrm{T}})P + P(-A)$ is negative definite. However the interesting point is that when A is an M-matrix then a *diagonal* matrix satisfying (11.66) can be found.

(M7) There exists a column vector $x > 0$ such that $Ax > 0$.

If the requirement that A be non-singular is removed then each condition below is necessary and sufficient for $A \in Z$ to be an M-matrix in this more general sense.

$(M1)_g$ Every real eigenvalue of A is non-negative.

$(M2)_g$ All principal minors of A are non-negative.

$(M3)_g$ There exists a matrix $B \geq 0$ and a scalar $c \geq \rho(B)$ such that $A = cI - B$.

$(M4)_g$ There exists a permutation matrix P such that $P^T A P = LU$, with the lower and upper triangular matrices L and U having non-negative diagonal entries.

$(M5)_g$ The real part of every eigenvalue of A is non-negative.

If A satisfies any one of $(M1)_g$ to $(M5)_g$ and is also non-singular, then it satisfies (M1) to (M7), and so can be called a *non-singular* M-matrix. The situation is analogous to the relationship between positive semidefinite and positive definite hermitian (or real symmetric) matrices, defined in Section 7.4.1: if such a matrix satisfies a condition for positive semidefiniteness (e.g. all its eigenvalues ≥ 0) and in addition is non-singular, then it is in fact positive definite.

Two special cases are of interest. When $A \in Z$ is symmetric then it is a non-singular M-matrix if and only if A is positive definite, in which case it is called a *Stieltjes* matrix; when $A \in Z$ is irreducible then it is a non-singular M-matrix if and only if there exists an $x > 0$ such that $Ax \geq 0$.

There are important applications to iterative methods for solving a standard set of linear equations

$$Ax = b \tag{14.17}$$

in Section 6.7.1. We saw that when A is split up as $A = B - C$, where B is non-singular, then convergence of the procedures requires that the spectral radius of $L = B^{-1}C$ must be less than unity (i.e. L must be a convergent matrix). The splitting of A is called *regular* if $B^{-1} \geq 0$ and $C \geq 0$, and *weak regular* if $B^{-1} \geq 0$ and $L \geq 0$. In fact when A in (14.17) is a non-singular M-matrix then every regular or weak regular splitting has L convergent. In particular, both the Jacobi and Gauss–Seidel methods described in Section 6.7.1 converge when A is an M-matrix. More generally, in (14.17) let $A = [a_{ij}]$ be a complex $n \times n$ matrix with non-zero diagonal elements, and construct the real *comparison matrix* $C_A = [\alpha_{ij}]$ according to

$$\alpha_{ii} = |a_{ii}|, \qquad \alpha_{ij} = -|a_{ij}|, \qquad i \neq j, \qquad \text{for all } i \text{ and } j.$$

Then Jacobi's method for (14.17) converges if C_A is an M-matrix.

There is also a connection with the concept of a diagonal dominant $n \times n$ complex matrix $A = [a_{ij}]$, which was defined in (6.131). We saw in Problem 6.5 that any diagonal dominant matrix is non-singular. It follows from the dominance property (6.131) that

$$C_A[1, 1, \ldots, 1]^\mathsf{T} > 0.$$

Since the off-diagonal elements of C_A are all non-positive, we conclude by property (M7) that C_A is a non-singular M-matrix whenever A is diagonal dominant.

14.2.3 Stochastic matrices

A non-negative matrix A each of whose row sums defined in (14.14) is equal to one is called (row) *stochastic*, and similarly (column) stochastic if the column sums are one (an illustration of the latter was given in Example 1.9). When A is both row and column stochastic it is called *doubly stochastic*. Such matrices arise in the theory of Markov chains, and the element a_{ij} represents the probability of a transition from state i to state j.

Notice that a square permutation matrix (defined in Section 13.4.5) is a doubly stochastic matrix. In fact, any doubly stochastic matrix can always be expressed as a linear combination $\sum \alpha_i P_i$ of permutation matrices P_i, where the α_i are positive scalars such that $\sum \alpha_i = 1$.

The Perron root of a row stochastic matrix is equal to 1, and a non-negative matrix A is row stochastic if and only if

$$e = [1, 1, \ldots, 1, 1]^\mathsf{T} \qquad (14.18)$$

is a right eigenvector corresponding to the eigenvalue 1. Furthermore, for any non-negative A with Perron root p and a positive Perron eigenvector, there exists a diagonal matrix $D > 0$ such that $(1/p)D^{-1}AD$ is row stochastic.

In applications to Markov chains the equations describing the process take the form of the linear difference eqns (9.51), where A is a stochastic matrix. Because of the form of the solution in (9.52), we are interested in the behaviour of A^k as $k \to \infty$. Since $\rho(A) = 1$, we know from Chapter 9 that $A^k \nrightarrow 0$ as $k \to \infty$. However, if A is primitive then $A^k \to ey^\mathsf{T}$ as $k \to \infty$, where e is defined in (14.18) and y is the right eigenvector of A^T corresponding to the eigenvalue 1, normalized so that $y^\mathsf{T}e = 1$.

We remarked earlier that if $A > 0$ then it is primitive, but when $A \geqslant 0$ a necessary and sufficient condition for A to be primitive is the existence of a positive integer k such that $A^k > 0$ (see Problem 14.2). If $A \geqslant 0$ is irreducible then a sufficient condition for it to be primitive is that all the elements on the principal diagonal are positive.

14.2.4 Other forms

Some other classes of non-negative matrices are now briefly discussed.

A real square matrix A is *totally non-negative* (*totally positive*) if all of its minors of all possible orders are non-negative (positive). This implies that $A \geqslant 0$ (>0) since the elements of A are its first-order minors. A totally positive symmetric matrix is positive definite, since all its leading principal minors are positive (see the remark at the end of Section 7.4.2), but the converse will certainly not hold in general. If A is totally non-negative, and there exists a positive integer k such that A^k is totally positive, then A is called *oscillatory*. The name arises from applications to vibration problems. A necessary and sufficient condition for a totally non-negative matrix A to be oscillatory is that $\det A > 0$, and all the elements on the tridiagonal band including the principal diagonal are positive. An interesting application is to the Hurwitz matrix H associated with a real polynomial $a(\lambda)$, defined in (11.44). We saw that all the roots of $a(\lambda)$ have negative real parts (i.e. $a(\lambda)$ is asymptotically stable) if and only if all the leading principal minors of H are positive. In fact, provided no root of $a(\lambda)$ has a positive real part (in which case $a(\lambda)$ is called *quasi-stable*) then H is totally non-negative. Furthermore, when $a(\lambda)$ is asymptotically stable, then the matrix obtained by deleting the last row and column in (11.44) is oscillatory.

The eigenvalue properties are also interesting: all the eigenvalues of a square totally non-negative matrix are real and non-negative; if the matrix is oscillatory the eigenvalues are all distinct and positive.

In many combinatorial applications the elements of the matrices involved are either 0 or 1. For example, the *adjacency matrix* $A = [a_{ij}]$ associated with a graph has $a_{ij} = 1$ if there is an arc from vertex i to vertex j, and $a_{ij} = 0$ otherwise. Permutation matrices, introduced in Section 13.4.5 are also $(0, 1)$ matrices. It is appropriate here to mention a more general class of matrices which, whilst not actually non-negative, are somewhat related. These arise in economic models where only the signs of the elements (including zeros) are known. If $A = [a_{ij}]$ is a real square matrix then $B = [b_{ij}]$ is *sign similar* to A if $\mathrm{sgn}(b_{ij}) = \mathrm{sgn}(a_{ij})$ when $a_{ij} \neq 0$, and $b_{ij} = 0$ when $a_{ij} = 0$, where the sign function was defined in Section 9.4. If at least one such matrix B is a stability matrix then A is called *potentially stable*, and A is *sign stable* if every matrix B is a stability matrix. In the latter case, A is a stability matrix whatever the values of its elements provided the pattern of positive, negative and zero entries is preserved. For further details see Barnett (1984).

General references on non-negative matrices are Berman and Plemmons (1979), Fiedler (1986), Gantmacher (Vol. 2, 1959), Horn and Johnson (1985) and Minc (1988).

14.3 Norms and conditioning

It is now useful to formalize and extend the concepts of norms of vectors and matrices which have been introduced earlier in the book, and to indicate their application to computational problems.

14.3.1 Vector norms

The simple idea of the euclidean norm of a column n-vector $x = [x_1, x_2, \ldots, x_n]^\mathsf{T}$ as the length of x was defined in (6.8). We use here the notation

$$\|x\|_2 = (|x_1|^2 + |x_2|^2 + \cdots + |x_n|^2)^{1/2}, \qquad (14.19)$$

since it is a special case of the *p-norm*

$$\|x\|_p = (|x_1|^p + |x_2|^p + \cdots + |x_n|^p)^{1/p}, \qquad (14.20)$$

where p is a positive integer. In particular, the *1-norm* (or *sum norm*) is

$$\|x\|_1 = \sum_{i=1}^n |x_i|. \qquad (14.21)$$

The *infinity-norm* is defined by

$$\|x\|_\infty = \max(|x_1|, |x_2|, \ldots, |x_n|). \tag{14.22}$$

All norms satisfy the properties

$$\left.\begin{array}{l} \|x\| \geqslant 0, \qquad \text{with } \|x\| = 0 \text{ only if } x = 0 \\ \|x + y\| \leqslant \|x\| + \|y\| \ (\textit{triangle inequality}) \\ \|cx\| = |c| \, \|x\|, \qquad \text{for all scalars } c \end{array}\right\}. \tag{14.23}$$

Any two p-norms $\|x\|_q$, $\|x\|_r$ are *equivalent*, which means that there exist positive constants c_1, c_2 such that

$$c_1 \|x\|_q \leqslant \|x\|_r \leqslant c_2 \|x\|_q. \tag{14.24}$$

Examples of (14.24) are

$$\|x\|_2 \leqslant \|x\|_1 \leqslant n^{1/2} \|x\|_2$$
$$\|x\|_\infty \leqslant \|x\|_p \leqslant n^{1/p} \|x\|_\infty,$$

the latter showing that $\|x\|_p \to \|x\|_\infty$ as $p \to \infty$. The *Hölder inequality* states that

$$|x^*y| = |\bar{x}_1 y_1 + \bar{x}_2 y_2 + \cdots + \bar{x}_n y_n|$$
$$\leqslant \|x\|_p \|y\|_q, \qquad \text{where } p^{-1} + q^{-1} = 1. \tag{14.25}$$

In particular take $p = q = 2$, assume x and y are real, and square both sides in (14.25) to obtain

$$(x_1 y_1 + x_2 y_2 + \cdots + x_n y_n)^2 \leqslant (x_1^2 + \cdots + x_n^2)(y_1^2 + \cdots + y_n^2), \tag{14.26}$$

which is the classical *Cauchy–Schwarz inequality*.

14.3.2 Matrix norms

In Section 8.5 (Exercise 8.32) we extended the definition of euclidean norm to an $m \times n$ matrix $A = [a_{ij}]$ in an obvious fashion:

$$\|A\|_e = \left(\sum_{i=1}^m \sum_{j=1}^n |a_{ij}|^2 \right)^{1/2}, \tag{14.27}$$

and this gives a measure of the 'size' of A (the case $m = n$ was first used in (6.117)). The terminology *Frobenius norm* or *Schur norm* is also used for (14.27). All matrix norms satisfy the same properties as displayed in (14.23) for the vector case, so

$$\left. \begin{array}{l} \|A\| \geq 0, \quad \text{with } \|A\| = 0 \text{ only if } A = 0 \\ \|A + B\| \leq \|A\| + \|B\| \\ \|cA\| = |c| \, \|A\|, \quad \text{for all scalars } c \end{array} \right\}, \qquad (14.28)$$

and there is an additional property

$$\|AB\| \leq \|A\| \, \|B\| \qquad (14.29)$$

for any $n \times r$ matrix B. The *p-norm* of A is

$$\|A\|_p = \max_{x \neq 0} \frac{\|Ax\|_p}{\|x\|_p}, \qquad (14.30)$$

where the vector p-norm is defined in (14.20). Notice that $\|A\|_2$ is *not* the euclidean norm in (14.27), although when $p = 2$ in (14.30) the vector norm on the right side is the euclidean norm. This explains why we use the notation $\|A\|_e$ in (14.27); in fact

$$\|A\|_2 \leq \|A\|_e \leq n^{1/2} \|A\|_2 .$$

In Section 8.5 we called the 2-norm of A the *spectral norm*, since $\|A\|_2$ is equal to the largest singular value of A (i.e. the square root of the largest eigenvalue of A^*A). It was also noted (Exercise 8.33) that $\|A\|_e^2$ is equal to the sum of the squares of the singular values of A. The matrix and vector p-norms have the property that

$$\|Ax\|_p \leq \|A\|_p \, \|x\|_p .$$

More generally, if a matrix norm $\|A\|_\alpha$ and a vector norm $\|x\|_\beta$ satisfy

$$\|Ax\|_\beta \leq \|A\|_\alpha \, \|x\|_\beta ,$$

then the two norms are *consistent* (or *compatible*). For example, the euclidean vector norm (14.19) and matrix norm (14.27) are consistent. If for any other vector norm $\|x\|_\gamma$ we define

$$\|A\|_\delta = \max_{\|x\|_\gamma = 1} \|Ax\|_\beta ,$$

then the matrix norm $\| \ \|_\sigma$ is *subordinate* to the vector norms $\| \ \|_\beta$ and $\| \ \|_\gamma$. Explicit formulae for the 1- and ∞-norms are

$$\|A\|_1 = \max_{1 \leq j \leq n} \sum_{i=1}^{m} |a_{ij}|, \qquad \|A\|_\infty = \max_{1 \leq i \leq m} \sum_{j=1}^{n} |a_{ij}|$$

and they are related to the 2-norm by

$$m^{-1/2} \|A\|_1 \leqslant \|A\|_2 \leqslant n^{1/2} \|A\|_1,$$
$$n^{-1/2} \|A\|_\infty \leqslant \|A\|_2 \leqslant m^{1/2} \|A\|_\infty.$$

When A is square with eigenvalues $\lambda_1, \ldots, \lambda_n$ then the spectral radius $\rho(A) = \max(|\lambda_i|)$ satisfies $\rho(A) \leqslant \|A\|$ for any matrix norm (see Exercise 14.19).

14.3.3 Conditioning

Roughly speaking, a problem is said to be *well-conditioned* if small changes in parameters produce small changes in the solution, and *ill-conditioned* if relatively large changes are produced (an illustration of the latter was given in Example 3.10).

Consider first the usual linear equations $Ax = b$ with A $n \times n$ and non-singular. Suppose that when b changes to $b + \delta b$ the corresponding solution becomes $x + \delta x$, so that

$$A(x + \delta x) = b + \delta b.$$

Hence $A \delta x = \delta b$, or $\delta x = A^{-1} \delta b$, and the ratio of the norms of the *relative* changes is

$$\frac{\|\delta x\|/\|x\|}{\|\delta b\|/\|b\|} = \frac{\|\delta x\| \, \|b\|}{\|\delta b\| \, \|x\|}. \tag{14.31}$$

However, by (14.29)

$$\|b\| = \|Ax\| \leqslant \|A\| \, \|x\|$$
$$\|\delta x\| = \|A^{-1} \delta b\| \leqslant \|A^{-1}\| \, \|\delta b\|,$$

which when substituted into (14.31) give

$$\frac{\|\delta x\|/\|x\|}{\|\delta b\|/\|b\|} \leqslant \|A\| \, \|A^{-1}\|. \tag{14.32}$$

The *condition number* $c(A)$ of A is defined to be

$$c(A) = \|A\| \, \|A^{-1}\|, \tag{14.33}$$

so that (14.32) can be rearranged as

$$\frac{\|\delta x\|}{\|x\|} \leqslant c(A) \frac{\|\delta b\|}{\|b\|},$$

which expresses the relative change in the solution in terms of the relative change in the right side of the equations. Similarly, if A changes to $A + \delta A$ but b is unaltered, we obtain

$$\frac{\|\delta x\|}{\|x\|} \leqslant \frac{c(A)\,\|\delta A\|/\|A\|}{1 - \|\delta A\|\,\|A^{-1}\|}$$

$$\simeq c(A)\,\frac{\|\delta A\|}{\|A\|},$$

provided $\|\delta A\|\,\|A^{-1}\|$ is small compared to 1. The matrix is called *ill-* (*well-*) *conditioned* according as $c(A)$ is large or small. The value of $c(A)$ depends upon the particular norm used, but if it is large in one norm then it is large in all others, and in any case $c \geqslant 1$ (see Exercise 14.17). For example, if $c_p(A)$ denotes the condition number relative to the p-norm (14.30) then

$$\frac{1}{n}c_j \leqslant c_i \leqslant nc_j, \qquad i = 1, j = 2; \qquad i = 2, j = \infty,$$

$$\frac{1}{n^2}c_1 \leqslant c_\infty \leqslant n^2 c_1.$$

For the spectral norm we have $c_2(A) = \sigma_1/\sigma_n$, where σ_1 and σ_n are the largest and smallest singular values of A (defined in Section 8.5).

Turning to the eigenvalue problem, suppose A is diagonalizable with eigenvalues $\lambda_1, \ldots, \lambda_n$, so that $T^{-1}AT = \mathrm{diag}[\lambda_1, \ldots, \lambda_n]$. If μ is any eigenvalue of the perturbed matrix $A + \delta A$ then the *Bauer–Fike theorem* states:

$$\min_{1 \leqslant i \leqslant n} (|\lambda_i - \mu|) \leqslant c_p(T)\,\|\delta A\|_p. \qquad (14.34)$$

If A is normal then it was noted in Section 6.4.3 that T can be taken to be unitary, in which case (14.34) simplifies when $p = 2$ since $c_2(T) = 1$.

Finally, consider the conditioning of a single specific eigenvalue λ_i, assumed to be a simple root of the characteristic polynomial of A, with corresponding left and right normalized eigenvectors y_i^T and x_i. Suppose $\delta A = \varepsilon K$ where K is a constant matrix and ε is a small parameter. The perturbation in A produces a first order change $\varepsilon y_i^T K x_i / s_i$ in λ_i, where $s_i = y_i^T x_i$, and

$$\|y_i^T K x_i\|_2 \leqslant \|y_i\|_2\,\|K\|_2\,\|x_i\|_2 \leqslant \|K\|_2.$$

If s_i is very small then λ_i will be very sensitive to this change δA in A. The numbers s_i^{-1}, $i = 1, 2, \ldots, n$ are the *condition numbers* of A with respect to the eigenvalue problem.

For an authoritative treatment of conditioning, see Wilkinson (1965).

14.4 Further material

14.4.1 Hadamard matrices

An $N \times N$ *Hadamard matrix* H_N has all its elements ± 1 and satisfies

$$H_N H_N^\mathsf{T} = H_N^\mathsf{T} H_N = N I_N. \qquad (14.35)$$

It follows immediately from (14.35) that $N^{-1/2} H_N$ is orthogonal. It can be shown that Hadamard matrices exist only when N is 1, 2 or a multiple of 4. It is easy to see that multiplying the elements in any row or column of H_N by -1 produces another Hadamard matrix. In this way, all the elements in the first row and first column of H_N can be made equal to $+1$, in which case the Hadamard matrix is called *normalized*. The smallest normalized Hadamard matrix is

$$H_2 = \begin{bmatrix} 1 & 1 \\ 1 & -1 \end{bmatrix} \qquad (14.36)$$

and a family of symmetric normalized matrices can be constructed from the formula

$$H_N = H_2 \otimes H_{N/2}, \qquad N = 2^k, \qquad k = 2, 3, 4, \ldots . \qquad (14.37)$$

Using the notation in Problem 2.13, we can write $H_N = H_2^{[k]}$.

Example 14.7 Taking $k = 2$, $N = 4$ in (14.37) gives

$$H_4 = H_2 \otimes H_2 = H_2^{[2]}$$

$$= \begin{bmatrix} 1 & 1 & 1 & 1 \\ 1 & -1 & 1 & -1 \\ 1 & 1 & -1 & -1 \\ 1 & -1 & -1 & 1 \end{bmatrix}. \qquad (14.38)$$

Regard the rows of the matrix in (14.38) as defining four piecewise linear functions $\varphi_i(t)$ on the interval $0 < t < 1$ divided into four equal

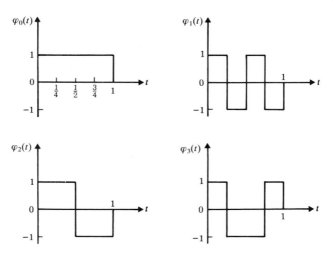

Fig. 14.1 The functions $\varphi_i(t)$.

parts, as shown in Fig. 14.1. For example, the second row of (14.38) gives

$$\varphi_1(t) = \begin{cases} 1, & 0 < t < \frac{1}{4} \\ -1, & \frac{1}{4} < t < \frac{1}{2} \\ 1, & \frac{1}{2} < t < \frac{3}{4} \\ -1, & \frac{3}{4} < t < 1 \end{cases}.$$

It is easy to see from Fig. 14.1 that the four rectangular pulse functions $\varphi_i(t)$ are orthonormal, i.e.

$$\int_0^1 \varphi_i(t)\varphi_j(t)\,\mathrm{d}t = \begin{cases} 0, & i \neq j \\ 1, & i = j \end{cases}. \tag{14.39}$$

They are an example of *Walsh functions*, which have many applications in information theory and signal processing (Ahmed and Rao 1975; Beauchamp 1975).

In general, the rows of H_N generated by (14.37) correspond to a set of Walsh functions wal(i, t) in *lexicographic* (or *Kronecker* or *Hadamard*) *order*, which is defined as follows: Each function wal(i, t), $i = 0, 1, 2, \ldots, N - 1$ takes the value ± 1 over the interval $0 < t < 1$ divided into $N = 2^k$ equal parts. Write i as a

k-bit binary number $(b_k b_{k-1} \cdots b_1)_2$, i.e.

$$i = b_k 2^{k-1} + b_{k-1} 2^{k-2} + \cdots + b_1 2^0$$

and define the *bit-reversal* $b(i)$ of i as the decimal number $b(i) = (b_1 b_2 \cdots b_k)_2$, i.e.

$$b(i) = b_1 2^{k-1} + b_2 2^{k-2} + \cdots + b_k 2^0.$$

The required set of Walsh functions is given by

$$\text{wal}(i, t) = \text{pal}[b(i), t], \qquad i = 0, 1, 2, \ldots, N-1, \quad (14.40)$$

where the $\text{pal}(j, t)$ represents the Walsh functions in *Paley* (or *dyadic*) *order*. These are defined in turn by

$$\text{pal}(j, t) = \prod_{n=1}^{m} [r(n, t)]^{\beta_n}, \qquad j = 0, 1, 2, \ldots, N-1, \quad (14.41)$$

where the binary representation of j is $(\beta_m \beta_{m-1} \cdots \beta_1)_2$ and

$$m = 1 + (\text{greatest integer} \leqslant \log_2 j).$$

Finally, the *Rademacher functions* $r(n, t)$ in (14.41) are defined by

$$r(n, t) = \text{sgn}[\sin(2^n \pi t)], \qquad 0 < t < 1, \qquad (14.42)$$

and can be regarded as 'squared-off' sine waves.

Example 14.8 When $k = 5$ (so $N = 2^5$) the five-bit binary representation of $i = 6$ is 00110, so the bit reversal of i is

$$b(6) = (01100)_2 = 1 \cdot 2^3 + 1 \cdot 2^2 = 12.$$

From (14.40) we then have

$$\text{wal}(6, t) = \text{pal}(12, t).$$

The binary representation of $j = 12$ has $m = 4$, $\beta_4 = \beta_3 = 1$, $\beta_2 = \beta_1 = 0$, so that in (14.41)

$$\text{pal}(12, t) = r(3, t) r(4, t)$$
$$= \text{sgn}(\sin 8\pi t) \, \text{sgn}(\sin 16\pi t), \quad \text{by (14.42)}.$$

Example 14.9 When $k = 2$ and $N = 4$ then (14.41) and (14.42) give:

$j = 0$, $\beta_1 = 0$: $\quad \text{pal}(0, t) = r(0, t) = 1$, $\quad 0 < t < 1$

$j = 1$, $m = 1$, $\beta_1 = 1$: $\quad \text{pal}(1, t) = r(1, t) = \text{sgn}(\sin 2\pi t)$

$j = 2$, $m = 2$, $\beta_2 = 1$, $\beta_1 = 0$: $\quad \text{pal}(2, t) = r(2, t) = \text{sgn}(\sin 4\pi t)$

$j = 3$, $m = 2$, $\beta_2 = 1$, $\beta_1 = 1$: $\quad \text{pal}(3, t) = r(1, t) r(2, t)$
$$= \text{sgn}(\sin 2\pi t) \, \text{sgn}(\sin 4\pi t).$$

Comparing these expressions with Fig. 14.1, it is easy to see that

$$\text{pal}(0, t) = \varphi_0(t), \qquad \text{pal}(1, t) = \varphi_2(t),$$
$$\text{pal}(2, t) = \varphi_1(t), \qquad \text{pal}(3, t) = \varphi_3(t). \tag{14.43}$$

In order to apply (14.40) we need the bit-reversals of the integers 0 to 3: clearly $b(0) = 0$, and

$$1 = (01)_2, \qquad b(1) = (10)_2 = 2$$
$$2 = (10)_2, \qquad b(2) = (01)_2 = 1$$
$$3 = (11)_2, \qquad b(3) = (11)_2 = 3.$$

Hence from (14.40) and (14.43) we obtain

$$\text{wal}(0, t) = \text{pal}(0, t) = \varphi_0(t) \qquad \text{wal}(2, t) = \text{pal}(1, t) = \varphi_2(t)$$
$$\text{wal}(1, t) = \text{pal}(2, t) = \varphi_1(t) \qquad \text{wal}(3, t) = \text{pal}(3, t) = \varphi_3(t),$$

where the $\varphi_i(t)$ were obtained in Example 14.7. This illustrates for the case $k = 2$, $N = 4$ that the Walsh functions in lexicographic order correspond to the rows of the Hadamard matrix H_4 in (14.38).

In general, the N values of $\text{wal}(i, t)$ on the N intervals $s/N < t < (s + 1)/N$, $s = 0, 1, 2, \ldots, N - 1$ are given by the $(i + 1)$th row of H_N in (14.37).

The scheme for constructing Hadamard matrices via (14.37) is sometimes named after *Sylvester*; a second, so-called *Paley* construction produces Hadamard matrices of order $p + 1$ where p is any odd prime such that $p + 1$ is a multiple of 4. Some preliminary definitions are needed:

(i)   The *quadratic residues mod p* are the integers $i^2 \bmod p$, $i = 1, 2, 3, \ldots, \frac{1}{2}(p - 1)$. The remaining numbers mod p are called the *non-residues*.

(ii) The *Legendre symbol $\chi(i)$* satisfies $\chi(0) = 0$ and

$$\chi(i) = \begin{cases} 1, & \text{if } i \text{ is a quadratic residue mod } p \\ -1, & \text{if } i \text{ is a nonresidue mod } p. \end{cases}$$

(iii) The $p \times p$ *Jacobsthal matrix Q* is a circulant matrix whose first row is

$$[\chi(0), \chi(1), \chi(2), \ldots, \chi(p - 1)].$$

It is easy to check that Q is skew symmetric.

The required normalized Hadamard matrix is then given by

$$H_{p+1} = \begin{bmatrix} 1 & e^{\mathsf{T}} \\ e & Q - I_p \end{bmatrix}, \tag{14.44}$$

where e is the column vector defined in (14.18).

Example 14.10 When $p = 7$ the quadratic residues mod 7 are

$$1^2 = 1, \qquad 2^2 = 4, \qquad 3^2 = 9 \equiv 2,$$

so the non-residues are 3, 5, 6. Hence

$$\chi(1) = \chi(2) = \chi(4) = 1, \qquad \chi(3) = \chi(5) = \chi(6) = -1,$$

and the first row of Q is

$$[0, 1, 1, -1, 1, -1, -1].$$

The normalized Hadamard matrix of order 8 in (14.44) is

$$H_8 = \begin{bmatrix} 1 & 1 \cdots 1 \\ 1 & \\ \vdots & Q - I_7 \\ 1 & \end{bmatrix}.$$

Hadamard matrices are used in the theory of error-correcting codes, briefly introduced in Section 5.8. Replacing all the 1's by zeros, and -1's by 1's in a normalized Hadamard matrix produces a *binary Hadamard matrix* which can be used to construct certain codes. For details see MacWilliams and Sloane (1977).

For further material on Hadamard matrices, see Agaian (1985), which includes applications to combinatorics; and Harwit and Sloane (1979), which covers applications to spectroscopy and imaging.

14.4.2 Inequalities

(1) *Cauchy-Schwarz inequality.* If A and B are two real $n \times n$ matrices, then

$$[\text{tr}(A^\mathsf{T}B)]^2 \leq [\text{tr}(A^\mathsf{T}A)][\text{tr}(B^\mathsf{T}B)]$$

with equality if and only if $A = tB$ for some scalar t. The expression still holds when A and B are column n-vectors a and b respectively, in which case it becomes the classical result (14.26).

(2) *Hadamard's inequalities.* For any complex $n \times n$ matrix $A = [a_{ij}]$,

$$|\det A|^2 \leq \prod_{j=1}^{n} \sum_{i=1}^{n} |a_{ij}|^2$$

and equality holds if and only if A^*A is diagonal, or A has a zero column.

If A is hermitian and positive definite then

$$\det A \leqslant \prod_{i=1}^{n} a_{ii} \qquad (14.45)$$

with equality if and only if A is diagonal.

(3) *Minkowski's inequalities.* If A and B are two $n \times n$ positive semidefinite hermitian matrices, then

$$[\det(A + B)]^{1/n} \geqslant (\det A)^{1/n} + (\det B)^{1/n} \qquad (14.46)$$

and

$$[\mathrm{tr}(A + B)^p]^{1/p} \leqslant (\mathrm{tr}\, A^p)^{1/p} + (\mathrm{tr}\, B^p)^{1/p}$$

for any integer $p > 1$. Equality holds in either expression if and only if $A = tB$ for some positive scalar t; in (14.46) an alternative condition for equality is $\det(A + B) = 0$.

(4) If A and B are two $n \times n$ positive definite (or semidefinite) hermitian matrices then their Hadamard product $A \circ B$ (defined in (2.69)) is also positive definite (respectively, semidefinite) and

$$a_{11} \cdots a_{nn} b_{11} \cdots b_{nn} \geqslant \det(A \circ B) \geqslant b_{11} b_{22} \cdots b_{nn} \det A \geqslant \det A \det B$$

$$\det(A \circ B) + (\det A)(\det B) \geqslant b_{11} \cdots b_{nn} \det A + a_{11} \cdots a_{nn} \det B.$$

Good sources for matrix inequalities are Magnus and Neudecker (1988) and Marcus and Minc (1964).

14.4.3 Interval matrices

Of increasing importance in applications is the situation where each element a_{ij} of a real $n \times n$ matrix A lies in a closed interval $[\alpha_{ij}, \beta_{ij}]$, so that

$$\alpha_{ij} \leqslant a_{ij} \leqslant \beta_{ij}.$$

It is convenient to write $a_{ij} = [\alpha_{ij}, \beta_{ij}]$ to denote the general element of an *interval matrix* A.

In order to define basic operations on two interval matrices A and $B = [b_{ij}]$ we must first define operations on two real intervals $a = [a_1, a_2]$ and $b = [b_1, b_2]$:

equality: $a = b \Leftrightarrow a_1 = b_1,\ a_2 = b_2;$

addition and subtraction:

$$a + b = [a_1 + b_1, a_2 + b_2], \qquad a - b = [a_1 - b_2, a_2 - b_1]; \quad (14.47)$$

multiplication:

$$a * b = [\min c, \max c], \qquad c = (a_1 b_1, a_1 b_2, a_2 b_1, a_2 b_2). \quad (14.48)$$

Then $A \pm B$ has i, j element $a_{ij} \pm b_{ij}$, and the general element in the product is

$$[AB]_{ij} = \sum_{k=1}^{n} a_{ik} * b_{kj}.$$

These expressions look just like the ones for the ordinary sum and product defined in (2.5) and (2.19), the difference being that the operations used obey (14.47) and (14.48).

Example 14.11 Consider the interval matrices

$$A = \begin{bmatrix} [1, 3] & [-1, 0] \\ [-2, 2] & [3, 6] \end{bmatrix}, \qquad B = \begin{bmatrix} [-2, -1] & [-3, 1] \\ [0, 4] & [-2, 4] \end{bmatrix}. \quad (14.49)$$

Then using (14.47)

$$A + B = \begin{bmatrix} [-1, 2] & [-4, 1] \\ [-2, 6] & [1, 10] \end{bmatrix}, \qquad A - B = \begin{bmatrix} [2, 5] & [-2, 3] \\ [-6, 2] & [-1, 8] \end{bmatrix}. \quad (14.50)$$

This means, for example, that

$$1 \leq a_{11} \leq 3, \qquad -2 \leq b_{11} \leq -1$$

and

$$-1 \leq (A + B)_{11} \leq 2, \qquad 2 \leq (A - B)_{11} \leq 5.$$

Using (14.48) gives

$$\begin{aligned}
a_{11} * b_{11} &= [1, 3] * [-2, -1] \\
&= [\min(-2, -1, -6, -3), \quad \max(-2, -1, -6, -3)] \\
&= [-6, -1] \\
a_{12} * b_{21} &= [-1, 0] * [0, 4] = [\min(0, -4, 0, 0), \quad \max(0, -4, 0, 0)] \\
&= [-4, 0],
\end{aligned}$$

so the 1, 1 element of AB is

$$a_{11} * b_{11} + a_{12} * b_{21} = [-6, -1] + [-4, 0] = [-10, -1].$$

Proceeding similarly for the other elements in the product, we obtain

$$AB = \begin{bmatrix} [-10, -1] & [-13, 5] \\ [-4, 28] & [-18, 30] \end{bmatrix}. \quad (14.51)$$

For any real interval $a = [a_1, a_2]$, we define the *absolute value* of the interval by

$$|a| = \max(|a_1|, |a_2|)$$

and the *width* of the interval by

$$d(a) = a_2 - a_1 \geq 0.$$

The *distance* between the intervals a and b is defined as

$$q(a, b) = \max(|a_1 - b_1|, |a_2 - b_2|).$$

When $a_1 = a_2$ the interval $[a_1, a_1]$ is called a *point interval*, and the definition of distance between two point intervals reduces to that of distance between two real numbers, i.e.

$$q([a_1, a_1], \quad [b_1, b_2]) = |a_1 - b_1|.$$

Three non-negative matrices can now be constructed using the above definitions: the *absolute value matrix* $|A|$ has i, j element $|a_{ij}|$; the *width matrix* $d(A)$ has i, j element $d(a_{ij})$; and the *distance matrix* $q(A, B)$ has i, j element $q(a_{ij}, b_{ij})$. Some simple properties are

$$|A + B| \leq |A| + |B| \tag{14.52}$$

$$|AB| \leq |A|\,|B| \tag{14.53}$$

$$d(A \pm B) = d(A) + d(B) \tag{14.54}$$

$$|A|\,d(B) \leq d(AB) \leq d(A)\,|B| + |A|\,d(B) \tag{14.55}$$

$$\left. \begin{array}{l} q(A, B) \leq q(A, C) + q(B, C) \\ q(AB, AC) \leq |A|\,q(B, C) \end{array} \right\} \quad \begin{array}{l} \text{for any interval} \\ \text{matrix } C, \end{array} \tag{14.56}$$

where the inequalities are in the sense defined for non-negative matrices at the beginning of Section 14.2.1.

Example 14.11 (continued)   For the matrices in (14.49) we have

$$|A| = \begin{bmatrix} 3 & 1 \\ 2 & 6 \end{bmatrix}, \qquad |B| = \begin{bmatrix} 2 & 3 \\ 4 & 4 \end{bmatrix},$$

$$|A + B| = \begin{bmatrix} 2 & 4 \\ 6 & 10 \end{bmatrix}, \qquad |A|\,|B| = \begin{bmatrix} 10 & 13 \\ 28 & 30 \end{bmatrix},$$

and from (14.51)

$$|AB| = \begin{bmatrix} 10 & 13 \\ 28 & 30 \end{bmatrix},$$

showing that in this example (14.52) and (14.53) become

$$|A + B| < |A| + |B|, \qquad |AB| = |A| \, |B| \, .$$

We also see that

$$d(A) = \begin{bmatrix} 2 & 1 \\ 4 & 3 \end{bmatrix}, \qquad d(B) = \begin{bmatrix} 1 & 4 \\ 4 & 6 \end{bmatrix}$$

and from (14.50)

$$d(A + B) = \begin{bmatrix} 3 & 5 \\ 8 & 9 \end{bmatrix} = d(A - B),$$

which verifies (14.54). Returning to (14.51) reveals that

$$d(AB) = \begin{bmatrix} 9 & 18 \\ 32 & 48 \end{bmatrix}$$

and the other expressions in (14.55) are

$$|A| \, d(B) = \begin{bmatrix} 7 & 18 \\ 26 & 44 \end{bmatrix},$$

$$d(A) \, |B| + |A| \, d(B) = \begin{bmatrix} 8 & 10 \\ 20 & 24 \end{bmatrix} + \begin{bmatrix} 7 & 18 \\ 26 & 44 \end{bmatrix} = \begin{bmatrix} 15 & 28 \\ 46 & 68 \end{bmatrix},$$

confirming that (14.55) is satisfied.

It is possible to develop the gaussian elimination scheme of Chapter 3 for systems of linear equations with interval coefficients, and to devise iterative methods for finding an interval matrix which includes the inverse of a given fixed matrix; for details see Alefeld and Herzberger (1983).

14.4.4 Unimodular integer matrices

Linear programming models were introduced in Example 1.5, Problem 1.4 and Problem 5.2. In a general formulation, it is required to find the maximum (or minimum) of some linear function $\sum c_i x_i$ subject to a set of m linear equations

$$Ax = b \qquad (14.57)$$

in n unknowns x_i. Since the x_i represent real physical quantities they are required to be non-negative; furthermore, in many situations the elements of A and b in (14.57) are all integers, and the solution x must also have only integer components (this constitutes the *integer programming* problem). For example, in the transportation model in Problem 5.2 the x_i represent numbers of items delivered from warehouses to supermarkets.

We call A an *integer matrix* when all of its elements are integers, and such matrices have very similar properties to those of polynomial matrices (studied in Chapter 12), since the set of polynomials and the set of integers are each examples of a ring (i.e. the usual operations of addition, multiplication, etc., apply, but *not* division). In particular, if $R(A) = r$ and $m \leq n$ then A can be expressed as

$$A = PSQ$$

where the $m \times n$ matrix

$$S = [\text{diag}(i_1, i_2, \ldots, i_r), 0] \qquad (14.58)$$

is the *Smith normal form* of A, and P and Q are matrices whose inverses are also integer matrices. From the expression $P^{-1} = \text{adj } P/\det P$ it follows that both $\det P$ and $\det Q$ must be ± 1. The expression (14.58) is the direct analogue of (12.17) and (12.18) for polynomial matrices.

The importance of the Smith normal form for our present purposes is that the eqns (15.47) possess an integer solution for all possible integer vectors b if and only if $S = I_m$. For the integer programming problem we need to go further, since $R(A) = m$ and we require so-called *basic* integer solutions of (14.57) in which m of the x_i are non-zero. The necessary and sufficient condition for all basic solutions of (14.57) to be integer, for all integer vectors b, is that all non-zero $m \times m$ minors of A have value ± 1, in which case A is *unimodular*. In particular, when $m = n$ then $\det A = \pm 1$, and our definition agrees with that used for square polynomial matrices in Chapter 12.

Often in integer programming models the variables are constrained not by a set of linear equations (14.57) but by a set of linear inequalities in the form

$$Ax \leq b, \qquad (14.59)$$

where again A and b have all their elements integers. For all basic solutions of (14.59) to be integer for all possible b, A must be *totally unimodular*: that is, all its minors of all possible orders must have value 0, 1 or -1 (this includes the elements themselves). An important class of totally unimodular matrices $B = [b_{ij}]$, which includes those arising in transportation problems, can be characterized as follows:

(i) $b_{ij} = 0$, 1 or -1, all i and j;

(ii)   every column of B contains at most two nonzero entries;

(iii) the rows of B can be separated into two disjoint sets R_1 and R_2 such that if in column j we have $b_{ij} \neq 0$, $b_{kj} \neq 0$ then

(a) $\text{sgn}(b_{ij}) = \text{sgn}(b_{kj}) \Rightarrow i \in R_1, j \in R_2$

(b) $\text{sgn}(b_{ij}) \neq \text{sgn}(b_{jk}) \Rightarrow$ either $i, k \in R_1$ or $i, k \in R_2$.

Example 14.12   The matrix of coefficients in the transportation model in Problem 5.2 is found to be

$$B = \begin{bmatrix} 1 & 0 & 0 & 1 & 0 & 0 \\ 0 & 1 & 0 & 0 & 1 & 0 \\ 0 & 0 & 1 & 0 & 0 & 1 \\ 1 & 1 & 1 & 0 & 0 & 0 \\ 0 & 0 & 0 & 1 & 1 & 1 \end{bmatrix} \begin{matrix} \left.\vphantom{\begin{matrix}1\\0\\0\end{matrix}}\right\} R_1 \\ \left.\vphantom{\begin{matrix}1\\0\end{matrix}}\right\} R_2. \end{matrix} \tag{14.60}$$

Divide the rows of B into the two indicated sets. It is easy to see that the above conditions for B to be totally unimodular are satisfied; in particular (iii)(a) holds because in each column of B there is a single 1 amongst the first three rows (R_1) and a single 1 amongst the last two rows (R_2). The reader can check that all the minors of B in (14.60) of orders 2, 3, 4, 5 take the values 0, $+1$ or -1.

Total unimodularity of a given $m \times n$ matrix $A = [a_{ij}]$ is preserved if any of the following operations is performed on it:

(i)   permutation of rows or columns;

(ii)   transposition;

(iii) multiplication of any row or column by -1;

(iv) adding an extra zero row or column, or a row or column containing a single element (± 1);

(v)   *pivoting* on the element $a_{ij} = \varepsilon \in \{\pm 1\}$, by which after the row and column interchanges $(Ri) \leftrightarrow (R1)$, $(Cj) \leftrightarrow (C1)$, we apply the transformation

$$\begin{matrix} & \begin{matrix} 1 & n-1 \end{matrix} \\ \begin{matrix} 1 \\ m-1 \end{matrix} & \begin{bmatrix} \varepsilon & r \\ c & A' \end{bmatrix} \end{matrix} \rightarrow \begin{bmatrix} -\varepsilon & \varepsilon r \\ \varepsilon c & A' - \varepsilon cr \end{bmatrix}.$$

The theory of integer matrices is covered by Newman (1972), further results on unimodular matrices are given by Barnett (1971, 1984), and a comprehensive treatment is contained in Schrijver (1986), including relationships to networks.

Exercises

Section 14.1.1

Exercise 14.1 Determine the general form of the solution of (14.3) when the Jordan forms in (14.4) are defined by

$$\lambda_1 = -1, \quad k_1 = 5, \quad \lambda_2 = 0, \quad k_2 = 2, \quad \lambda_3 = 1, \quad k_3 = 2, \quad \lambda_4 = 3, \quad k_4 = 3,$$

$$\mu_1 = 0, \quad l_1 = 2, \quad \mu_2 = 2, \quad l_2 = 3, \quad \mu_3 = 1, \quad l_3 = 4.$$

Section 14.1.2

Exercise 14.2 Determine the general form of matrices which commute with

$$\text{diag}\left\{ \begin{bmatrix} 3 & 1 & 0 \\ 0 & 3 & 1 \\ 0 & 0 & 3 \end{bmatrix}, \begin{bmatrix} 3 & 1 \\ 0 & 3 \end{bmatrix}, -1, -1 \right\}.$$

Exercise 14.3 Show that when the Jordan form in (14.9) is diagonal, then the general solution Y is block diagonal. Deduce that if λ_i has multiplicity r_i, $i = 1, 2, \ldots, t$, where $\lambda_1, \ldots, \lambda_t$ are the distinct eigenvalues, then Y contains $r_1^2 + r_2^2 + \cdots + r_t^2$ arbitrary parameters.

Section 14.1.3

Exercise 14.4 If A is non-singular, show that $(A + \lambda B)^{-1}$ is linear in λ provided $B = AX$ or XA, where $X^2 = 0$.

This arose in a problem of parametric linear programming (Barnett 1971).

Exercise 14.5 Determine the general form of an $n \times n$ matrix which is nilpotent with index k, i.e. $X^k = 0$.

Exercise 14.6 Use the result in Problem 12.1 to prove the statement about right and left solvents of $A(\lambda)$ at the end of Section 14.1.3.

Section 14.1.4

Exercise 14.7 Use the result in Problem 9.12 with $X(0) = C$ to show that if A and B are stability matrices then the solution of $AY + YB = C$ is

$$Y = -\int_0^\infty e^{At}Ce^{Bt}\,dt.$$

Section 14.2.1

Exercise 14.8 Find the Perron root and a Perron eigenvector of

$$\begin{bmatrix} 3 & 1 & 2 \\ 1 & 1 & 0 \\ 3 & 2 & 1 \end{bmatrix}.$$

Exercise 14.9 Test whether the non-negative matrix

$$\begin{bmatrix} 0 & 0 & 0 & 1 \\ 1 & 0 & 0 & 0 \\ 0 & 0 & 1 & 1 \\ 0 & 1 & 0 & 0 \end{bmatrix}$$

is irreducible.

Section 14.2.2

Exercise 14.10 Verify that

$$A = \begin{bmatrix} 5 & -2 & 0 \\ 0 & 4 & -1 \\ -1 & 0 & 5 \end{bmatrix}$$

is an M-matrix. Check that

$$A = \begin{bmatrix} 5 & -1 & 0 \\ 0 & 4 & 0 \\ 0 & 0 & 5 \end{bmatrix} - \begin{bmatrix} 0 & 1 & 0 \\ 0 & 0 & 1 \\ 1 & 0 & 0 \end{bmatrix} = B - C$$

is a regular splitting. Use the tabular form of the Schur–Cohn theorem (Section 11.4.2) to show that $B^{-1}C$ is convergent.

Exercise 14.11 Prove that if A is an M-matrix then so is $A + D$ for any non-negative diagonal matrix D.

Section 14.2.3

Exercise 14.12 In a certain genetics problem the matrix of transition pobabilities is

$$A = \begin{bmatrix} 1 & 0 & 0 \\ \dfrac{1}{4} & \dfrac{1}{2} & \dfrac{1}{4} \\ \dfrac{1}{18} & \dfrac{4}{9} & \dfrac{1}{2} \end{bmatrix}.$$

Use Sylvester's formula (9.32) to show that as $k \to \infty$,

$$A^k \to \begin{bmatrix} 1 & 0 & 0 \\ 1 & 0 & 0 \\ 1 & 0 & 0 \end{bmatrix}.$$

Section 14.2.4

Exercise 14.13 Construct the Hurwitz matrix H defined in (11.44) for the polynomial in Exercise 11.19(b). Verify that H is totally non-negative. If H_r is the matrix obtained by deleting the last row and column of H, show that H_r^2 is totally positive.

Section 14.3.2

Exercise 14.14 Use (14.29) to show that $\|A^k\| \le \|A\|^k$ for any positive integer k.

Exercise 14.15 Use (14.29) to prove that

$$\|XAY\|_p \le \|X\|_p \, \|Y\|_p \, \|A\|_p$$

for A, X, Y having appropriate dimensions. By taking X and Y to be vectors having a single non-zero entry, deduce that $|a_{ij}| \le \|A\|_p$.

Exercise 14.16 By using the results in the preceding two exercises, together with the ∞-norm, deduce that if

$$A = \begin{bmatrix} 0.2 & 0.6 \\ 0.5 & 0.4 \end{bmatrix},$$

then the elements $\hat{a}_{ij}$ of A^k satisfy $|\hat{a}_{ij}| \le (0.9)^k$ for any positive integer k.

Section 14.3.3

Exercise 14.17 Prove that for any p-norm, $\|I_n\|_p = 1$. Hence by considering $AA^{-1} = I$, show that the condition number in (14.33) satisfies $c(A) \geqslant 1$.

Exercise 14.18 Deduce that if A is positive definite hermitian then the condition number $c_2(A)$ is equal to the ratio of the largest to the smallest eigenvalues of A (use Problem 8.5(d)).

Exercise 14.19 Deduce from the eigenvalue–eigenvector equation $Ax = \lambda x$ that $|\lambda| \leqslant \|A\|$.

Section 14.4.1

Exercise 14.20 Verify that H_2 in (14.36) and H_4 in (14.38) satisfy the condition (14.35). Show that the 4×4 matrix with -1's on the secondary diagonal and 1's elsewhere is a different fourth-order Hadamard matrix.

Exercise 14.21 Verify that circ$(1, 1, 1, -1)$ is a Hadamard matrix.

Exercise 14.22 Use (14.44) to construct Hadamard matrices of orders 4 and 8.

Section 14.4.2

Exercise 14.23 Deduce from Minkowski's inequality (14.46) that if A and B are positive definite hermitian matrices then

$$\det(A + B) > \det A + \det B.$$

Section 14.4.3

Exercise 14.24 Complete the remaining elements of the product AB in Example 14.11, and hence verify (14.51).

Exercise 14.25 In addition to those in (14.55) the inequality $d(AB) \geqslant |B| \, d(A)$ also applies. Verify that this holds for the interval matrices A, B in (14.49).

Exercise 14.26 Verify that (14.56) holds for the interval matrices A, B in (14.29) and

$$C = \begin{bmatrix} [-3, 1] & [2, 5] \\ [-7, 9] & [-11, -1] \end{bmatrix}.$$

Problems

14.1   Consider the equation

$$\sum_{j=0}^{N} \sum_{i=0}^{N} \alpha_{ij}(A^{\mathrm{T}})^i X A^j = Q, \tag{14.61}$$

where A is a real $n \times n$ matrix, Q is a real symmetric matrix and the α_{ij} are scalars. Equation (14.61) represents $\frac{1}{2}n(n+1)$ linear equations for the $\frac{1}{2}n(n+1)$ unknown elements of the real symmetric matrix X. The Liapunov equations (11.66) and (11.69) are special cases of (14.61).

Show that (14.61) can be written as

$$\sum_{i=0}^{N} \sum_{j=0}^{N} \alpha_{ij}(A^{\mathrm{T}})^i S A^j = R,$$

where

$$S = A^{\mathrm{T}}X - XA, \qquad R = A^{\mathrm{T}}Q - QA$$

are both real skew symmetric matrices. This represents a reduction of n in the number of equations and unknowns.

By showing that

$$(A^{\mathrm{T}})^i X = \sum_{k=1}^{i} (A^{\mathrm{T}})^{i-k} S A^{k-1} + X A^i, \qquad i = 1, 2, 3, \ldots,$$

deduce that the solution of (14.61) is

$$X = (Q - F)E^{-1},$$

where

$$E = \sum_{i=0}^{N} \sum_{j=0}^{N} \alpha_{ij} A^{i+j}$$

$$F = \sum_{j=0}^{N} \sum_{i=1}^{N} \alpha_{ij} \sum_{k=1}^{i} (A^{\mathrm{T}})^{i-k} S A^{j+k-1}.$$

The result in Problem 11.7 is a special case of the above.

14.2   Let A be a non-negative square matrix such that $A^k > 0$ for some positive integer k. Show that A is irreducible. By considering the eigenvalues of A^k, deduce that A has a single eigenvalue of maximal modulus (i.e. A is primitive).

The converse also holds: if A is primitive then $A^k > 0$ for some integer k.

14.3 Prove that a non-negative $n \times n$ matrix A is row stochastic if and only if $AE = E$, where E is the $n \times n$ matrix all of whose elements are equal to 1. Hence prove that the product of two row stochastic matrices is also row stochastic.

14.4 Use the result in Problem 6.5(b) to prove that a real square matrix is potentially stable provided all its diagonal elements are negative.

14.5 A matrix A is called *D-stable* if DA is a stability matrix for any positive diagonal matrix D. Use the Liapunov eqn (11.66) with $P = D$ to show that $S - kI$, where S is a real skew symmetric matrix and k is a positive scalar, is *D*-stable.

14.6 An $n \times n$ positive matrix $A = [a_{ij}]$ is called *equitable* if $a_{ij}a_{jk} = a_{ik}$ for all i, j and k. There are applications in economics and group theory.

(a) Prove that $A^2 = nA$.

(b) If B is a second such matrix show that the Hadamard product $A \circ B$ (defined in (2.69)) is also equitable.

(c) Show that the $n \times n$ matrix E all of whose elements are equal to one is equitable.

(d) Obtain an expression for A in terms of the elements in its first column. Show that

$$R^{-1}AR = \text{diag}[n, 0, 0, \ldots, 0],$$

where

$$R = \begin{bmatrix} 1 & 0 & 0 & \cdot\cdot & 0 & -1 \\ a_{21} & a_{21} & 0 & \cdot\cdot & 0 & 0 \\ a_{31} & -a_{31} & a_{31} & \cdot\cdot & 0 & 0 \\ a_{41} & 0 & -a_{41} & \cdot\cdot & \cdot & \cdot \\ \vdots & \vdots & \vdots & & & \\ a_{n1} & 0 & 0 & \cdot\cdot & -a_{n1} & a_{n1} \end{bmatrix}.$$

Hence deduce that any equitable matrix is similar to E.

14.7 An $n \times n$ matrix A is called *quasi diagonal dominant* if there exists a positive diagonal matrix D such that AD is diagonal dominant in the sense of (6.131). Show that in this case the comparison matrix C_{AD} defined in Section 14.2.2 is an M-matrix.

It can be shown that if in addition A is a so-called *Metzler matrix* (i.e. $a_{ii} < 0$, $a_{ij} \geq 0$ for all i and j) then A is a stability matrix.

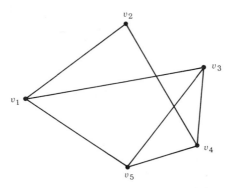

Fig. 14.2   Graph for Problem 14.8.

14.8   For a graph with n vertices $v_1, v_2, \ldots, v_n$, denote an arc
connecting v_i to v_j by (v_i, v_j). A *chain* of *length k* which connects
v_x to v_y is a sequence of k arcs $(v_x, v_{i_1}), (v_{i_1}, v_{i_2}), \ldots, (v_{i_{k-1}}, v_y)$;
notice that the sequence can include $(v_\alpha, v_\beta), (v_\beta, v_\alpha)$—i.e.
'retracing steps' is allowed.

   If A is the $n \times n$ adjacency matrix for the graph then the (i, j)
entry of A^k is the number of chains of length k connecting v_i to v_j.
For the graph shown in Fig. 14.2, write down A and interpret A^2
and A^3.

   Furthermore, the graph is said to be *connected* if every pair of
vertices is connected by a chain. A necessary and sufficient
condition for a graph to be connected is that no entry of

$$A + A^2 + A^3 + \cdots + A^{n-1}$$

is zero. Test whether the graph in Fig. 14.2 is connected.

14.9   Let A be an $n \times n$ positive definite real symmetric matrix. By
using the fact (Problem 7.2) that $A = B^{\mathsf{T}}B$ with B non-singular,
deduce that

$$\|x\| = (x^{\mathsf{T}}Ax)^{1/2}$$

satisfies the axioms (14.23) for a vector norm.

14.10   Let $\bar{x}$ be a computed solution to the set of linear equations
$Ax = b$, and let $r = A\bar{x} - b$ denote the *residual*. Show that

$$\frac{\|\bar{x} - x\|}{\|x\|} = c(A) \frac{\|r\|}{\|b\|},$$

where $c(A)$ is the condition number in (14.33).

This shows that a relatively small residual does not ensure a relatively small error in the solution if $c(A)$ is large.

14.11 Consider a possible matrix norm for an $n \times n$ matrix $A = [a_{ij}]$ defined by max $|a_{ij}|$, $i, j = 1, \ldots, n$. By taking $A = \begin{bmatrix} 1 & 1 \\ 1 & 1 \end{bmatrix}$, show that this is not a matrix norm since (14.29) is not satisfied. Prove, however, that $n \max_{i,j} |a_{ij}|$ is a matrix norm by verifying that it satisfies (14.28) and (14.29).

This norm was used in Exercise 9.6 and the subsequent development.

14.12 If H_M and H_N are Hadamard matrices defined by (14.35) show that $H_M \otimes H_N$ is a Hadamard matrix of order MN.

14.13 Use the formulae (14.40) and (14.41) with $k = 3$ to construct the eight Walsh functions in lexicographic order. Confirm that they agree with the rows of H_8 obtained from (14.37).

14.14 Let

$$H_{N+1} = \begin{bmatrix} 1 & e^{\mathsf{T}} \\ e & G \end{bmatrix} \begin{matrix} 1 \\ N \end{matrix}$$

$$\begin{matrix} 1 & N \end{matrix}$$

be a normalized Hadamard matrix of order $N + 1$, where $e^{\mathsf{T}} = [1, 1, \ldots, 1]$. Prove that if $E = ee^{\mathsf{T}}$ then

(a) $GG^{\mathsf{T}} = (N + 1)I_N - E$;

(b) $EG^{\mathsf{T}} = GE = -E$;

(c) $G^{-1} = \dfrac{1}{N + 1}(G^{\mathsf{T}} - E)$.

14.15 Define S to be the $(0, 1)$ matrix obtained from a normalized Hadamard matrix H_{N+1} of order $N + 1$ by deleting the first row and column of H_{N+1} and changing 1's to zeros and -1's to 1's. Prove:

(a) $S = \frac{1}{2}(E - G)$, where E and G are defined in Problem 14.14;

(b) $SS^{\mathsf{T}} = \frac{1}{4}(N + 1)(I_N + E)$;

(c) $SE = \frac{1}{2}(N + 1)E$;

(d) $S^{-1} = \dfrac{2}{N + 1}(2S^{\mathsf{T}} - E)$.

14.16 For an $m \times m$ matrix A and an $n \times n$ matrix B having rows $b_1, \ldots, b_n$, define the *Paley product* $A \bar{\otimes} B$ as the matrix having rows $A \otimes b_1, \ldots, A \otimes b_n$, where $\otimes$ denotes the usual Kronecker product. Prove that if A and B are Hadamard matrices then so is $A \otimes B$.

14.17 If A and B are real $n \times n$ matrices such that $A \geqslant B$, $A^{-1} \geqslant 0$, $B^{-1} \geqslant 0$, prove that $A^{-1} \leqslant B^{-1}$. Hence deduce that the matrices in Exercise 14.11 satisfy $(A + D)^{-1} \leqslant A^{-1}$.

14.8 The definition (14.48) of the product of two intervals still applies when $a = [a_1, a_1]$ is a point interval. Use this to show that

$$d(a_1 B) = |a_1| \, d(B).$$

14.19 Let an $n \times n$ integer matrix A be such that each row sum r_i defined in (14.14) is even. By considering the equation

$$Ax = \tfrac{1}{2}[r_1, r_2, \ldots, r_n]^{\mathrm{T}}$$

and using Cramer's rule (Section 4.5), deduce that $\det A$ is even.

References

Agaian (1985), Ahmed and Rao (1975), Alefeld and Herzberger (1983), Barnett (1971, 1984), Beauchamp (1975), Berman and Plemmons (1979), Fiedler (1986), Gantmacher (1959, Vol. 2), Harwit and Sloane (1979), Horn and Johnson (1985), MacWilliams and Sloane (1977), Magnus and Neudecker (1988), Marcus and Minc (1964), Minc (1988), Newman (1972), Schrijver (1986).

Bibliography

References for chapters

Agaian, S. S. (1975). *Hadamard matrices and their applications.* Springer-Verlag, Berlin.

Ahmed, N. and Rao, K. R. (1975). *Orthogonal transforms for digital signal processing.* Springer-Verlag, Berlin.

Alefeld, G. and Herzberger, J. (1983). *Introduction to interval computations.* Academic Press, New York.

Barnett, S. (1971). *Matrices in control theory with applications to linear programming.* Van Nostrand Reinhold, London.

Barnett, S. (1979). *Matrix methods for engineers and scientists.* McGraw-Hill, London.

Barnett, S. (1983). *Polynomials and linear control systems.* Marcel Dekker, New York.

Barnett, S. (1984). *Matrices in control theory,* (2nd edn). Krieger, Florida.

Barnett, S. and Cameron, R. G. (1985). *Introduction to mathematical control theory,* (2nd edn). Clarendon Press, Oxford.

Barnett, S. and Cronin, T. M. (1986). *Mathematical formulae for engineering and science students,* (4th edn). Longman, London.

Barnett, S. and Storey, C. (1970). *Matrix methods in stability theory.* Nelson, London.

Basilevsky, A. (1983). *Applied matrix algebra in the statistical sciences.* North-Holland, New York.

Beauchamp, K. G. (1985). *Walsh functions and their applications.* Academic Press, London.

Ben-Israel, A. and Greville, T. N. E. (1980). *Generalized inverses: theory and applications,* (2nd edn). Krieger, Florida.

Berman, A. and Plemmons, R. J. (1979). *Non-negative matrices in the mathematical sciences.* Academic Press, New York.

Boullion, T. L. and Odell, P. L. (1971). *Generalized inverse matrices.* Wiley, New York.

Bradley, I. and Meek, R. L. (1986). *Matrices and society.* Penguin, Harmondsworth, Middlesex.

Brigham, E. O. (1988). *The fast Fourier transform and its applications.* Prentice-Hall, Englewood Cliffs, New Jersey.

Campbell, S. L. and Meyer, C. D. Jr. (1979). *Generalized inverses of linear transformations.* Pitman, London.

Campbell, S. L. (1982) (ed.). *Recent applications of generalized inverses.* Pitman, London.

Chen, C.-T. (1984). *Linear system theory and design.* Holt, Rinehart and Winston, New York.

Dahlquist, G. and Björck, A. (1974). *Numerical methods.* Prentice-Hall, Englewood Cliffs, New Jersey.

Davis, P. J. (1979). *Circulant matrices.* Wiley, New York.

Fiedler, M. (1986). *Special matrices and their applications in numerical mathematics.* Martinus Nijhoff, Dordrecht.

Fletcher, T. J. (1972). *Linear algebra through its applications.* Van Nostrand Reinhold, London.

Gantmacher, F. R. (1959). *The theory of matrices,* Vols. 1 and 2. Chelsea, New York.

Gohberg, I., Lancaster, P., and Rodman, L. (1982). *Matrix polynomials.* Academic Press, New York.

Golub, G. H. and Van Loan, C. F. (1989). *Matrix computations,* (2nd edn). Johns Hopkins University Press, Baltimore.

Gregory, R. T. and Karney, D. L. (1969). *A collection of matrices for testing computational algorithms.* Wiley, New York.

Gregory, R. T. and Krishnamurthy, E. V. (1984). *Methods and applications of error-free computation.* Springer-Verlag, New York.

Hager, W. W. (1988). *Applied numerical linear algebra.* Prentice-Hall, Englewood Cliffs, New Jersey.

Harwit, M. and Sloane, N. J. A. (1979). *Hadamard transform optics.* Academic Press, New York.

Heinig, G. and Rost, R. (1984). *Algebraic methods for Toeplitz-like matrices and operators.* Akademie-Verlag, Berlin.

Hill, R. (1986). *A first course in coding theory.* Clarendon Press, Oxford.

Horn, R. A. and Johnson, C. R. (1985). *Matrix analysis.* Cambridge University Press.

Householder, A. S. (1964). *The theory of matrices in numerical analysis.* Blaisdell, New York.

Jury, E. I. (1982). *Inners and stability of dynamic systems,* (2nd edn). Krieger, Florida.

Knuth, D. E. (1969). *The art of computer programming,* Vol. 2. Addison-Wesley, Reading, Mass.

Kučera, V. (1979). *Discrete linear control.* Wiley, Chichester.

Lancaster, P. (1966). *Lambda-matrices and vibrating systems.* Pergamon, Oxford.

MacWilliams, F. J. and Sloane, N. J. A. (1977). *The theory of error-correcting codes.* North-Holland, Amsterdam.

Magnus, J. R. and Neudecker, H. (1988). *Matrix differential calculus with applications in statistics and econometrics.* Wiley, Chichester.

Marcus, M. and Minc, H. (1964). *A survey of matrix theory and matrix inequalities.* Allyn and Bacon, Boston.

Miller, K. S. (1987). *Some eclectic matrix theory.* Krieger, Florida.

Minc, H. (1988). *Nonnegative matrices.* Wiley, New York.

Mirsky, L. (1963). *An introduction to linear algebra.* Clarendon Press, Oxford.

Nashed, M. Z. (ed.) (1976). *Generalized inverses and applications.* Academic Press, New York.

Newman, M. (1972). *Integral matrices.* Academic Press, New York.

Ogata, K. (1987). *Discrete-time control systems.* Prentice-Hall, Englewood Cliffs, New Jersey.

Pringle, R. M. and Rayner, A. A. (1971). *Generalized inverse matrices with applications to statistics.* Griffin, London.

Rao, C. R. and Mitra, S. K. (1971). *Generalized inverse of matrices and its applications.* Wiley, New York.

Rorres, C. and Anton, H. (1984). *Applications of linear algebra,* (3rd edn). Wiley, New York.

Rosenbrock, H. H. (1970). *State-space and multivariable theory,* Nelson, London.

Schrijver, A. (1986). *Theory of linear and integer programming.* Wiley, Chichester.

Searle, S. R. (1982). *Matrix algebra useful for statistics.* Wiley, New York.

Stewart, G. W. (1973). *Introduction to matrix computations.* Academic Press, New York.

Strang, G. (1988). *Linear algebra and its applications,* (3rd edn). Harcourt, Brace, Jovanovich, San Diego.

Wilkinson, J. H. (1965). *The algebraic eigenvalue problem.* Clarendon Press, Oxford.

Willems, J. L. (1970). *Stability theory of dynamical systems.* Nelson, London.

Some additional references on applications

Berman, A., Neumann, M., and Stern, R. J. (1989). *Non-negative matrices in systems theory.* Wiley, New York.

Björck, A., Plemmons, R. J., and Schneider, H. (eds) (1981). *Large scale matrix problems.* North Holland, Amsterdam.

Brualdi, R. A., Carlson, D. H., Datta, B. N., Johnson, C. R., and Plemmons, R. J. (eds) (1984). *Linear algebra and its role in systems theory.* American Mathematical Society, Providence, Rhode Island.

Bunch, J. R. and Rose, D. J. (eds) (1976). *Sparse matrix computations.* Academic Press, New York.

Bibliography

Collar, A. R. and Simpson, A. (1987). *Matrices and engineering dynamics.* Ellis Horwood, Chichester.

Datta, B. N., Johnson, C. R., Kaashoek, M. A., Plemmons, R. J., and Sontag, E. D. (eds) (1988). *Linear algebra in signals, systems and control.* SIAM, Philadelphia.

Duff, I. S., Erisman, A. M., and Reid, J. K. (1987). *Direct methods for sparse matrices.* Clarendon Press, Oxford.

Gill, P. E. and Wright, M. H. (1989). *Numerical linear algebra and optimization.* Addison-Wesley, Reading, Mass.

Graybill, F. A. (1983). *Matrices with applications in statistics,* (2nd edn). Wadsworth, Belmont, California.

Kaczorek, T. (1985). *Two-dimensional linear systems.* Springer-Verlag, Berlin.

Kailath, T. (1980). *Linear systems.* Prentice-Hall, Englewood Cliffs, New Jersey.

Kim, K. H. (1982). *Boolean matrix theory and applications.* Marcel Dekker, New York.

Krishnamurthy, E. V. (1985). *Error-free polynomial matrix computations.* Springer-Verlag, New York.

Noble, B. and Daniel, J. W. (1988). *Applied linear algebra,* (3rd edn). Prentice-Hall, Englewood Cliffs, New Jersey.

Ortega, J. M. (1987). *Matrix theory, a second course.* Plenum Press, New York.

Patel, R. V. and Munro, N. (1982). *Multivariable system theory and design.* Pergamon, Oxford.

Pissanetsky, S. (1984). *Sparse matrix technology.* Academic Press, London.

Tsay, Y. T., Shieh, L.-S., and Barnett, S. (1988). *Structural analysis and design of multivariable control systems. An algebraic approach.* Springer-Verlag, Berlin.

Answers to exercises

Chapter 2

2.1   (a) 4×2; (b) $3, 5$, undefined.

2.2   (a) $\begin{bmatrix} 1 & 0 & -1 \\ 3 & 2 & 1 \end{bmatrix}$;   (b) $\begin{bmatrix} 0 & 1 & 2 \\ 1 & 0 & 1 \\ 2 & 1 & 0 \end{bmatrix}$.

2.3   (a) $\begin{bmatrix} 2 & 2 \\ 7 & 3 \\ 9 & 6 \\ 5 & -14 \end{bmatrix}$;   (b) $\begin{bmatrix} 0 & -4 \\ -7 & 5 \\ -1 & -2 \\ 1 & -2 \end{bmatrix}$;

(c) $\begin{bmatrix} 2 & 6 \\ 14 & -2 \\ 10 & 8 \\ 4 & -12 \end{bmatrix}$;   (d) $\begin{bmatrix} -2 & 10 \\ 14 & -18 \\ -6 & 0 \\ -8 & 20 \end{bmatrix}$.

2.5   49 units vit. A, 109 units vit. B, 1205 units.

2.6   (a) $A + B = \begin{bmatrix} 0 & 0 \\ 2 & 1 \end{bmatrix}$, $AB = \begin{bmatrix} 1 & 0 \\ 1 & -1 \end{bmatrix}$, $BA = \begin{bmatrix} -1 & -1 \\ 0 & 1 \end{bmatrix}$.

(b) $A + B = \begin{bmatrix} 2 & 3 & 4 \\ 2 & 5 & 9 \end{bmatrix}$.

(c) $AB = \begin{bmatrix} -1 & -8 & -10 \\ 1 & -2 & -5 \\ 9 & 22 & 15 \end{bmatrix}$, $BA = \begin{bmatrix} 15 & -21 \\ 10 & -3 \end{bmatrix}$.

2.7 $m = 3$, $n = 8$

2.10 $a_2 = b_1$, $b_2 = c_1$, $\ldots$, $y_2 = z_1$; $a_1 \times z_2$.

2.19 ith row, column of A.

2.20 $M = \begin{bmatrix} 3 & 0 & 0 \\ 0 & 1 & -1 \\ 0 & -1 & 1 \end{bmatrix}$, $S = \begin{bmatrix} 0 & -1 & -1 \\ 1 & 0 & 0 \\ 1 & 0 & 0 \end{bmatrix}$.

2.23 $\frac{1}{2}n(n - 1)$.

2.25 (b) $A_1A_2 = -A_2A_1$.

2.27 $M = \frac{1}{2}(A + A^*)$, $S = \frac{1}{2}(A - A^*)$.

2.28 (R1), (C1); (R2), (C2).

2.31 $A^\mathsf{T} = \text{diag}[A_1^\mathsf{T}, A_2^\mathsf{T}]$, $A^2 = \text{diag}[A_1^2, A_2^2]$.

2.38 $\dot{A}A + A\dot{A}$, $\dot{A}A^2 + A\dot{A}A + A^2\dot{A}$.

Chapter 3

3.1 $x_1 = (a_{22}b_1 - a_{12}b_2)/d$, $x_2 = (a_{11}b_2 - a_{21}b_1)/d$, $d = a_{11}a_{22} - a_{12}a_{21}$;
(a) $d \neq 0$; (b) $d = 0$, $a_{11}/a_{21} = b_1/b_2$; (c) $d = 0$, $a_{11}/a_{21} \neq b_1/b_2$.

3.2 $(1, 13/5, 11/5)$.

3.3 $(3, 4, 5)$.

3.6 $(1, -2, -1, 1)$.

3.7 $\begin{bmatrix} 1 & 0 & 0 \\ 2 & 1 & 0 \\ 3 & 2 & 1 \end{bmatrix} \begin{bmatrix} 2 & 3 & 4 \\ 0 & 4 & 1 \\ 0 & 0 & 6 \end{bmatrix}$, $(5, 3, 1)$.

3.9 $(1, 2, 3)$, $(3, -1, 2)$.

3.11 $U = \frac{1}{2} \begin{bmatrix} 4 & 1 & 3 \\ 0 & 3 & -1 \\ 0 & 0 & 5\sqrt{2} \end{bmatrix}$.

3.12 (a) $(0.609, -0.739)$; (b) $(-28.913, -43.891)$.

Chapter 4

4.2 -55.

4.5 -45.

4.12 $2\mathbf{i} - 3\mathbf{j} - 5\mathbf{k}$, $\frac{1}{2}\sqrt{38}$.

4.14 $R_1R_2C \neq L$.

4.15 25, -27.

4.16 (a) 7; (b) 63.

4.19 $a_{1n}a_{2,n-1}\cdots a_{n1}d$; $\det K_n = d = (-1)^{n/2}$, n even;
$d = (-1)^{(n-1)/2}$, n odd.

4.21 $\dfrac{1}{17}\begin{bmatrix} 6 & -3 & 14 \\ -5 & 11 & -23 \\ -2 & 1 & 1 \end{bmatrix}$.

4.22 $(2, -3, 0)$.

4.25 A and B commute.

4.29 $\dfrac{1}{289}\begin{bmatrix} 65 & -75 & 197 \\ -75 & 131 & -294 \\ 197 & -294 & 726 \end{bmatrix}$.

4.31 $\dfrac{1}{25}\begin{bmatrix} 55 & -25 & 5 \\ 24 & -10 & -1 \\ -2 & 5 & -2 \end{bmatrix}$.

4.32 $\dfrac{1}{14}\begin{bmatrix} 0 & 4 & 2 & -6 \\ 21 & -9 & -8 & 3 \\ 7 & -1 & -4 & 5 \\ -14 & 6 & 10 & -2 \end{bmatrix}$.

Chapter 5

5.2 (a) $k = 8$; (b) $k \neq 8$; (c) none.

5.4 2.

5.5 (a) 2; (b) 2; (c) 3.

5.10 (a) yes in all cases; (b) and (c) yes if non-singular, no if singular.

5.11 (a) $x_1 = \frac{7}{2}t_1 - \frac{5}{2}t_2$, $x_2 = -3t_1 + 2t_2$, $x_3 = t_1$, $x_4 = t_2$;
(b) $x_1 = 2t_1$, $x_2 = -4t_1$, $x_3 = t_1$;
(c) $x_1 = t_1 - t_3$, $x_2 = -4t_1 - 4t_2 - 2t_3$, $x_3 = t_1$, $x_4 = 0$, $x_5 = t_2$, $x_6 = t_3$.

5.12 (a) $x_1 = -2t_1$, $x_2 = t_1$, $x_3 = 0$; (b) no.

5.13 consistent.

5.14 $k = 6$.

5.15 $x_1 = 22 - 7t_1 + 5t_2$, $x_2 = 6 - t_1 + t_2$, $x_3 = t_1$, $x_4 = t_2$.

5.16 $x_1 = -9 - 2t_1$, $x_2 = 15 + t_1$, $x_3 = t_1$.

5.17 (a) $k = 3$; (b) $k = 1$.

5.18 $y = \frac{1}{2} + \frac{8}{5}x$.

5.20 $X = \dfrac{1}{238} \begin{bmatrix} 95 & -33 \\ -7 & 35 \end{bmatrix}.$

5.22 (a) $7(C1) - 6(C2) + 2(C3) \equiv 0$, $(R1) + 2(R2) - (R3) \equiv 0$;
    (b) $2(C1) - 4(C2) + (C3) \equiv 0$, $5(R1) - (R3) - 2(R4) \equiv 0$.

5.23 $b = \frac{1}{3}(4b_1 + b_2 - 2b_3)g_1 + \frac{1}{3}(-b_1 - b_2 + 2b_3)g_2$
    $+ \frac{1}{3}(-2b_1 + b_2 + b_3)g_3.$

5.27 (a) yes; (b) no; (c) yes.

5.28 $[1\,1\,1\cdots1].$

5.29 (a) 000000, 011100, 101010, 111001, 110110, 100101, 010011,
       001111;
    (b) 0000000, 1111001, 1110010, 1010100, 0100110, 0001011,
       0101101, 1011111.

5.30 $\begin{bmatrix} 1 & 1 & 0 & 0 & 1 \\ 0 & 0 & 0 & 1 & 1 \\ 0 & 1 & 1 & 0 & 0 \end{bmatrix}$, 00000, 11100, 10011, 01111.

5.31 (a) 10110; (b) 11101; (c) more than one error.

5.32 000000, 111000, 100110, 010101, 011110, 101101, 110011, 001011.

5.33 $\begin{bmatrix} 0 & 0 & 0 & 1 & 1 & 1 & 1 \\ 0 & 1 & 1 & 0 & 0 & 1 & 1 \\ 1 & 0 & 1 & 0 & 1 & 0 & 1 \end{bmatrix}$, 1101001.

5.34 (a) $\begin{bmatrix} 0 & 1 & 1 & 1 & 0 & 0 \\ 1 & 0 & 1 & 0 & 1 & 0 \\ 1 & 1 & 1 & 0 & 0 & 1 \end{bmatrix}$; (b) $\begin{bmatrix} 1 & 0 & 1 & 0 & 1 & 0 & 0 \\ 1 & 1 & 1 & 0 & 0 & 1 & 0 \\ 1 & 1 & 1 & 1 & 0 & 0 & 1 \end{bmatrix}.$

Chapter 6

6.1 $4, (1/\sqrt{2})[1, 1]^{\mathsf{T}}; -2, (1/\sqrt{2})[1, -1]^{\mathsf{T}}.$

6.6 (a) $-2 \pm i\sqrt{(21)}$; (b) 4, $\pm 2i$.

6.7 (a) $\frac{1}{3}[1, 2, 2]^{\mathsf{T}}, \frac{1}{3}[2, 1, -1]^{\mathsf{T}}.$
    (b) $(1/\sqrt{59})[-1, 7, 3]^{\mathsf{T}}, (1/\sqrt{3})[i, 1, -1]^{\mathsf{T}}, (1/\sqrt{3})[i, -1, 1]^{\mathsf{T}}.$

6.8 (a) $0, 3 \pm 3\sqrt{3}$; (b) 1, 2, 3.

6.9 $1, (1/\sqrt{6})[-(1 + i), 2]^{\mathsf{T}}; 4, (1/\sqrt{3})[(1 + i), 1]^{\mathsf{T}}.$

6.18 $\lambda^3 - 5\lambda^2 + 8\lambda - 4.$

6.19 $A^4 = \begin{bmatrix} -49 & -50 & -40 \\ 65 & 66 & 40 \\ 130 & 130 & 81 \end{bmatrix}$, $A^{-1} = \dfrac{1}{6}\begin{bmatrix} 4 & -2 & 2 \\ -1 & 5 & -2 \\ -2 & -2 & 2 \end{bmatrix}$.

6.20 $\begin{bmatrix} -15 & -8 & 0 \\ 8 & 17 & 24 \\ 32 & 40 & 49 \end{bmatrix}$.

6.22 $\begin{bmatrix} -2 & -1 & 1 \\ -6 & -13 & -7 \\ 42 & 71 & 29 \end{bmatrix}$, yes.

6.23 $A^2 \otimes I_m + I_n \otimes B^2$.

6.24 $\lambda_i \mu_j \nu_k$, $u_i \otimes y_j \otimes z_k$.

6.27 similar in each case.

6.28 $[1, -1, 0]^{\mathrm{T}}$, $[2, -1, -2]^{\mathrm{T}}$, $[1, -1, -2]^{\mathrm{T}}$.

6.29 $A^3 = \begin{bmatrix} -11 & -12 & -13 \\ 19 & 20 & 13 \\ 38 & 38 & 27 \end{bmatrix}$.

6.32 9, 18.

6.33 n.

6.36 $x_1 = \alpha_1 e^t + 2\alpha_2 e^{2t} + \alpha_3 e^{3t}$, $x_2 = -\alpha_1 e^t - \alpha_2 e^{2t} - \alpha_3 e^{3t}$,
$x_3 = -2\alpha_2 e^{2t} - 2\alpha_3 e^{3t}$.

6.39 $\lambda_1 = 6$, $u_1 = [\tfrac{1}{2}, 1]^{\mathrm{T}}$, $\lambda_2 = 3$.

6.42 (a) $x^{(3)} = [2.120, 1.414, -1.236]^{\mathrm{T}}$; (b) $x^{(3)} = [2.055, 1.469, -1.318]^{\mathrm{T}}$;
Exact solution is $x = [2, 3/2, -4/3]^{\mathrm{T}}$.

6.43 $-2 < a < 2$, $-4 < b < 2$.

6.44 $X_1 = \begin{bmatrix} 0.09 & 0.18 \\ 0.27 & -0.46 \end{bmatrix}$, $X_2 = \begin{bmatrix} 0.091 & 0.182 \\ 0.273 & -0.455 \end{bmatrix}$.

Chapter 7

7.2 $2x_1^2 + 4x_1x_2 - x_2^2 + 5x_1x_3 - x_2x_3 + 4x_3^2$.

7.3 $2|x_1|^2 + (1 + i)\bar{x}_1 x_2 + (1 - i)x_1\bar{x}_2 + 7|x_2|^2 + (5 - i)\bar{x}_1 x_3$
$+ (5 + i)x_1\bar{x}_3 + i\bar{x}_2 x_3 - ix_2\bar{x}_3 - |x_3|^2$.

7.7 3.

7.8 $2y_1^2 + y_2^2 + \tfrac{5}{8}y_3^2$; $x_1 = y_1 + y_2 + \tfrac{15}{4}y_3$, $x_2 = y_2 + \tfrac{5}{2}y_3$, $x_3 = y_3$.

7.9 (a) $2(x_3 + \frac{5}{4}x_2)^2 - \frac{25}{8}(x_2 - \frac{12}{25}x_1)^2 + \frac{18}{25}x_1^2$;

(b) $(z_1 + \frac{1}{2}z_2)^2 - \frac{1}{4}(z_2 - 2z_3)^2 + z_3^2$, $x_1 = z_1$, $x_2 = z_1 + z_2$, $x_3 = z_3$.

7.10 3, 1.

7.11 positive semidefinite.

7.14 positive definite, indefinite, indefinite.

7.20 (a) negative semidefinite; (b) indefinite.

7.21 (a) indefinite; (b) positive semidefinite; (c) positive definite.

7.30 $P = \dfrac{1}{30}\begin{bmatrix} 13 & -1 \\ -1 & 4 \end{bmatrix}$, yes.

7.31 $\lambda_i + \lambda_j \neq 0$, for all eigenvalues λ_i of A.

Chapter 8

8.2 $\begin{bmatrix} 3 & 1 & & & \\ & 3 & & & \\ & & 2 & 1 & \\ & & & 2 & 1 \\ & & & & 2 \end{bmatrix}$, $\begin{bmatrix} 3 & 1 & & & \\ & 3 & & & \\ & & 2 & & \\ & & & 2 & 1 \\ & & & & 2 \end{bmatrix}$, $\begin{bmatrix} 3 & 1 & & & \\ & 3 & & & \\ & & 2 & & \\ & & & 2 & \\ & & & & 2 \end{bmatrix}$

$\begin{bmatrix} 3 & & & & \\ & 3 & & & \\ & & 2 & 1 & \\ & & & 2 & 1 \\ & & & & 2 \end{bmatrix}$, $\begin{bmatrix} 3 & & & & \\ & 3 & & & \\ & & 2 & & \\ & & & 2 & 1 \\ & & & & 2 \end{bmatrix}$, $\begin{bmatrix} 3 & & & & \\ & 3 & & & \\ & & 2 & & \\ & & & 2 & \\ & & & & 2 \end{bmatrix}$.

8.7 $T = \begin{bmatrix} 1 & -1 & 1 \\ 1 & 0 & 0 \\ 1 & 1 & 0 \end{bmatrix}$, $\mathcal{J} = \begin{bmatrix} 1 & 1 & 0 \\ 0 & 1 & 1 \\ 0 & 0 & 1 \end{bmatrix}$.

8.8 $x_1(t) = 2\alpha_1 e^{3t} + 2t\alpha_2 e^{3t} - 3\alpha_3 e^t$,
$x_2(t) = \alpha_2 e^{3t} + \alpha_3 e^t$, $x_3(t) = \alpha_3 e^t$.

8.9 $X_1(n) = \alpha_1 \lambda^n + n\alpha_2 \lambda^{n-1} + \frac{1}{2}n(n-1)\alpha_3 \lambda^{n-2}$,
$X_2(n) = \alpha_2 \lambda^n + n\alpha_3 \lambda^{n-1}$, $X_3(n) = \alpha_3 \lambda^n$.

8.14 diag$\left\{ -3, \begin{bmatrix} 0 & 1 \\ 9 & 0 \end{bmatrix}, \begin{bmatrix} 0 & 1 & 0 \\ 0 & 0 & 1 \\ 45 & 9 & -5 \end{bmatrix} \right\}$,   diag$[-3, -3, -3, 3, 3, -5]$.

8.17 $S = \begin{bmatrix} 1 & 0 & 0 \\ 0 & 1 & -\sqrt{2} \\ 0 & 0 & 5 \end{bmatrix}$, $U = \begin{bmatrix} 1/\sqrt{2} & -1/\sqrt{3} & 1/\sqrt{6} \\ 0 & 1/\sqrt{3} & 2/\sqrt{6} \\ -1/\sqrt{2} & -1/\sqrt{3} & 1/\sqrt{6} \end{bmatrix}$.

8.18 $\begin{bmatrix} \dfrac{1}{\sqrt{3}} & \dfrac{1}{\sqrt{6}} & \dfrac{1}{\sqrt{2}} \\ \dfrac{1}{\sqrt{3}} & \dfrac{1}{\sqrt{6}} & -\dfrac{1}{\sqrt{2}} \\ \dfrac{1}{\sqrt{3}} & -\dfrac{2}{\sqrt{6}} & 0 \end{bmatrix} \begin{bmatrix} \sqrt{3} & \dfrac{2}{\sqrt{3}} & \dfrac{1}{\sqrt{3}} \\ 0 & \dfrac{\sqrt{2}}{\sqrt{3}} & \dfrac{\sqrt{2}}{2\sqrt{3}} \\ 0 & 0 & \dfrac{1}{\sqrt{2}} \end{bmatrix}$.

8.22 $\dfrac{1}{6+\sqrt{6}} \begin{bmatrix} -(1+\sqrt{6}) & -(1+\sqrt{6}) & -2(1+\sqrt{6}) \\ -(1+\sqrt{6}) & 5+\sqrt{6} & -2 \\ -2(1+\sqrt{6}) & -2 & 2+\sqrt{6} \end{bmatrix}$.

8.27 $\lambda^3 - 11\lambda^2 + 59\lambda - 97$, $\lambda^5 - 12\lambda^4 + 72\lambda^3 - 178\lambda^2 + 215\lambda - 194$.

8.31 $\dfrac{1}{\sqrt{5}} \begin{bmatrix} 2 & 1 \\ 1 & -2 \end{bmatrix} \begin{bmatrix} 3\sqrt{5} & 0 \\ 0 & 2\sqrt{5} \end{bmatrix} \dfrac{1}{5} \begin{bmatrix} 4 & 3 \\ 3 & -4 \end{bmatrix}$.

8.35 $Q = \dfrac{1}{5\sqrt{5}} \begin{bmatrix} 11 & 2 \\ -2 & 11 \end{bmatrix}$, $H = \dfrac{\sqrt{5}}{25} \begin{bmatrix} 66 & 12 \\ 12 & 59 \end{bmatrix}$.

Chapter 9

9.2 $\begin{bmatrix} 1 & 0 & -3 \\ 0 & 1 & 0 \\ 0 & 0 & 1 \end{bmatrix}$.

9.3 $\begin{bmatrix} (2\sin\frac{1}{4} - \sin\frac{1}{2}) & 4(\sin\frac{1}{2} - \sin\frac{1}{4}) \\ \frac{1}{2}(\sin\frac{1}{4} - \sin\frac{1}{2}) & (2\sin\frac{1}{2} - \sin\frac{1}{4}) \end{bmatrix}$.

9.9 $e^A = \begin{bmatrix} e & (e-1) \\ 0 & 1 \end{bmatrix}$, $e^B = \begin{bmatrix} e & (1-e) \\ 0 & 1 \end{bmatrix}$, $e^{A+B} = \begin{bmatrix} e^2 & 0 \\ 0 & 1 \end{bmatrix}$.

9.15 $\sin A = \dfrac{1}{2} \begin{bmatrix} (\sin 5 + \sin 1) & (\sin 5 - \sin 1) \\ (\sin 5 - \sin 1) & (\sin 5 + \sin 1) \end{bmatrix}$,

$A^{100} = \dfrac{1}{2} \begin{bmatrix} (5^{100} + 1) & (5^{100} - 1) \\ (5^{100} - 1) & (5^{100} + 1) \end{bmatrix}$.

9.16 $\begin{bmatrix} (2e^2 - e^3) & (2e^2 - e - e^3) & \frac{1}{2}(e - e^3) \\ (e^3 - e^2) & (e - e^2 + e^3) & \frac{1}{2}(e^3 - e) \\ (2e^3 - 2e^2) & (2e^3 - 2e^2) & e^3 \end{bmatrix}$.

9.20 $\begin{bmatrix} e^{2t} & (e^t - e^{2t}) & (-e^t + e^{2t}) \\ te^{2t} & (e^t - te^{2t}) & (-e^t + e^{2t} + te^{2t}) \\ te^{2t} & -te^{2t} & (e^{2t} + te^{2t}) \end{bmatrix}$.

9.21 $\begin{bmatrix} 1176 & -2400 & 1225 \\ 1225 & -2499 & 1275 \\ 1275 & -2600 & 1326 \end{bmatrix}$.

9.22 $(\lambda - 2)^2$

$\begin{bmatrix} \sin 2t & 0 & 0 \\ -11t \cos 2t & (-2t \cos 2t + \sin 2t) & t \cos 2t \\ -22t \cos 2t & -4t \cos 2t & (2t \cos 2t + \sin 2t) \end{bmatrix}$.

9.24 $\begin{bmatrix} 1 & 2t & (t + 3t^2) \\ 0 & 1 & 3t \\ 0 & 0 & 1 \end{bmatrix}$.

9.30 $\dfrac{1}{9} \begin{bmatrix} 7 & 4 & 4 \\ 4 & 1 & -8 \\ 4 & -8 & 1 \end{bmatrix}$.

9.35 $\begin{bmatrix} -0.186 & 1.220 \\ 0.814 & 0.220 \end{bmatrix}$.

Chapter 10

10.4 $\dfrac{1}{62} \begin{bmatrix} 22 & 23 & 17 \\ 4 & 7 & -11 \end{bmatrix}$.

10.15 $\dfrac{1}{180} \begin{bmatrix} 1 & 4 & 1 \\ 3 & 12 & 3 \end{bmatrix}$.

10.17 $\dfrac{1}{\alpha} \begin{bmatrix} \bar{a}_{11} & \bar{a}_{21} \\ \bar{a}_{12} & \bar{a}_{22} \end{bmatrix}$, $\alpha = |a_{11}|^2 + |a_{12}|^2 + |a_{21}|^2 + |a_{22}|^2$.

10.18 $\dfrac{1}{2745} \begin{bmatrix} -20 & -40 & 215 \\ -40 & -80 & 430 \\ 150 & 300 & -240 \\ 49 & 98 & -115 \end{bmatrix}$.

10.20 $\dfrac{1}{18} \begin{bmatrix} 1 & 2 & 3 & 1 \\ -2 & 2 & 0 & 4 \\ 5 & -2 & 3 & -7 \end{bmatrix}$.

10.25 $\begin{bmatrix} 1+\alpha & -1+\beta \\ -3-5\alpha & 4-5\beta \\ \alpha & \beta \end{bmatrix}$.

10.35 $\dfrac{1}{2}\begin{bmatrix} 2 & 5 & 0 \\ 0 & 1 & 0 \end{bmatrix}$.

10.51 $\begin{bmatrix} (1-4y_2-y_3-3y_4-12y_5-3y_6) & y_2 & y_3 \\ y_4 & & \\ & y_5 & y_6 \end{bmatrix}$,

$x_1 = 49/180$, $x_2 = 147/180$.

Chapter 11

11.6 $\begin{bmatrix} -16 & -18 & -2 \\ -18 & -19 & -1 \\ -2 & -1 & 1 \end{bmatrix}$.

11.9 $\begin{bmatrix} 16 & 6 & 2 \\ 6 & -3 & 1 \\ 2 & 1 & 1 \end{bmatrix}$, $Z^{-1} = \dfrac{1}{16}\begin{bmatrix} 1 & 1 & -3 \\ 1 & -3 & 1 \\ -3 & 1 & 21 \end{bmatrix}$.

11.11 $\lambda + 1$.

11.12 $\lambda + 7$.

11.14 $f_2 = \frac{73}{4}\lambda^2 + \frac{33}{4}\lambda + \frac{41}{4}$, $f_3 = \frac{4}{5329}(-2631\lambda + 932)$,
$f_4 = 80341.73^2/4.2631^2$.

11.18 $x(\lambda) = -\lambda + 6$, $y(\lambda) = \lambda^3 - 4\lambda^2 - 7\lambda - 26$.

11.19 (a) yes; (b) yes; (c) no; (d) no; (e) yes.

11.20 One on left, two on right of imaginary axis.

11.21 Two on left, two on right of imaginary axis.

11.22 One inside, three outside.

11.23 $22\mu^3 + 16\mu^2 - 14\mu - 16$.

11.24 $15\mu^4 - 20\mu^3 + 30\mu^2 - 12\mu + 3$; two inside, two outside.

11.25 $\frac{1}{4}(2\lambda^3 + 3\lambda^2 + 8\lambda - 5)$.

11.26 One eigenvalue on left, two on right of imaginary axis.

11.27 $P = \dfrac{1}{7}\begin{bmatrix} -27 & 19 & 7 \\ 19 & -42 & -13 \\ 7 & -13 & -2 \end{bmatrix}$.

11.29 $P = \begin{bmatrix} 32 & 12 \\ 12 & 2 \end{bmatrix}$, one inside, one outside.

11.31 12.

11.32 $P = \begin{bmatrix} 0.171 & 0.316 \\ 0.316 & 0.856 \end{bmatrix}$; $u = -1.71x_1 - 3.16x_2$.

11.39 $P = \begin{bmatrix} 2 & 0 \\ 0 & 1 \end{bmatrix}$.

11.40 $P = [v_1^T \cdots v_n^T]X \begin{bmatrix} v_1 \\ \vdots \\ v_n \end{bmatrix}$.

11.42 Exact solution $= \dfrac{1}{6} \begin{bmatrix} 3 & 1 \\ 1 & 1 \end{bmatrix}$.

11.43 $P = \begin{bmatrix} 6 & 2 \\ 2 & 1 \end{bmatrix}$.

11.44 $P = \begin{bmatrix} -6 & 2 \\ 2 & -5 \end{bmatrix}$.

Chapter 12

12.3 $-\dfrac{1}{8} \begin{bmatrix} (5\lambda^2 + 10\lambda + 7) & -(\lambda^2 + 2\lambda + 3) \\ -(5\lambda^3 + 15\lambda^2 + 17\lambda + 12) & (\lambda^3 + 3\lambda^2 + 5\lambda + 4) \end{bmatrix}$.

12.4 $Q_2 = \begin{bmatrix} 2\lambda + 3 & \lambda - 2 \\ -9 & 2 \end{bmatrix}$, $R_2 = \begin{bmatrix} -17 & 8 \\ 27 & -2 \end{bmatrix}$.

12.9 $\begin{bmatrix} 1 & 0 & 0 \\ 0 & 1 & 0 \\ 0 & 0 & \lambda^2 + \lambda \end{bmatrix}$.

12.10 (a) λ, λ^2, λ^4, $\lambda + 2$, $\lambda + 2$, $(\lambda + 2)^2$;
       (b) $\lambda + 2$, $\lambda + 2$, $(\lambda + 2)^2$, $(\lambda + 2)^3$, $\lambda - 2$, $(\lambda - 2)^2$, $(\lambda - 2)^3$, $\lambda + 3$, $(\lambda + 3)^2$.

12.11 (a) $i_1 = i_2 = i_3 = i_4 = i_5 = 1$, $i_6 = \lambda$, $i_7 = \lambda(\lambda^2 - 1)(\lambda^2 - 4)$;
       (b) $i_1 = i_2 = 1$, $i_3 = \lambda + 2$, $i_4 = i_5 = (\lambda + 2)^2$, $i_6 = (\lambda + 2)^3$,
       $i_7 = (\lambda + 2)^4(\lambda - 1)^2$.

12.13 (a) yes;   (b) no.

12.14 $X = \dfrac{1}{9}\begin{bmatrix} 7 & 11 \\ 6 & 3 \end{bmatrix}$, $Y = -\dfrac{1}{9}\begin{bmatrix} 7 & 11 \\ 6 & 3 \end{bmatrix}$.

12.15 (a) $\begin{bmatrix} \dfrac{1}{(\lambda+2)^2(\lambda+3)} & 0 \end{bmatrix}$; (b) $\begin{bmatrix} \dfrac{1}{(\lambda+3)^2(\lambda+5)^2} & 0 \\ 0 & (\lambda+2)(\lambda+4) \end{bmatrix}$;

(c) $\begin{bmatrix} \dfrac{1}{(\lambda+2)^2(\lambda+3)^2} & 0 \\ 0 & \lambda+3 \end{bmatrix}$.

12.17 $G(s) = \begin{bmatrix} \dfrac{3s-4}{(s-1)(s-2)} & \dfrac{1}{s-2} \\ \dfrac{-7s+6}{(s-1)(s-2)} & \dfrac{3}{s-2} \end{bmatrix}$, $g(s) = (s-1)(s-2) \neq \det(sI - A)$.

12.18 $G(s) = \begin{bmatrix} (s+3)^2(s+4) & 0 \\ 0 & (s+3)(s+4)^2 \end{bmatrix}^{-1} \begin{bmatrix} s+4 & s+3 \\ -6 & (s+3)(s-1) \end{bmatrix}$.

Chapter 13

13.1 $\begin{bmatrix} 1 & 0 & 0 & 0 \\ \dfrac{5}{2} & 1 & 0 & 0 \\ 0 & \dfrac{6}{7} & 1 & 0 \\ 0 & 0 & -\dfrac{49}{4} & 1 \end{bmatrix} \begin{bmatrix} 2 & 1 & 0 & 0 \\ 0 & -\dfrac{7}{2} & 3 & 0 \\ 0 & 0 & -\dfrac{4}{7} & 2 \\ 0 & 0 & 0 & \dfrac{61}{2} \end{bmatrix}$.

13.2 $\begin{bmatrix} 1 & 0 & 0 & 0 & 0 \\ \dfrac{5}{2} & 1 & 0 & 0 & 0 \\ 1 & \dfrac{8}{7} & 1 & 0 & 0 \\ 4 & \dfrac{18}{7} & 4 & 1 & 0 \\ 0 & \dfrac{4}{7} & \dfrac{5}{3} & -\dfrac{2}{15} & 1 \end{bmatrix} \begin{bmatrix} 2 & 1 & 1 & 0 & 0 \\ 0 & -\dfrac{7}{2} & \dfrac{1}{2} & -4 & 0 \\ 0 & 0 & \dfrac{3}{7} & \dfrac{46}{7} & 11 \\ 0 & 0 & 0 & -10 & -35 \\ 0 & 0 & 0 & 0 & -23 \end{bmatrix}$.

13.3 122.

13.6 $C^{-1} = -\dfrac{1}{2}\begin{bmatrix} 1 & 1 & 1 & \cdots & 1 \\ 1 & 3 & 3 & \cdots & 3 \\ 1 & 3 & 5 & \cdots & 5 \\ . & . & . & \cdots & . \\ 1 & 3 & 5 & \cdots & 2n-1 \end{bmatrix}.$

13.7 $a = r,\ b = r,\ \lambda = -(1 + r^2)/r.$

13.19 $X = \alpha\begin{bmatrix} 1 & -1 & 1 \\ -1 & 1 & -1 \\ 1 & -1 & 1 \end{bmatrix},\ \alpha \neq 0.$

13.21 $T^{-1} = \dfrac{1}{350}\begin{bmatrix} -2 & 2 & 5 \\ -6 & 24 & 2 \\ 8 & -6 & -2 \end{bmatrix}.$

13.22 Inverse $= \dfrac{1}{5}\begin{bmatrix} 5 & 5 & 5 & -20 & 30 \\ -2 & -3 & -3 & 13 & -20 \\ 0 & 2 & 1 & -3 & 5 \\ 1 & 0 & 2 & -3 & 5 \\ 1 & 1 & 0 & -2 & 5 \end{bmatrix}.$

13.23 Determinant $= -61250.$

13.25 $x_3 = [-\tfrac{2}{5}, 1, -\tfrac{3}{5}]^{\mathsf{T}}.$

13.26 $\dfrac{1}{19}\begin{bmatrix} -11 & 15 & 9 & -27 \\ 10 & -5 & -3 & 9 \\ 4 & -2 & -5 & 15 \\ -8 & 4 & 10 & -11 \end{bmatrix}.$

13.36 det $H = 1/6048000,$

$$H^{-1} = \begin{bmatrix} 16 & & & \text{symmetric} \\ -120 & 1200 & & \\ 240 & -2700 & 6480 & \\ -140 & 1680 & -4200 & 2800 \end{bmatrix}.$$

Chapter 14

14.1 $Y = \begin{bmatrix} 0 & 0 & 0 \\ Y_{21} & 0 & 0 \\ 0 & 0 & Y_{33} \\ 0 & 0 & 0 \end{bmatrix}$, $\quad Y_{21} = \begin{bmatrix} y_1 & y_2 \\ 0 & y_1 \end{bmatrix}$, $\quad Y_{33} = \begin{bmatrix} 0 & 0 & y_3 & y_4 \\ 0 & 0 & 0 & y_3 \end{bmatrix}$.

14.2 $Y = \begin{bmatrix} Y_{11} & Y_{12} & 0 & 0 \\ Y_{21} & Y_{22} & 0 & 0 \\ 0 & 0 & Y_{33} & Y_{34} \\ 0 & 0 & Y_{43} & Y_{44} \end{bmatrix}$,

$Y_{11} = \begin{bmatrix} y_1 & y_2 & y_3 \\ 0 & y_1 & y_2 \\ 0 & 0 & y_1 \end{bmatrix}$, $\quad Y_{12} = \begin{bmatrix} y_4 & y_5 \\ 0 & y_4 \\ 0 & 0 \end{bmatrix}$,

$Y_{21} = \begin{bmatrix} 0 & y_6 & y_7 \\ 0 & 0 & y_6 \end{bmatrix}$, $\quad Y_{22} = \begin{bmatrix} y_8 & y_9 \\ 0 & y_8 \end{bmatrix}$,

$Y_{33} = [y_{10}]$, $\quad Y_{34} = [y_{11}]$, $Y_{43} = [y_{12}]$, $\quad Y_{44} = [y_{13}]$.

14.5 $X = T \operatorname{diag}\{\underbrace{H_1, \ldots, H_1}_{t_1}, \underbrace{H_2, \ldots, H_2}_{t_2}, \cdots, \underbrace{H_k, \ldots, H_k}_{t_k}\} T^{-1}$,

$t_1 + 2t_2 + \cdots + kt_k = n$.

$H_i = \begin{bmatrix} 0 & 1 & 0 & \cdots & 0 \\ 0 & 0 & 1 & \cdots & 0 \\ \cdot & \cdot & \cdot & \cdots & \cdot \\ \cdot & \cdot & \cdot & \cdots & 1 \\ \cdot & \cdot & \cdot & \cdots & 0 \end{bmatrix}_{i \times i}$

14.8 $5, [8, 2, 7]^{\mathsf{T}}$.

14.9 no.

14.22 $H_4 = \begin{bmatrix} 1 & 1 & 1 & 1 \\ 1 & -1 & 1 & -1 \\ 1 & -1 & -1 & 1 \\ 1 & 1 & -1 & -1 \end{bmatrix}$,

$H_8 = \begin{bmatrix} 1 & 1 & 1 & 1 & \cdots\cdots\cdots & 1 \\ 1 & & & & & \\ \cdot & & \operatorname{circ}(0, 1, 1, -1, 1, -1, -1) & & \\ 1 & & & -I_7 & & \end{bmatrix}$.

Answers to problems

Chapter 1

1.1 $\begin{array}{c} \\ e_1 \\ e_2 \\ e_3 \end{array} \begin{array}{cc} f_1 & f_2 \\ \left[\begin{array}{cc} 1 & 0 \\ 1 & 1 \\ 0 & 1 \end{array}\right] \end{array}$.

1.2 (a) $\begin{array}{c} \\ B_1 \\ B_2 \\ B_3 \end{array} \begin{array}{ccc} C_1 & C_2 & C_3 \\ \left[\begin{array}{ccc} 0 & 1 & 2 \\ 2 & 3 & 0 \\ 0 & 0 & 1 \end{array}\right] \end{array}$; (b) $\begin{array}{c} \\ A_1 \\ A_2 \end{array} \begin{array}{ccc} C_1 & C_2 & C_3 \\ \left[\begin{array}{ccc} 2 & 7 & 11 \\ 4 & 6 & 2 \end{array}\right] \end{array}$.

1.3 $A \begin{array}{c} H \\ T \end{array} \begin{array}{cc} H & T \\ \left[\begin{array}{cc} 5 & -2 \\ 3 & -7 \end{array}\right] \end{array}$, $+$ = money from A to B.

1.4 $\left[\begin{array}{ccccccc} 2 & 8 & 0 & 2 & -1 & 0 & 0 \\ 1 & 2 & 28 & 9 & 0 & -1 & 0 \\ 16 & 0 & 0 & 1 & 0 & 0 & -1 \end{array}\right] \left[\begin{array}{c} x_1 \\ x_2 \\ \vdots \\ x_7 \end{array}\right] = \left[\begin{array}{c} 2000 \\ 2500 \\ 8500 \end{array}\right]$, each $x_i \geqslant 0$.

1.5 (a) reflection in x_1-axis; (b) stretch; (c) rotation about x_3-axis.

1.6 $\left[\begin{array}{c} x_{n+1} \\ y_{n+1} \\ z_{n+1} \end{array}\right] = \left[\begin{array}{ccc} 0 & \frac{1}{3} & 0 \\ 0 & 0 & \frac{1}{2} \\ 6 & 0 & 0 \end{array}\right] \left[\begin{array}{c} x_n \\ y_n \\ z_n \end{array}\right]$.

1.7 picked up in city

$\begin{array}{cc} B & L \\ \left[\begin{array}{cc} 0.4 & 0.3 \\ 0.6 & 0.7 \end{array}\right] \begin{array}{c} B \\ L \end{array} \end{array}$ dropped off in city.

Chapter 2

2.2 $v_1 = (-31 + 16i)v_3 + (-28 + 24i)i_3,$
$i_1 = (-32 + 8i)v_3 + (-31 + 16i)i_3.$

2.5 $\begin{bmatrix} 2 & 3 & 0 \\ 0 & 0 & 4 \\ 1 & -7 & 3 \end{bmatrix}$.

2.11 21, 34.

2.16 $A_u = \begin{bmatrix} 0 & 0 & 0 \\ * & 0 & 0 \\ 0 & * & 0 \\ 0 & 0 & * \end{bmatrix}$, $A_k = \begin{bmatrix} 9 & 1 & 3 \\ 0 & 2 & 4 \\ 5 & 0 & 7 \\ 1 & 6 & 0 \end{bmatrix}$.

Chapter 3

3.1 $(1.26, 0.05, -0.33)$.

3.2 $a_1 \neq 0$, $d_1 = a_1 a_2 - b_1 c_1 \neq 0$, $d_2 = a_3 d_1 - a_1 b_2 c_2 \neq 0$, $a_4 d_2 - b_3 c_3 d_1 \neq 0$.

3.3 $l_1 = 2$, $l_2 = -3/5$, $l_3 = -35/12$, $u_1 = 1$, $u_2 = -5$, $u_3 = -12/5$, $u_4 = 13/12$, $v_1 = 2$, $v_2 = 1$, $v_3 = -1$, $x = (2, 3, -2, 5)$.

3.4 $x_1^2 + x_2^2 - 6x_1 - 4x_2 - 12 = 0$.

3.5 $\begin{bmatrix} 2 & 5/2 \\ -4 & -9/2 \\ -1 & -2 \end{bmatrix}$.

3.8 $a \neq 1/3$.

3.9 $\begin{bmatrix} 2 & 1 \\ -1 & 3 \end{bmatrix}$.

Chapter 4

4.2 $k < 2$.

4.6 $\begin{bmatrix} A^{-1} & 0 \\ -D^{-1}CA^{-1} & D^{-1} \end{bmatrix}$.

4.13 $A^5 = \begin{bmatrix} 122 & 121 \\ 121 & 122 \end{bmatrix}$.

Chapter 5

5.1 not controllable.

5.2 $x_{11} = -2 + t_1 + t_2$, $x_{12} = 12 - t_1$, $x_{13} = 7 - t_2$, $x_{21} = 11 - t_1 - t_2$, $x_{22} = t_1$, $x_{23} = t_2$.

5.3 $x_1 = 2 - t_1 - t_2$, $x_2 = -t_1 + 2t_2$, $x_3 = t_1$, $x_4 = t_2$.

5.4 $k = 2$, $x_1 = -t_1$, $x_2 = t_1$, $x_3 = t_1$; $k = 5$, $x_1 = \frac{1}{2}t_2$, $x_2 = \frac{1}{4}t_2$, $x_3 = t_2$.

5.5 $\lambda = -1$, $x_1 = -1/11$, $x_2 = -15/11$; $\lambda = 1$, $x_1 = -5$, $x_2 = 1$; $\lambda = 12$, $x_1 = \frac{1}{2}$, $x_2 = 1$.

5.6 $a = 0$, $b = \frac{7}{2}$, $R(A) = 2$, $x_1 = -3t_1 - \frac{5}{2}t_2$, $x_2 = 2t_1 + \frac{3}{2}t_2$, $x_3 = t_1$, $x_4 = t_2$; $R(A) = 3$ if $a \neq 0$ (any b) or $b \neq \frac{7}{2}$ (any a), and $x_1 = -3t_1$, $x_2 = 2t_1$, $x_3 = t_1$, $x_4 = 0$.

Chapter 6

6.6 $\lambda^4 - \lambda^3 - 13\lambda^2 - 20\lambda + 13$.

6.7 $\frac{1}{2}(3 \pm \sqrt{5})$, $\frac{1}{2}(5 \pm \sqrt{5})$.

6.13 $k_n\lambda^n + k_{n-1}\lambda^{n-1} + \cdots + k_1\lambda + 1 = 0$.

6.18 $\dfrac{1}{9}\begin{bmatrix} 14 & -2 & -14 \\ -2 & -1 & -16 \\ -14 & -16 & 5 \end{bmatrix}$.

6.29 $\lambda^3 - 4\lambda^2 + 4\lambda - 16$,

$$\begin{bmatrix} \lambda^2 - 4\lambda & 2 - \lambda & \lambda - 6 \\ 2\lambda - 8 & \lambda^2 - \lambda + 2 & 3\lambda + 2 \\ 8 - 2\lambda & \lambda + 2 & \lambda^2 - 3\lambda + 2 \end{bmatrix}.$$

6.30 $\lambda_1 = 16$, $u_1 = [\frac{1}{2}, 1, -\frac{1}{2}]^T$.

6.31 4, 2.

6.33 -3.56.

Chapter 7

7.2 positive semidefinite if $R(A) < n$, positive definite if $R(A) = n$.

7.4 (a) $3 > k > 2$; (b) $k > 25$.

7.7 $[1, -2, -2]^T$

7.13 $(13\alpha^2 - 2\alpha\beta + 4\beta^2)/30$.

Chapter 9

9.4 $z(0) \cos \omega t + [\dot{z}(0) \sin \omega t]/\omega$.

9.8 $|c| < 1$.

9.16 $\begin{bmatrix} (2e^{2t} - e^{3t}) & (-e^t + 2e^{2t} - e^{3t}) & \frac{1}{2}(e^t - e^{3t}) \\ (e^{3t} - e^{2t}) & (e^t - e^{2t} + e^{3t}) & \frac{1}{2}(e^{3t} - e^t) \\ (2e^{3t} - 2e^{2t}) & 2(e^{3t} - e^{2t}) & e^{3t} \end{bmatrix}$.

9.18 $\begin{bmatrix} 0 & \frac{1}{3} & 0 \\ 0 & 0 & \frac{1}{2} \\ 3 & 1 & 0 \end{bmatrix}$.

Chapter 10

10.3 $A^+ = \dfrac{1}{552} \begin{bmatrix} 0 & 0 & 0 & 0 \\ -26 & -6 & 14 & 20 \\ 10 & 9 & 8 & -1 \\ 30 & 27 & 24 & -3 \\ -124 & -42 & 40 & 82 \end{bmatrix}$.

10.16 $u = \begin{bmatrix} 3 & -1 \\ -1 & 0 \end{bmatrix} y$.

Chapter 11

11.5 $\det \Gamma = \pm 2^{n(n+1)/2}$, $\Gamma^{-1} = \Gamma/2^n$.

11.7 n

11.14   (a) $\dfrac{7}{16} > k > 0$;   (b) $4 > k > 2$.

11.17 $u = -2x_1 - x_2$.

Chapter 12

12.6 $\dfrac{1}{s^2 - 4s + 5} \begin{bmatrix} s - 1 \\ s - 5 \end{bmatrix}$.

12.7 $\begin{bmatrix} \dfrac{1}{(\lambda+3)(\lambda+4)} & 0 \\ 0 & \dfrac{\lambda+5}{\lambda+4} \end{bmatrix}$, 3.

12.8 $A^{-1} = \begin{bmatrix} -\lambda & -(\lambda^2+3\lambda+2) & 3\lambda^2+8\lambda+5 \\ \lambda-1 & \lambda^2+2\lambda-1 & -(3\lambda^2+5\lambda-3) \\ 1 & \lambda+2 & -(3\lambda+5) \end{bmatrix}$.

Chapter 13

13.19 $P = \begin{bmatrix} 1 & 0 & 0 & 0 & 0 \\ 0 & 1 & 0 & 0 & 0 \\ 0 & \dfrac{1}{2} & 1 & 0 & 0 \\ 0 & 0 & -1 & 1 & 0 \\ 0 & 0 & 0 & \dfrac{1}{7} & 1 \end{bmatrix}$, $D = \begin{bmatrix} 6 & 2 & 0 & 0 & 0 \\ & 6 & -1 & 0 & 0 \\ & & \dfrac{11}{2} & 7 & 0 \\ & & & \dfrac{29}{2} & \dfrac{13}{14} \\ & \text{symmetric} & & & \dfrac{533}{98} \end{bmatrix}$.

13.20 $R = \begin{bmatrix} 1 & 2 & 3 \\ 0 & 1 & 7 \\ 0 & 0 & 1 \end{bmatrix}$, $D = \text{diag}[1, -1, 50]$.

13.26 determinant $= (\alpha - \beta)^{n-1}[\alpha + (n-1)\beta]$;

inverse $= \dfrac{1}{\alpha - \beta}\left[I - \dfrac{\beta E}{\alpha + (n-1)\beta}\right]$.

Chapter 14

14.6 (d) $a_{ii} = 1$, $a_{ij} = a_{i1}/a_{j1}$.

14.8 yes.

Index